D0075942

Atmospheric Chemistry

Atmospheric Chemistry

Ann M. Holloway and Richard P. Wayne
Department of Chemistry, University of Oxford, Oxford, UK

RSC Publishing

ISBN: 978-1-84755-807-7

A catalogue record for this book is available from the British Library

Published by The Royal Society of Chemistry,
Thomas Graham House, Science Park, Milton Road,
Cambridge CB4 0WF, UK

Registered Charity Number 207890

For further information see our website at www.rsc.org

Preface

To the poet, the atmosphere ought to have been "this brave o'erhanging firmament, this majestical roof fretted with golden fire" but Hamlet, in a bad mood, found that "it appears no other thing to me but a foul and pestilent congregation of vapours". Environmental scientists investigating air pollution might tend to agree with Hamlet, while atmospheric physicists studying the intriguing phenomenon of the airglow undoubtedly feel that Shakespeare's introductory phrases much better describe the magic and beauty of the firmament from which comes the glowing light that they capture by day and by night. Our atmosphere most certainly offers many different facets to its human observers. Allowing ourselves for the moment the indulgence of an artificial 19th-century classification of the sciences, and confining ourselves just to these disciplines, we could pursue these thoughts a little further by drawing some caricatures. To the physicist, the atmosphere displays highly complex fluid dynamics in which temperature, gravitation and rotation all play a part. The meteorologist builds on this information to describe 'weather' and 'climate'. Biologists see the atmosphere as in part a support system for living organisms that gain their energy by respiration. Hydrologists and geologists are concerned with precipitation of atmospheric water to form the oceans, lakes and rivers, and the effects of so-called weathering on the composition and morphology of the exposed surface of the planet.

What about the chemist? One quite useful concept likens the atmosphere to a giant photochemical reactor, with the atmospheric chemist challenged to interpret what is going on inside the reactor. The challenge is rather a serious one because, as we shall see as this book develops, there are a surprisingly large number of reactants inside the 'vessel', and some of the chemical interactions are far from simple. As the atmospheric sciences have developed over the past decades, it has become increasingly evident that chemists are often central to the enterprise of discovering how the atmosphere works, despite our earlier suggestion that the formal divisions into the individual sciences are unprofitable and perhaps meaningless. Identification and measurement of the concentrations of the species within the atmosphere, and how they got there, are clearly matters in which chemists will play key parts. Even more obviously, finding out which chemical transformations can occur and how fast they take place are jobs for chemists to do, perhaps largely in laboratory experimentation. That information then needs to be combined with the physics of atmospheric motions, as one example, to see how this chemistry works in the atmosphere. What is unprofitable is to regard chemistry and physics as operating separately in the atmosphere, but what is very profitable is to deploy the special and different skills of people who have been trained as

Atmospheric Chemistry
By Ann M. Holloway and Richard P. Wayne

Published by the Royal Society of Chemistry, www.rsc.org

chemists or physicists in looking together at how we can continue to gain a better and more detailed understanding of our atmospheric environment.

It is in the light of these remarks that we can set out the purpose of this book. Our aim is to inform and stimulate a person who has studied (or is studying) chemistry in a conventional academic course; we hope to provide the additional key information that will unlock the gates to understanding how chemistry works in the atmosphere and what it does there. Let there be no doubt about the importance of chemists possessing at least a modest familiarity with the topics that will be laid out in the book. To gain the 'better and more detailed understanding' that we suggested at the end of the preceding paragraph is, for sure, an intellectually stimulating task at the level of fundamental science. But it may be that the improved understanding will play a vital rôle in maintaining our atmosphere as one that is a healthy part of the land–water–air environment in which we live. Man's activities are certainly providing challenges for the atmosphere. There can be few scientists who have not seen at least some of the phrases *air pollution*, *photochemical smog*, *acid rain*, *ozone holes*, and, above all, *the greenhouse effect* and *climate change*. These are all phenomena largely or entirely brought about by Man. A knowledge of chemistry, and specifically of atmospheric chemistry, must be at the heart of attempts to formulate measures to reduce or reverse the adverse environmental impacts. What is more, the control measures have to be implemented by policy decisions whose effects may range from the purely local to the truly global. Policy makers thus need to be advised by well informed chemists if they are to make sensible decisions (even if the advice alone may not ensure good policy). Chemists who know about atmospheric chemistry can thus help at several different levels, ranging from fundamental research through to guiding scientifically valid legislation. Incidentally, it may not be only the threats to the atmosphere posed by Man that require action, but those of Nature itself. Volcanic eruptions, solar flares, incoming solar system material (especially in a collision with an asteroid!), and long-term changes in solar luminosity and of the Earth's orbit can all be expected to affect the composition and behaviour of the atmosphere, quite probably adversely. That chemists could offer ways of avoiding some of the worst atmospheric effects of cataclysmic events may be wishful speculation, but an optimistic view is that the knowledge and intelligence possessed by such chemists would make the outlook less bleak. At least these thoughts highlight how important atmospheric chemistry and atmospheric chemists are now, and will be in the future. We hope that with this encouragement our readers will find much to fascinate and to stimulate in what will unfold here.

This book relies quite heavily on the content of a much more extensive book written by one of the authors (R.P. Wayne, *Chemistry of Atmospheres* 3rd edn, OUP, 2000). The topics selected for that book were carefully chosen, and have found favour with many readers. There are, however, significant differences between the earlier work and the one presented here. First, this book is considerably shorter than its predecessor; secondly, it is addressed to a somewhat different audience. Here, we have aimed at a level that will be suitable for (advanced) undergraduate courses or for postgraduates setting out on the first steps of their research into atmospheric chemistry. We believe that, at the same time, this level will also prove attractive to the general, but scientifically literate, reader who wishes to be better informed about atmospheric chemistry and the several pressing environmental problems that involve the atmosphere.

The structure of the book and the development of our subject deviate somewhat from those usually encountered. Our idea is to bring together important and recurring concepts, such as the sources and sinks of trace gases, and their budgets and lifetimes, before discussing in more detail the atmospheric behaviour of specific chemical species. That is, the emphasis is initially on the principles of the subject, with the finer points emerging at later points in the book, sometimes in several successive chapters. In this way, some of the core material gets repeated exposure, but in new ways and new contexts. We hope that our approach will serve a valuable pedagogic purpose.

The constraints of space mean that it is not possible to examine every exception to a general principle or every twist to a particular hypothesis, however entertaining or instructive those exceptions and twists might be. Rigorous scholarship might demand that these matters be properly explored and assessed, but experience tells us that our intended audiences can easily lose track of the main themes and ultimately lose interest in the subject if there is too much emphasis on rigour. To the best of our ability, though, we have attempted to ensure that all the statements made are scientifically correct as far as they go. Our hope is that we have struck a good balance between the competing claims of rigour and delivering a smooth flow of ideas to build up the reader's understanding of our subject. Similar reasoning lies behind our decision not to provide bibliographic citations within the text. This is not a work of reference, and we know that citations disrupt the narrative flow in a way that disturbs our target audiences. Instead, we have made suggestions for further reading of books or articles that will allow our readers to build on what they have learned from the present book. Again in view of the needs of many undergraduates studying chemistry, we have tried to make what mathematical material there is as approachable as possible.

We now come to that pleasant part of writing a book where we can at last thank the many people who made it possible. A number of good friends and colleagues went out of their way to provide advice, material and encouragement. Of the Oxford team, there are those still here (Pete Biggs, Carlos Canosa-Mas), those who are former members (Dudley Shallcross, Paul Monks), and one who is just visiting (John Burrows). Pete redrew the latest incarnation of Figure 1.1. John and his colleague Mark Weber have made available new and reworked GOME and SCIAMACHY satellite data from their home base, the Institute for Environmental Physics, Bremen, and they constructed new figures to include in this book. Anne Douglass and Charley Jackman, both of NASA's Goddard Space Flight Center, each most generously gave of their own time and expertise to perform entirely new model calculations to help illustrate important points. RPW also recognizes that, in view of the relationship between this new text and his *Chemistry of Atmospheres*, he remains deeply indebted to the many dozens of people who over more than a quarter of a century since the first edition was being written have kept him supplied with good science to write about. Their names are recorded in the prefaces to the three editions, and I hope that they will not mind that they are not listed again here.

Finally, we have the greatest pleasure in recording our gratitude to our respective spouses. Dan and Brenda have provided much succour and sustenance, both spiritual and physical, without which the task of writing would have been well-nigh impossible. We are fortunate that each of them is not only an excellent cook, but also possesses a sharp editorial eye and editorial skills to match. As a result, there are many fewer errors than there would otherwise have been (and, of course, there will be some remaining for which the authors take full responsibility). Heartfelt thanks are due to both of you!

AMH and RPW
Oxford

Contents

Atmospheric Chemistry
By Ann M. Holloway and Richard P. Wayne

Published by the Royal Society of Chemistry, www.rsc.org

CHAPTER 1

Earth's Atmosphere

1.1 CHEMISTRY IN THE ATMOSPHERE

Our atmosphere is an extraordinary mixture of gases and suspended particles, some inert and some highly reactive, some present in large quantities while others are found only in the minutest traces. New material is being added continuously to the atmosphere at the surface boundary, with materials trapped within the Earth, perhaps when the planet was formed, being liberated, sometimes slowly and gently, and sometimes violently in volcanic eruptions. So far, this description would also fit the atmospheres of our near neighbours Venus and Mars. But on Earth, the living organisms (the *biota*) make a quite dramatic contribution of their own to the supply of chemicals to the atmosphere. Humans are part of the biota, and are making an impact on the atmosphere out of all proportion to human life's biological importance, as we shall see time and again in this book. The study of the chemistry of the 'natural' atmosphere is truly fascinating in its own right. But it is also clear that the chemist who has a good understanding of how this natural atmosphere works is one of the most likely candidates to make rational and informed suggestions about ways to offset Man's depredations. Policy makers (and politicians) need chemists!

The mixture of gases and particles naturally contains many substances that can react with others that are present, and one of the primary tasks of the atmospheric chemist is to interpret the composition of atmosphere in terms of the pathways and rates of the reactions that occur. A knowledge, preferably obtained from experimental observations, of the possible *mechanisms* and of the *chemical kinetics* of the reactions thus lies at the core of this aspect of atmospheric chemistry. It is often helpful to imagine the atmosphere as a giant chemical reactor in which the chemical soup interacts to remove some components and to replace them with new species.

Of the biogenic gases, the most notable both in abundance and in properties is oxygen, O_2, produced by photosynthesis. The existence of life allows oxygen to make up about one-fifth (21 per cent) of Earth's atmosphere, while oxygen is almost absent from the atmospheres of Venus and Mars. However, many of the other gases released are reduced (CH_4, H_2S, $(CH_3)_2S$, for example), or at least only partially oxidized (CO and NO will serve as examples). Much of the chemical change in the lower part of the atmosphere thus consists of oxidation steps that lead ultimately to CO_2, H_2O, SO_3, NO_2 and N_2O_5 for the examples presented. We will look again at these products quite soon. For the moment, the significant feature to note is that the atmospheric conversion steps are driven, directly or indirectly, by solar ultraviolet radiation, so that the atmosphere is not just a reactor, but

Atmospheric Chemistry
By Ann M. Holloway and Richard P. Wayne

Published by the Royal Society of Chemistry, www.rsc.org

a *photochemical reactor*. To be sure, chemistry continues at night: charged particles from the Sun and lightning discharges provide energy for some reactions, while others can be promoted thermally near volcanoes. But by far the greatest proportion of the chemistry that we have outlined so far is ultimately dependent on photons emitted by the Sun.

This apparently simple picture of chemistry in the lower part of the atmosphere not surprisingly reveals considerable complexity when looked at more closely. It is at this point that the very close interactions between chemical, physical, biological, and geological processes become evident. Chemical change is greatly influenced by temperature and pressure; not only the rates of reactions, but even the pathways open for reaction and thus the products formed may depend on these two factors. As a result, in the atmosphere altitude, latitude and longitude may all play a part in determining which processes occur, or at least compete successfully with alternative pathways. Pressures and temperatures in the atmosphere fairly obviously change with altitude, although the behaviour turns out to be far from obvious. It is here that we encounter the concept of different regions of the atmosphere, such as the *troposphere*, the *stratosphere* and the *mesosphere*, in which the classification is based on temperature structure. Figure 1.1 shows where the regions are located. Altitudes z are shown on the left-hand vertical scales, and temperatures T are plotted horizontally; the line shows how T varies with z. Atmospheric motions transport substances that live long enough to other heights and places. Reactions take place on the surfaces of suspended solids (aerosols) such as ice, soot and dusts, and within droplets (liquid H_2O clouds, for example), as well as just within the gas phase, so that physics and meteorology are necessarily again involved in a detailed description of atmospheric chemistry. But chemical composition, and chemistry itself, may have an effect on temperatures, so there is a reverse interaction. Biology makes many of the chemicals released to the atmosphere, but life is sustained or shielded in one way or another by the atmosphere. Geology and geochemistry influence the chemical composition of the atmosphere, and in turn are influenced by it. It is worth noting, too, how the interactions often run in two directions: temperature affects chemistry, but chemistry influences temperatures. These so-called *feedbacks* thus need to be taken into account when attempting to understand atmospheric behaviour. Indeed, we do not get to a very deep understanding without considering the feedbacks between all the systems taken as a whole. It's a complex and daunting task, but a fascinating and scientifically stimulating one that provides insights at all sorts of levels.

1.2 EARTH'S ATMOSPHERE IN PERSPECTIVE

Although we shall consider many more of the effects that life has had on Earth's atmosphere later in the book, it is interesting to examine the pie charts in Figure 1.2, which indicate the actual compositions of the atmospheres of Venus, Earth and Mars. Comparison of the chart for Earth with the charts for Venus and Mars, the planets on either side of us, will show just how dramatically the existence of life on Earth has altered the composition of our atmosphere, particularly with respect to the proportion of carbon dioxide in the atmosphere.

At this point, the unexpected nature of the Earth's atmosphere becomes apparent. Since Earth lies in the solar system between Venus and Mars, Earth's atmosphere might have been expected to consist primarily of the *oxidized* compound, carbon dioxide. But CO_2 is only a minor (although very important) constituent. The presence of elemental oxygen as a major constituent is obviously one of the most significant features of our atmosphere and has some of the greatest impacts on its chemistry. Earth's atmosphere appears to be a combustible mixture, since there is too much oxygen in the presence of too many gases that react with oxygen. Oxygen reacts with hydrogen to form water, with nitrogen to form nitrates, with methane to form carbon dioxide and water, and so on. Biological processes are dominant in the production of the oxidizable components of our atmosphere, and Earth's atmosphere consequently maintains a steady-state disequilibrium composition.

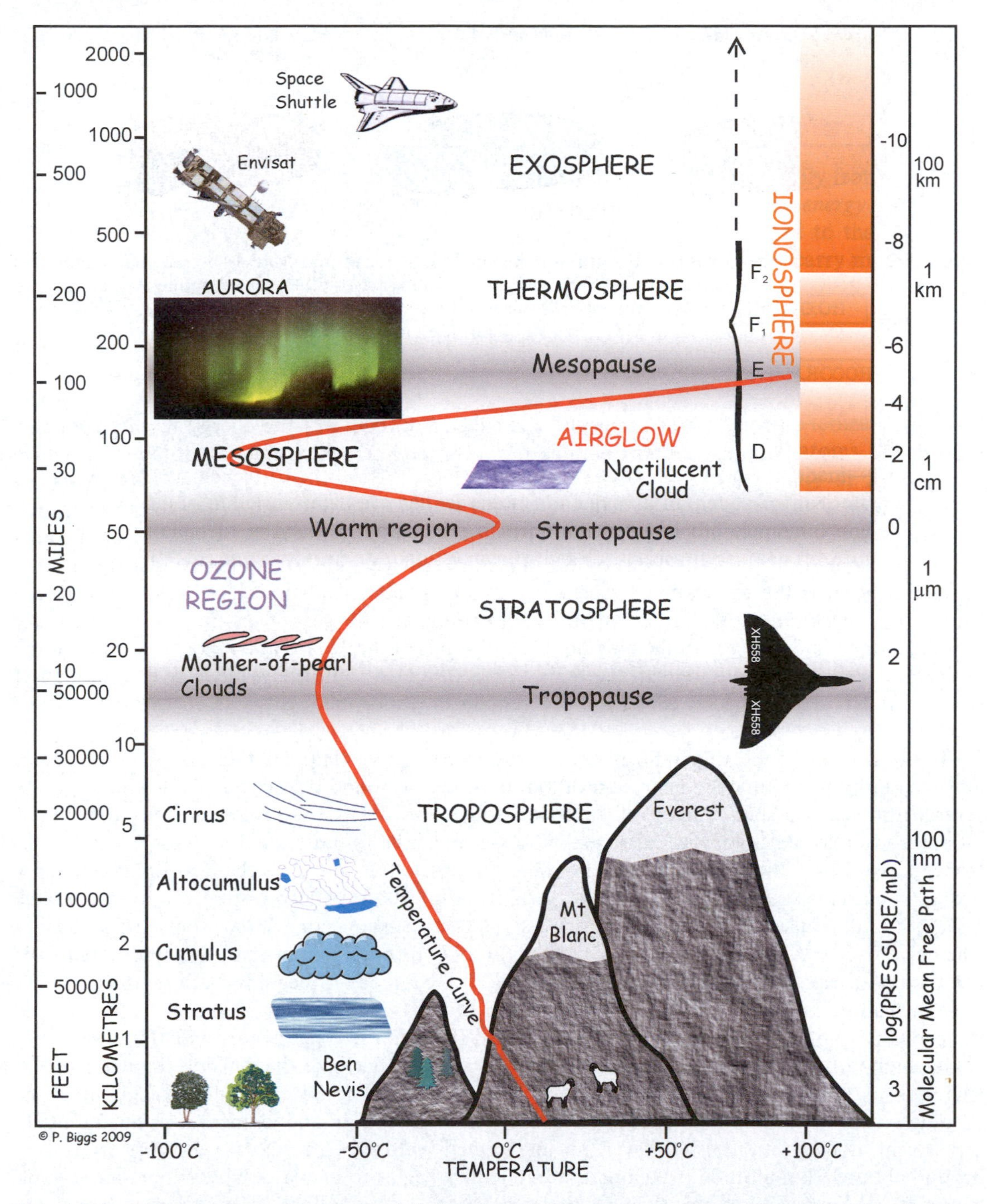

Figure 1.1 The temperature structure of the atmosphere. Temperatures show a complex dependence on altitude, decreasing with altitude at some heights but increasing at others. The turning points of the temperature gradient mark the boundaries between regions of the atmosphere. The diagram indicates the clouds and other features found at different altitudes. The right-hand ordinate scale shows both the pressure and the mean free path (λ) corresponding to the left-hand altitude scales. This version of the figure was constructed in 2009 by Dr P. Biggs, who kindly gave permission for its use here.

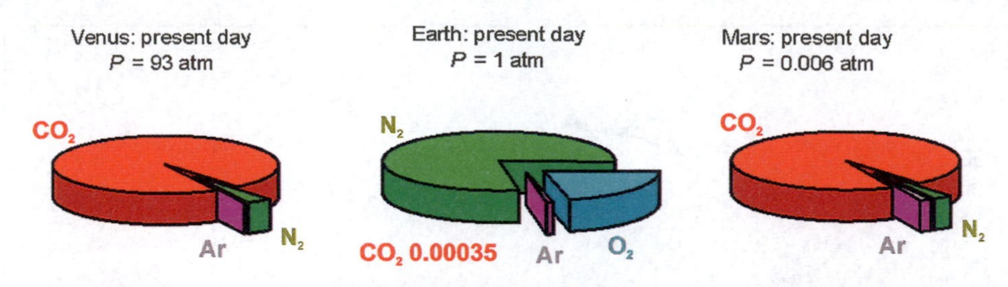

Figure 1.2 Abundances of gases in the atmospheres of Venus, Earth and Mars. The pie charts show the fractional abundances of the dominant gases, while those of other key components are shown in parentheses. Data for contemporary atmospheres are taken from R.P. Wayne, *Chemistry of Atmospheres,* 3rd edn, OUP, 2000, who cites the original references.

Furthermore, the entropy of the atmosphere is effectively reduced—the energy required for this reduction is supplied almost entirely by radiation from the Sun, which enables biology to produce both oxidizable components and, of course, oxygen.

As any high-altitude mountaineer will attest, most of this oxygen is found close to the surface of the Earth, and the atmosphere loses density very quickly as an ascent is made. Gaseous components in the atmosphere do not settle down on the planetary surface under the influence of gravitational attraction because the translational kinetic energy of the atoms or molecules competes with the forces of *sedimentation*. As a result of this competition, the density of gas falls with increasing altitude in the atmosphere. For the bulk liquids and solids of the oceans and land, the masses of the individual 'particles' are so great that this competition is completely ineffective, although if the material is finely divided enough it can remain in atmospheric suspension (an *aerosol*: see Section 1.3).

The total mass of Earth's atmosphere is around 5×10^{18} kg. Half of this mass lies below an altitude of about 5.5 km and 99 per cent of that mass can be found below roughly 30 km, although several atmospheric species may still be found some 160 km or so above sea level, and ions of atomic and molecular hydrogen, nitrogen and oxygen can be found many hundreds of kilometres from the surface of the Earth. However, where the outer 'boundary' of the atmosphere actually occurs is extremely difficult to define; at an altitude of 100 km the pressure is only about one millionth of that at sea-level and the average distance between collisions of the atoms and molecules (the *mean free path*, λ), has reached 1 mm, several million times greater than the diameters of the colliding partners. At this altitude, chemical reactions are already rather slow because the reactants have such a low chance of colliding with each other. By the time we reach 160 km above the surface, the probability of collisions is very low indeed, and the mean free path is around 100 m.

Our high-altitude mountaineer will have also discovered that, like the density, the temperature decreases as we climb higher. In fact, it drops by about 6 K for every additional kilometre of altitude even for the highest mountains (Everest's peak is at 8848 m and temperatures of 230–240 K are typical at that altitude). However, an unexpected feature of the Earth's atmosphere is that beyond about 15 km altitude the temperature in reality begins to increase. Now is the time to look at Figure 1.1 again. Below this level lies the *troposphere* (with the final kilometre or so before the surface making up a *boundary layer*). This is the atmospheric region in which we live: we will discuss it in detail in Chapter 8. For now, we observe that since this lowest part of the atmosphere has higher temperatures towards the bottom, it is subject to convection currents and is characterized by rather strong vertical mixing. It is this behaviour that gives rise to the name of the region, since the Greek for 'turning' is 'tropos'. Above about 15 km altitude, temperatures start to rise, with the consequence that there is very little convection or vertical mixing in this region. The atmosphere becomes 'layered', and is called the *stratosphere* (Greek 'stratos' = 'layered').

To illustrate just how different these rates of mixing are, we may like to note here that individual molecules usually take several years to travel through the stratosphere, but that they can traverse the entire depth of the troposphere in a few days, or even a few minutes in the updraughts of large thunderstorms. The height at which the boundary between the troposphere and stratosphere (the *tropopause*) occurs is also variable and depends on season and latitude.

In fact, the temperature structure of the atmosphere as a whole is even more complex than this brief discussion would suggest. As illustrated in Figure 1.1, beyond roughly 50 km, atmospheric temperature once again decreases, the turning point (the *stratopause*) marking the limit of the stratosphere. We are now entering the *mesosphere*; temperature reaches its minimum at the top of that region before rising again in the *thermosphere*, as shown in Figure 1.1.

We shall return later to the questions of how and why the temperatures vary in this way and why the pressures drop with increasing altitude. For the time being, it is just useful to introduce these key concepts, and to remind ourselves that most living organisms are found near the surface. The lowest parts of the atmosphere (the troposphere) receive the greatest variety of chemical species arriving geologically and biologically from the planet's surface; but, because solar radiation has had to pass through the atmosphere that lies above, contribute the lowest-energy photochemistry. In general, then, the chemistry becomes both more energetic and involves simpler species the higher the altitude. At the greatest altitudes, neutral species can be split into atoms, and the atoms and molecules can become ionized.

1.3 AEROSOLS, PARTICLES AND DROPLETS

Most of the mass of the atmosphere is made up of gaseous elements and compounds, and a large part of the chemistry that we shall be exploring in this book naturally concerns gas-phase reactants and reactions. However, there is a surprisingly large amount of non-gaseous material in the atmosphere. Such material includes sea-spray from the ocean, wind-blown dust from surface erosion, particles from forest fires and volcanic emissions, and meteoric debris. Some chemical reactions that occur between gaseous components in the atmosphere can also lead to particle formation. Even in the unpolluted atmosphere, these processes lead to the haze seen in places such as the Smoky Mountains of the USA. In addition, human activity generates further particulate material through combustion of fuels, and this material has a significant impact on the chemistry of the atmosphere, as we shall see later in the book. At present, mankind probably generates about 20 per cent of all solid particles currently being added to the atmospheric mix. Chemical reactions can occur on the surface of non-gaseous particles or within the body of liquid droplets, and these processes thus add to the rich chemistry of the atmosphere. Although to a large extent these non-gaseous species are to be found in the lower part of the atmosphere (say up to an altitude of 10 km), some are found much higher up. Polar stratospheric clouds, which we will meet in Section 9.4.4, are found in the stratosphere (obviously), and meteors, fragments of asteroids, cometary dust and the like enter the atmosphere at its outermost, indistinct, boundary. All these particles have been found to exert important influences on chemical behaviour. This may be a good point to say that the purely gas-phase processes are called *homogeneous* and those involving some phase (liquid, solid) in addition to gas are called *heterogeneous*.

Several different types of non-gaseous atmospheric particles exist. Material may be in liquid or solid forms and varies from the extremely small (around 0.05 μm—although, as we shall see, this end of the scale is very difficult to define) to the rather sizeable (around 1000 μm ≡ 1 mm). Suspensions of particles in a gas are called *aerosols*. In principle, if the particles are liquid we refer to the aerosol as a *cloud* or *mist*, while if they are solid, the aerosol is called *smoke* or *dust*. However, we need, at this point, to define what we mean by 'suspension'. An aerosol particle in the atmosphere is subject to the competing physical influences of gravity on the one hand and the kinetic energy of surrounding particles on the other, just as are the atoms or molecules of a gas. Gravity

Table 1.1 Size and terminal velocities of water droplets in the atmosphere.

Radius (μm)	*Name of droplet*	*Approximate number of H_2O molecules per drop*	*Terminal fall velocity ($m\,s^{-1}$)*
0.1	Condensation nucleus	10^8	10^{-6}
1	Cloud	10^{11}	10^{-4}
10	Cloud	10^{14}	10^{-2}
50	Large cloud	10^{16}	0.27
100	Drizzle	10^{17}	0.70
1000	Rain	10^{18}	6.50

Data of J.M. Wallace and P.V. Hobbs, *Atmospheric Science*, 2nd edn, Academic Press, London, 2006.

constantly pulls the particle downwards and tries to cause sedimentation, whereas the motion of the surrounding molecules means that the particle is subject to drag, which stops it falling. Whether or not the particle remains in the atmosphere (in 'suspension') or falls to the surface of the Earth is largely dependent upon its size. For example, a typical cloud droplet of 10 μm diameter would take a day to fall through a cloud 1 km thick and is, for practical purposes, permanently suspended. However, a raindrop of 2000 μm diameter would fall through the same cloud in under three minutes and is certainly not in suspension, as anyone getting wet as a result of its fall would testify. Table 1.1 shows terminal fall velocities for several sizes of water droplet. When they grow large enough, particles of water in clouds become *precipitation* in the form of rain, hail, snow and so on. Condensation of water vapour to form droplets usually starts by nucleation on foreign solid-aerosol particles, which act as *cloud-condensation nuclei* (CCN). Hygroscopic or soluble nuclei are particularly effective CCNs, so their presence in the atmosphere must be regarded as an important component of the evaporation–condensation–precipitation cycle of the Earth's water.

At the other end of the scale there is an imperceptible merging between what is a small aerosol and what is a large molecule or cluster of molecules. The smallest solid aerosols (those with radii less than about 0.5 μm) are known as *Aitken nuclei* and their concentration in the atmosphere is significantly affected by human activity. Typical counts of Aitken nuclei near the Earth's surface are 10^5, 10^4, and 10^3 particles cm^{-3} over cities, rural areas and the sea, respectively. We shall discuss these particles further in Section 8.6 and in Chapter 11. For the moment, we simply note that these aerosols have a significant effect on the chemistry of the atmosphere. Clouds and other aerosols may also modify the atmospheric balance of incoming and outgoing radiation and alter atmospheric and surface temperatures by changing both the reflectivity and absorptivity of the atmosphere towards incoming solar radiation. The effects of the eruption of Mount Pinatubo in the Philippines in June 1991 provided an interesting demonstration of climatic effects. Vast quantities of dust and SO_2 were injected into the atmosphere, resulting in significantly increased aerosol concentrations, including those of droplets of H_2SO_4. Furthermore, solar radiation reaching the lower atmosphere was reduced by about two per cent and global average temperatures may have been reduced by as much as 0.5 °C during the year following the eruption. As chemists, we should be alerted to these effects of particulate material in the atmosphere and to the possibility of aerosols altering the course or rates of chemical change. We shall consider the sources of particles and their roles in chemical change throughout the book, and especially in Sections 8.6, 9.4.4 and 11.1. At this stage, we just summarize in Figure 1.3 the sources, lifetimes and effects of particles of different sizes.

1.4 MAJOR GASES

The major gaseous constituents of Earth's unique atmosphere are nitrogen (78 per cent), oxygen (21 per cent), argon (0.9 per cent), carbon dioxide (0.035 per cent) when the atmosphere is dry;

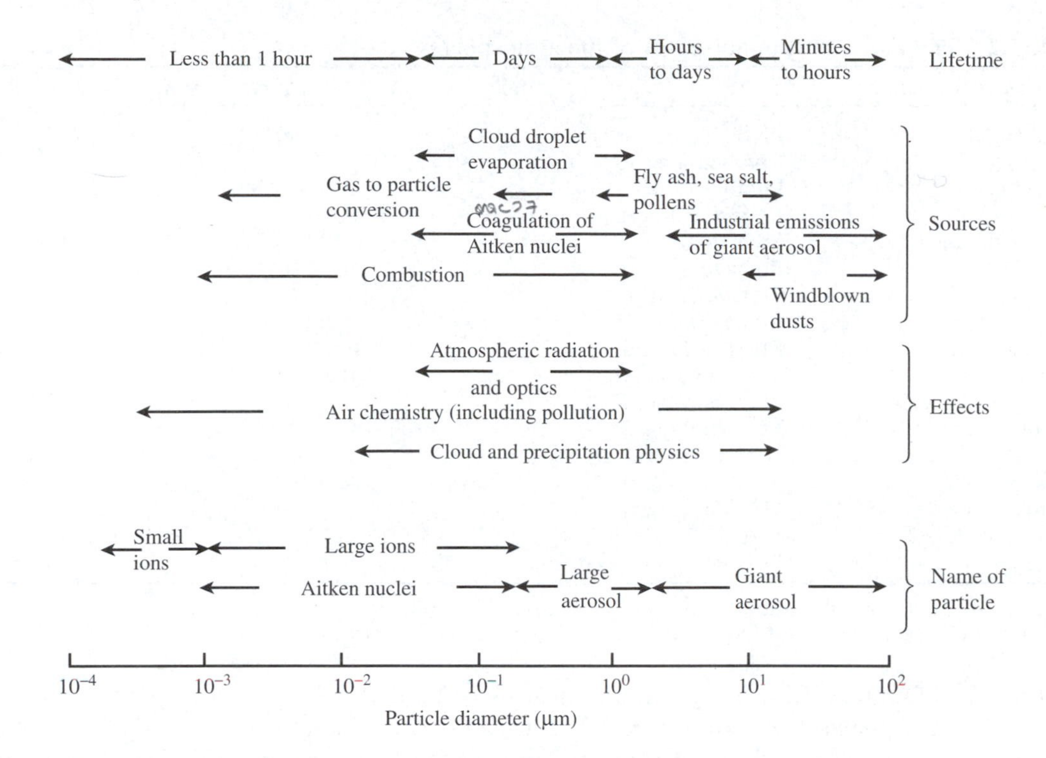

Figure 1.3 Names of atmospheric particles, together with effects, sources, and lifetimes. The lifetime of very small particles is short because they coagulate rapidly to form larger particles. Giant aerosol particles are short-lived because they precipitate out of the atmosphere. Figure drawn from data of J.M. Wallace and P.V. Hobbs, *Atmospheric Science*, 2nd edn, Academic Press, London, 2006.

water-vapour concentrations are highly variable, but up to 4 per cent can displace the other gases under very humid conditions.

Table 1.2 presents this information in the first entries. Subsequent entries refer to gases present at much smaller concentrations, and these *minor gases* are first discussed in Section 1.5 and then throughout the rest of the book. The concentrations are expressed in parts per million by volume (abbreviated to ppm or ppmv; in this book we shall use ppm for clarity, but it will always imply 'by volume'). Atmospheric scientists call this measure a *mixing ratio*: in several respects the mixing ratio is more significant than the absolute concentration (molecules per unit volume or *number density*, or moles per unit volume, *molarity*), since it shows the relative importance of an atmospheric component, whatever the absolute pressure at the altitude under consideration. Chemists will immediately recognize that the term represents a concept identical to that of a *mole fraction*. The values given in the table are typical global averages, but for some of the species there are enormous variations with geographical location and with altitude.

Although molecular nitrogen makes up the largest part of the atmosphere, it is chemically fairly inert owing to the large N–N bond energy of 745 kJ mol^{-1}. Most reactions involving N_2 are endothermic, or at least kinetically limited because of their large activation energies.

Oxygen is a much more active participant in the chemistry of the atmosphere. A great deal of chemical change in the Earth's atmosphere involves oxidation, and O_2 is the oxidant. For those living organisms that derive their energy by respiration, oxygen is evidently absolutely essential. Ozone is one of the most important minor gases in the atmosphere, and will be considered in the next section. It is made up only of O atoms, so without oxygen there would be no ozone. Atmospheric oxygen and its role will be explored in much greater detail throughout this book.

Table 1.2 The major and minor gases of the atmosphere.

Atmospheric constituent	*Source*	*Mixing ratio (ppm)*	*Percentage*
Nitrogen N_2	Prebiological Microbiological	780 840	78.08
Oxygen O_2	Photosynthesis	209 460	20.95
Argon Ar	Mainly radiogenic	9340	0.93
Carbon dioxide CO_2	Prebiological	350	0.035
Water H_2O	Prebiological	0–40 000[a]	
Neon Ne	Prebiological	18	
Helium He	Mainly radiogenic	5.2	
Methane CH_4	Microbiological	1.7	
Hydrogen H_2	Some volcanic (birth)	0.53	
Nitrous oxide N_2O	Microbiological	0.3	
Carbon monoxide CO	Microbiological Oxidation product	0.04–0.2	
Ammonia NH_3	Microbiological	<0.01	
Hydrogen sulfide H_2S	Microbiological	10^{-4}	
Sulfur dioxide SO_2	Volcanic	10^{-4}	
Ozone O_3	Photochemical production from O_2	0–10[b]	

[a]The amount of water varies considerably according to climatic conditions.
[b]The value for ozone varies principally with altitude (see Section 1.5 and Chapter 5).

Argon, is, of course, a 'noble' gas and is chemically inert. It does not participate in chemical cycles because of its inertness. However, it is a surprisingly abundant element in the atmosphere and, even more interesting, the most common isotope of atmospheric argon is ^{40}Ar, which is not the natural isotope. Because it is so inert, we might assume that the argon in the atmosphere was part of Earth's primordial atmosphere and that it has not altered over the lifetime of the planet; this assumption turns out to be a false one. In fact, the vast bulk of atmospheric argon has been produced as a result of radioactive decay of potassium, ^{40}K. The potassium in the Earth's crust, and deeper in the mantle, has decayed over the lifetime of the Earth to form ^{40}Ar, which has degassed from the solid, explaining why most of the argon in the atmosphere is ^{40}Ar, rather than the naturally occurring isotopes ^{36}Ar and ^{38}Ar, which are present at much smaller mixing ratios (see also Section 3.2.2).

Carbon dioxide plays a central role in atmospheric chemistry since it is converted by photosynthesis in living organisms into oxygen. Because of life, CO_2 is a relatively small constituent of the atmosphere, although we still classify it as a major atmospheric gas. We have already seen that Mars and Venus have atmospheres composed mainly of CO_2, and established that the situation on Earth is very different. The CO_2 concentration prior to the Industrial Revolution was about 280 ppm, although since then concentrations have been increasing owing to human activity. The increase in atmospheric carbon is possibly the most serious impact that Man may have on the atmosphere. Carbon that was originally removed from the atmosphere by life and buried as fossil fuels is now being dug up and burned by mankind, releasing it back into the atmosphere.

Water vapour is another interesting constituent of the atmosphere. It is very variable in concentration, which depends on location on the planet, vapour pressure, temperature, whether it is condensed or not, and how dry the climate is. Of course, water vapour is very significant because it drives much of the physics of the atmosphere, including the evaporation–condensation–precipitation cycle mentioned earlier.

1.5 MINOR GASES

As well as the major gases considered in the last section, there are numerous minor gases present in the atmosphere, many of which play significant roles in the chemistry of the atmosphere. Several of these minor gases are inorganic and they include the other noble gases: neon, krypton, xenon and radon, which are present in much smaller quantities than the argon discussed in Section 1.4. One of the more surprising features of the atmosphere is how little helium it contains. Helium is one of the main constituents of the solar environment, and indeed the universe as a whole, and we might imagine that it would have been present in large quantities when the Earth was being formed. Indeed, helium probably constituted up to 10 per cent of the primordial atmosphere but, since helium is so light, it has mainly escaped Earth's gravity and disappeared into space. However, not all helium now present in the atmosphere came from the original solar environment; like argon, helium has a radiogenic source and is continually generated, albeit in minuscule quantities. Hydrogen, the only element lighter than helium, is also almost absent from the atmosphere. Like helium, some of it has escaped to space, but, unlike helium, it has also entered into combination with oxygen to form water. Almost all the elemental hydrogen found in the present-day atmosphere is that produced by photolysis of water, perhaps with a small contribution from volcanic emissions. We say that H_2 has a very short chemical lifetime as it cannot remain in the atmosphere for very long before it is oxidized. We will look at the subject of chemical (and physical) lifetimes in Chapter 3.

Other inorganic gases in the atmosphere include sulfur compounds such as hydrogen sulfide (H_2S), carbonyl sulfide (COS) and sulfur dioxide (SO_2), all of which are released from volcanoes as well as being generated by living organisms. Dimethyl sulfide (DMS, $(CH_3)_2S$) is an organic sulfur-carrying gas that is produced in large quantities by algae in the oceans.

Nitrogen compounds in the atmosphere are represented by the oxides of nitrogen (nitrous oxide, N_2O; nitric oxide, NO; and nitrogen dioxide, NO_2). Nitrous oxide is liberated from soils as a result of microbiological activity, in which atmospheric N_2 is 'fixed'. It is chemically significant in the atmosphere since it is a precursor of the higher oxides NO and NO_2. All three oxides, especially NO and NO_2, are additionally formed by high-temperature reactions between N_2 and O_2, for example in lightning discharges or combustion. A further oxide, N_2O_5, is the product of oxidation of NO and NO_2, and since it is the acid anhydride of HNO_3 it is a source of the nitric acid found in 'acid rain' (see Section 11.1.6). The only significant sources of ammonia in the contemporary atmosphere are biological, and although it makes up only a tiny proportion of the atmosphere (<0.01 parts per million), ammonia is very important since it is one of only a very few trace gases that are alkaline.

As mentioned earlier, ozone is a particularly significant and interesting atmospheric gas, even though it is only a minor constituent of the atmosphere. The distribution of ozone in the atmosphere is of special interest because, unlike the situation for most minor gases whose concentration is highest nearest the ground, much of the ozone in the atmosphere is to be found in a layer in the stratosphere. The mixing ratio typically reaches a peak value of nearly 10 ppm at an altitude of about 35 km. In terms of absolute concentrations, roughly 10 per cent of the total ozone is found in the troposphere, reaching a peak of more than 10^{12} molecule cm^{-3} at 8 km, but the mixing ratio is only about 0.15 ppm. These statements illustrate nicely the difference between mixing ratio and concentration (number density), since the total atmospheric density at 8 km is very approximately 40 times that at 35 km. Chapter 5 will consider ozone, its chemistry, and its importance to life in some detail.

As well as the inorganic gases, there are literally thousands of organic gases present in the atmosphere. These gases have an extensive range of types, and, like ammonia, many of them are generated primarily by living species. Methane is one of the most abundant minor gases in the atmosphere with a concentration of nearly 1.8 ppm in today's atmosphere. Ice-core samples suggest that this figure was lower, about 0.7 ppm, for 800 years prior to the Industrial Revolution and that the increase in methane is therefore attributable to modern industrialized humans. Wetlands, termites and the oceans are responsible for the emission of 200 million tonnes of CH_4 annually, but nearly two-thirds of the methane released in the modern world is generated by Man's activities, ranging from farming practices to the extraction and use of fossil fuels, which release as much as 360 million tonnes each year.

In addition to methane, there are other biogenic *volatile organic compounds* (VOCs) present in the atmosphere. These include saturated and unsaturated hydrocarbons and aliphatic and aromatic compounds, including oxygenated species (OVOCs). Plants provide the main natural source for many of these compounds, which mostly have relatively short chemical lifetimes (they generally persist in the atmosphere for hours or days only). One of the most significant groups of VOCs is the terpenes. Terpenes are isoprenoid structures, based upon isoprene C_5H_8) and will be discussed in detail in Section 8.2.7.

The major and minor constituents of the atmosphere are present there (i) because they were trapped along with the material that formed the Earth, or (ii) because they have been emitted subsequently by some source, or (iii) because they have been generated as the result of some chemical transformation. The species considered so far all have relatively long lifetimes in the atmosphere, varying from minutes to centuries, which means that they could all, in principle, be stored in a bottle. However, there are many other species present in the atmosphere that are much more highly reactive, and thus short-lived. Lifetimes of these species can be as small as femtoseconds (10^{-15} s), but more usually range from microseconds to seconds: still short-lived enough that, when they are studied in the laboratory, they must be generated *in-situ*. Included in the category are free atoms, free radicals, ions and excited species. We will consider many of these *reaction intermediates* in different parts of the book as they are most appropriately examined during discussion of the chemical transformation processes with which they are involved.

CHAPTER 2

Physics of the Atmosphere

As noted briefly in Chapter 1, the chemical behaviour of the atmosphere at a particular element of latitude, longitude and altitude depends in part on the pressures and temperatures enjoyed by that element. The chemistry and chemical composition can in turn influence those pressures and temperatures. What is more, atmospheric motions can transport the chemicals into or out of the element, and the transport can be vertical, zonal (longitudinal) or latitudinal. Pressure, temperature, transport and winds are often regarded as the domain of meteorologists and atmospheric physicists, but an atmospheric chemist must certainly possess a rudimentary understanding of the principles that determine the physical behaviour of the atmosphere. Chapter 2 of our book is intended to provide the most relevant information at a fairly elementary level.

2.1 PRESSURES

Pressure is the force exerted per unit area. At the surface of a perfectly smooth, spherical and homogeneous planet, the force is the weight of the entire atmosphere above, equivalent to the atmospheric mass, M_A, multiplied by the acceleration due to gravity at the surface, g_0; the surface area is equal to 4π times the square of the planet's radius. Written as an equation,

$$p_0 = \frac{M_A g_0}{4\pi R_p^2} \tag{2.1}$$

where p_0 is the mean atmospheric pressure and R_p the planetary radius. For Earth, one standard atmosphere is defined as 101 325 $\mathrm{N\,m^{-2}}$; the units of $\mathrm{N\,m^{-2}}$ are frequently replaced by the Pa (Pascal), to which they are exactly equivalent. The equation can be rearranged in order to calculate M_A from the pressure and radius which for Earth gives $M_A = 5.3\times10^{18}$ kg.

The rather ungainly value of 101 325 Pa for a standard atmosphere arises in order to use the SI system of units properly. It is often convenient just to refer to 1 atm for the standard atmospheric pressure. Atmospheric scientists in general, and meteorologists in particular, often prefer to retain the units of bar and millibar (mbar). The bar, and hence millibar, are exactly defined in terms of the Pascal (1 mbar $= 10^2$ Pa), but they are not part of the SI. One millibar is thus equal to one hecto-Pascal (hPa), a unit that is finding increasing favour. For conformity with existing practice, we shall

Atmospheric Chemistry
By Ann M. Holloway and Richard P. Wayne

Published by the Royal Society of Chemistry, www.rsc.org

use the millibar throughout most of this book. We note from the definition that 1 atm is very roughly equivalent to 1 bar, or more exactly to 1013.25 mbar.

Gaseous components in the atmosphere are compressible. Unlike liquids, they therefore do not settle down on the planetary surface under the influence of gravitational attraction because the translational kinetic energy of the particles competes with the sedimentation forces. As a result of the competition, the density of gas falls with increasing altitude in the atmosphere. In fact, the gas density could be calculated exactly from the Boltzmann equation were the temperature and the acceleration due to gravity constant throughout the atmosphere. In such a hypothetical atmosphere, at temperature T

$$p = p_0 \exp(-mgz/kT), \tag{2.2}$$

where p is the pressure at any altitude z, g the acceleration due to gravity, m is the mass of an atom or molecule of gas, and k is Boltzmann's constant. Numbers of particles are proportional to pressures for a fixed temperature, so that this form is equivalent to the Boltzmann distribution, with mgz corresponding to (geopotential) energy. In reality, T varies with z (and g decreases somewhat, although for a planetary atmosphere, the acceleration due to gravity, g, is nearly constant at $\sim g_0$, since the atmospheric thickness is much less than the planetary radius). Equation (2.2) must thus be replaced by the more complex form

$$p = p_0 \exp\left\{-\int_0^z \frac{\mathrm{d}z}{kT/mg}\right\}, \tag{2.3}$$

which is known as the *hydrostatic equation*. Equation (2.2) is just a special case of equation (2.3) for constant T and g. The quantity (kT/mg) has the units of length, and represents a characteristic distance over which the pressure drops by a factor 1/e. It is given the symbol H_s, and is called the *scale height*.

Temperature is not independent of altitude in a real atmosphere, so that the scale height is not constant. For the Earth's lower atmosphere, the scale height, H_s, varies between 6 km at $T \sim 210$ K to 8.5 km at $T \sim 290$ K. Figure 2.1 shows pressure as a function of altitude for a 'standard' Earth atmosphere (one that represents the horizontal and time-averaged structure of the atmosphere). Since the pressure axis is logarithmic, the simplified expression (2.2) would predict a straight line, and the observed deviations reflect temperature variations in the real atmosphere.

2.2 MIXING RATIOS, MIXING, AND MEAN FREE PATHS

The relative concentration of an atmospheric constituent on a molecule-for-molecule basis is called the *mixing ratio*. For an ideal gas—and at pressures of one atmosphere and below, the atmosphere can be regarded as approaching ideality—the mixing ratio is the same as the volume-for-volume proportion of the component, and is often expressed in parts per million by volume, *ppmv*, or per billion (*ppbv*, a fraction of 10^{-9}) or per trillion (*pptv*, 10^{-12}). The mixing ratio expressed as a fraction is thus identical to the chemist's *mole fraction*.

Equations (2.2) and (2.3) contain the term m, so that the scale height seems to depend on the molecular or molar mass. Each component of a planetary atmosphere would apparently have its own scale height, and the pressure distribution would be specific to that species. In that case, mixing ratios even of unreactive gases would be a function of altitude. Yet, at least in the Earth's lower atmosphere, observation shows the gas composition, in the absence of sources and sinks, to be constant (water is an exception, because it can condense). In fact, the lower atmosphere behaves as though it is composed of a single species of relative molar mass $\sim(0.2\times32)+(0.8\times28)=28.8$.

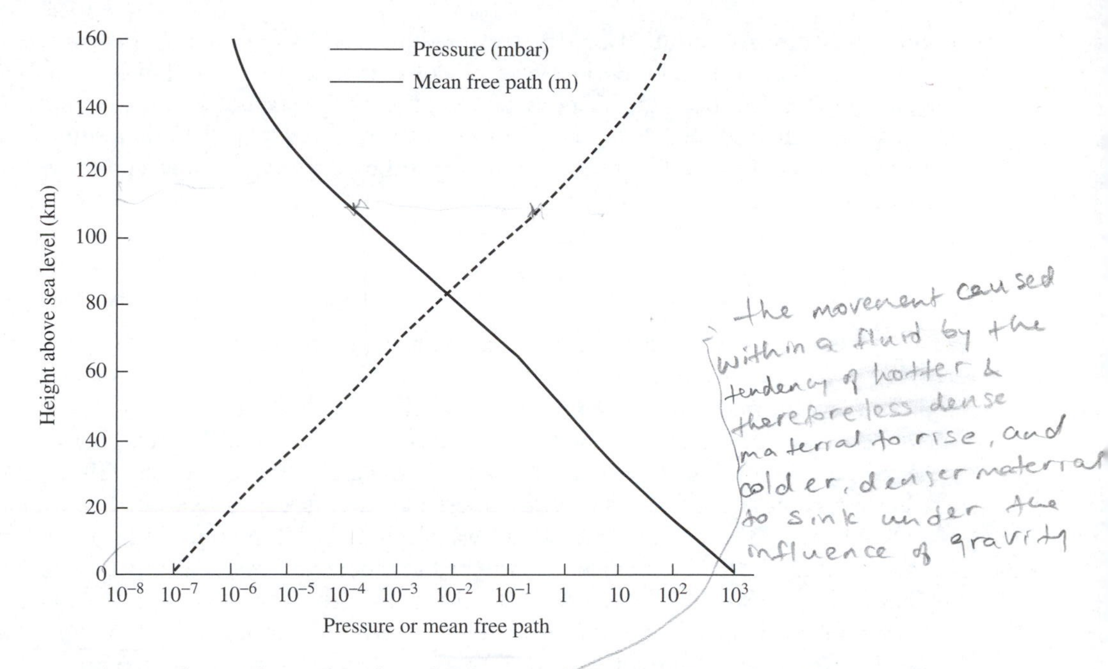

Figure 2.1 Pressure and mean free path as a function of altitude in the Earth's atmosphere. Redrawn from J.M. Wallace and P.V. Hobbs, *Atmospheric Science*, 2nd edn, Academic Press, London, 2006.

The homogeneity of lower atmospheres is a consequence of mixing due to fluid motions. Mixing on a macroscale, by convection, turbulence, or small eddies, does not discriminate according to molecular mass. It thus redistributes chemical species that gravitational attraction is trying to separate on a molecular scale by diffusion. The relative contributions of the molecular and bulk motions depend on the relative distances moved between transport events in each case. For molecular motions, that distance is clearly the *mean free path*, λ_m, the average distance a particle travels between collisions. The equivalent quantity for bulk fluid motions is the *mixing length*. A simple expression for mean free path in terms of pressure, p, and molecular diameter, d, can be derived from the kinetic theory of gases

$$\lambda_m = (1/\sqrt{2}\pi d^2)(kT/p). \tag{2.4}$$

The dotted line in Figure 2.1 shows mean free paths calculated for the Earth's atmosphere, and emphasizes the inverse pressure dependence suggested by equation (2.4). Fluid mixing lengths tend to decrease with increasing altitude, so that there is an altitude for which molecular diffusion and bulk mixing are of comparable importance. In the Earth's atmosphere, the mean free path and mixing lengths are both of the order of 0.1–1 m at $z = 100$–120 km. At higher altitudes, where the mean free path becomes larger than the mixing length, we may expect to see mass discrimination. Observations agree with these predictions. Above 100 km, molecular nitrogen begins to exceed its ground-level mixing ratio. Atomic fragments are favoured gravitationally (as well as chemically—see later) at high altitudes. Finally, at the highest altitudes, only the lightest species (H, H_2, He) are present.

2.3 TEMPERATURES

Figure 1.1 has already shown that the temperature structure of the Earth's atmosphere is by no means simple. Temperatures decrease at first with increasing altitude, and then begin to rise above about 15–20 km. At about 50 km, the temperature starts decreasing again, only to start increasing above about 90 km. In considering the temperatures to be found within the atmosphere, it is thus necessary to explain both the absolute temperature at the surface and the changes found as higher altitudes are probed. Since the temperature at any altitude will be determined by the balance between heating and cooling processes, it is useful to start by considering the two types of process separately.

2.3.1 Radiative Heating

Almost all the input of energy for the inner planets (Venus, Earth, Mars) is provided by solar heating. Radiation that reaches the planets is emitted from the Sun's *photosphere*. Physicists are fond of the term *black body* to denote an object that is a perfect emitter and absorber of radiation at all wavelengths. The photosphere behaves nearly as a black body of temperature ~5785 K in the spectral regions responsible for heat, visible light, and near ultraviolet (although there is much greater emission in the X-ray, far ultraviolet, and radio spectral regions than a black body would allow). The total amount of energy of all wavelengths intercepted in unit time by unit surface area at the top of the Earth's atmosphere, corrected to the Earth's mean distance from the Sun, is known as the *solar constant*. The degree of variability of this 'constant', which can now be investigated by satellites in orbit outside the atmosphere, is of great potential importance in assessing climatic changes. At present, the solar flux through a surface normal to the beam is approximately 1368 W m^{-2} near the Earth. Some of the radiation is reflected by the surface and by the atmosphere; the overall reflectivity, A, of a planet is called the *albedo*. For the Earth, the albedo is 0.31. The fraction of incident energy that is absorbed is thus $(1-A)=0.69$. A planet also radiates thermal energy itself. Let us make the simplifying assumption that the atmosphere does not itself absorb any radiation. An estimate of the effective or equilibrium temperature, T_e, of the planet can then be obtained by assuming that the planet is a black-body radiator. Radiation obeys the 'fourth power' (Stefan–Boltzmann) law, which states that the rate of radiation of energy per unit area (the *flux*) is proportional to T^4. If the overall temperature of the planet is to remain constant, then the inward and outward fluxes must be the same. A simple calculation based on balancing the absorbed solar flux against the Stefan–Boltzmann emission rate leads to a value of $T_e = 256$ K. For Earth, the real average surface temperature, T_s, is 288 K. This match might seem reasonable, but it should be remembered that a change of 32 K at the Earth's surface, up or down, would make the planet largely uninhabitable. That there is something missing in the calculations is clear when they are repeated for Venus, where surface temperatures (735 K) are more than 500 K higher than the calculated effective temperatures (227 K). Mars seems to behave more nearly as expected with $T_e = 217$ K and measured $T_s = 223$ K.

2.3.2 Radiation Trapping and the 'Greenhouse Effect'

Our calculation of T_e deliberately excluded any absorption of solar or planetary radiation by atmospheric gases, and it seems reasonable to look first at atmospheric absorption as a way of reconciling effective and surface temperatures. This idea is supported by the measured average infrared emission temperatures of the planets as seen from *outside* their atmospheres (230, 250, 220 K for Venus, Earth, and Mars), which are quite close to the calculated values of T_e. A more detailed look at the Earth's planetary emission gives a further indication of what is happening. Figure 2.2 shows a low-resolution spectrum obtained from a satellite in a cloud-free field of view, the dashed lines indicating the expected black-body radiance at different temperatures. Over some of the spectral region, the temperature corresponds to near-surface temperatures. However, there is a huge emission-temperature dip between 12 and 17 μm and smaller dips at 9.6 μm and at less

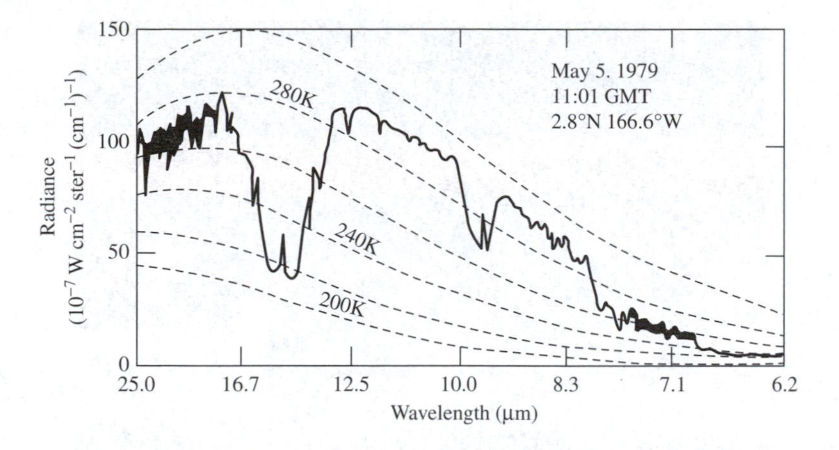

Figure 2.2 Spectrum of infrared emission escaping to space, as observed from outside the Earth's atmosphere by the Nimbus 4 satellite. Dashed lines represent the spectrum expected from a black body at different temperatures. Atmospheric absorbers cause the escaping radiation to come from different altitudes at different wavelengths, so that the effective temperature of the Earth's spectrum is wavelength dependent. Figure redrawn from R.E. Dickinson, in *Carbon Dioxide Review: 1982*, ed. W.C. Clark, Oxford University Press, Oxford, 1982 with permission from Oxford University Press.

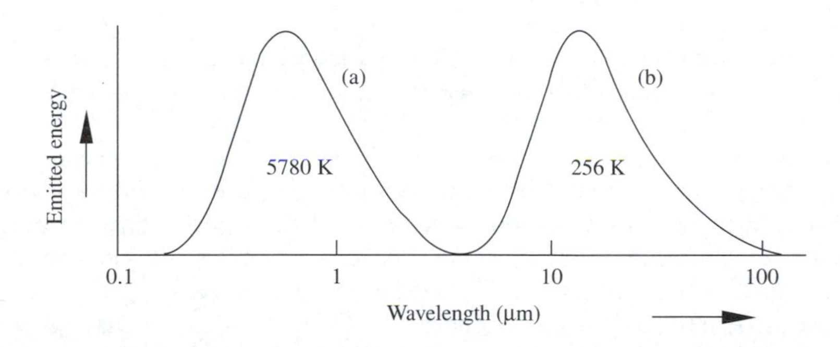

Figure 2.3 Curves of black-body emission plotted as a function of wavelength for (a) a temperature approximating to that of the Sun; and (b) the mean temperature of the Earth's atmosphere. The scales are relative, the absolute emission energy being much larger for curve (a) than for curve (b).

than 8 μm. The three spectral regions correspond to infrared active bands of CO_2, O_3, and H_2O, respectively, and the interpretation of emission temperatures is that where the atmospheric gases have a non-absorbing 'window', the satellite views the ground or layers near it. But at the wavelengths of absorption by atmospheric gases, the emission comes from higher, colder regions. Radiation from the Earth's surface has been *trapped* in the spectral regions of the absorption bands, and ultimately reradiated to space at lower temperatures than those of the surface. This *radiation trapping* is another key factor in determining atmospheric temperatures. Averaged over all wavelengths, the emitters are evidently about 32 K colder than the surface.

As we shall see soon, a temperature of 256 K is reached at about 6 km altitude, so that in this simplified picture one can visualize an equivalent radiating shell of the Earth lying 6 km above the surface. Why is the radiating shell not also the absorbing shell? The answer lies in the spectral distribution of black bodies at different temperatures. Figure 2.3 shows the Planck black-body curves for 5780 K (the Sun's photospheric temperature) and 256 K (T_e for Earth). The bulk of the Sun's radiation lies at wavelengths where the atmosphere absorbs only weakly. In contrast,

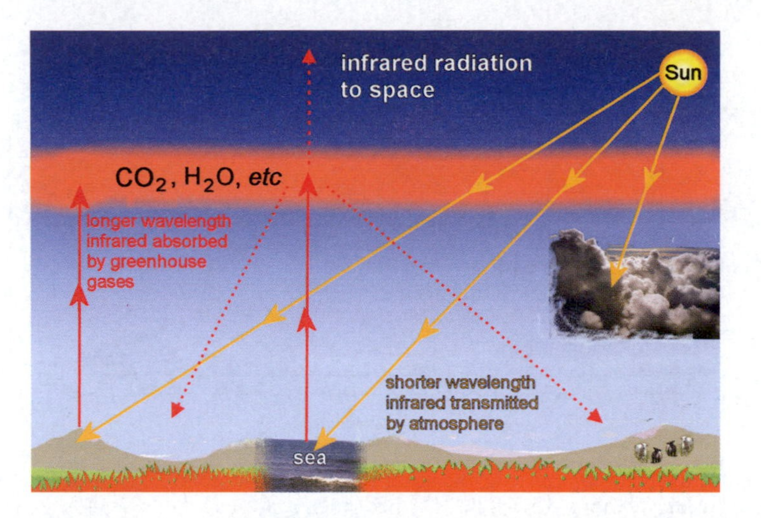

Figure 2.4 Radiation trapping or 'greenhouse' heating. Incoming solar infrared radiation (yellow line) passes through the atmosphere to warm the land and the oceans. The Earth emits radiation to balance the input, but at much longer wavelengths (red lines) that are absorbed by 'greenhouse gases' (GHGs) such as CO_2 and H_2O present in the atmosphere. This trapping of radiation means that the lower atmosphere acts as a blanket that keeps the surface warmer than it would otherwise be. N_2O, CH_4 and many other species are also GHGs.

throughout most of the wavelength region emitted by the low-temperature Earth, the atmosphere is opaque. The atmosphere lets shorter-wavelength radiation in, but does not let the longer-wavelength radiation out, as illustrated schematically in Figure 2.4.

By supposed analogy with the behaviour of panes of glass, the effect is often called the *greenhouse effect*. (In reality, greenhouses are effective almost entirely because they inhibit convection rather than because they trap radiation.) Absorption and emission for a particular transition occur at the same wavelengths, of course, so that one molecule can absorb radiation emitted by another molecule of the same chemical compound. Radiation is therefore passed back and forth between the infrared-active molecules, and escapes to space only from layers in the atmosphere high enough for the absorption to have become weak. The upper layers of the atmosphere suffer a net loss of energy that is relatively more than that for the lower ones, and they are therefore cooler.

The rates of upward and downward emission and absorption of radiation in a thin slice of the atmosphere can be calculated at any chosen altitude in the atmosphere. At a certain temperature, radiative equilibrium can be maintained. By calculating these temperatures for a large number of such slices stacked on top of each other, the altitude dependence of temperature can be obtained. Figure 2.5(a) shows the results of carrying out this procedure for a highly simplified and crude physical model of the atmosphere. Even at this limited degree of sophistication, the results bear a reasonable relation to what is actually observed. In comparison with the real atmosphere, the simple model predicts too steep a fall in temperature in the lowest few kilometres of the atmosphere, although it subsequently matches quite well for the next few kilometres. Convection in the lowest region reduces the *lapse rate* (rate of decrease of temperature with altitude), as we shall see in Section 2.3.3.

Several gases present in the atmosphere can contribute to radiation trapping. Each of these species is likely to be present with a different mixing ratio and possibly different altitude distribution. Each will possess spectral features (vibrational, rotational, isotopic, *etc.*), corresponding to its identity, and each feature will absorb or emit with a specific intensity. Spectral lines have widths determined by collisional (pressure-dependent) and Doppler (temperature-dependent) effects. Light scattering by molecules, and especially by particles, alters the optical behaviour of the

Table 2.1 Contribution of absorbers to atmospheric thermal trapping.

Species removed	*Percentage trapped radiation remaining*
None	100
O_3	97
CO_2	88
clouds	86
H_2O	64
H_2O, CO_2, O_3	50
H_2O, O_3, clouds	36
All	0

Data of V. Ramanathan and J.A Coakley, *Rev. Geophys. & Space Phys.* 1978, **16**, 465.

atmosphere. A proper study of radiation trapping must therefore accommodate at least these factors. The advent of high-speed computers with large memories has made it possible to investigate radiative transfer and trapping by direct *numerical modelling* (see Section 4.4).

Partly because the infrared bands of the various components overlap, the contributions of the individual absorbers do not add linearly. Table 2.1 shows the percentage of trapping that would remain if particular absorbers were removed from the atmosphere. We see that the clouds only contribute 14 per cent to the trapping with all other species present, but would trap 50 per cent if the other absorbers were removed. Carbon dioxide adds 12 per cent to the trapping of the present atmosphere: that is, it is a less important trapping agent than water vapour or clouds. On the other hand, on its own CO_2 would trap three times as much as it actually does in the Earth's atmosphere. The point is of importance in seeing how far increases in the greenhouse effect could provoke climatic response to changed carbon dioxide concentrations (Section 11.4). In this context, it is interesting to note that the upper layers of the atmosphere leak relatively more radiation to space than they trap. Consequently, increased atmospheric carbon dioxide leads to atmospheric cooling rather than warming for atmospheric layers above about 20 km. Many of the trace atmospheric gases, such as CH_4, N_2O, and O_3, have infrared modes active in the trapping region. These gases may therefore have a direct effect on global temperatures quite distinct from that exercised through their possible modification of concentration of major absorbers. The influence of minor constituents is particularly marked if they absorb where there is otherwise an atmospheric window.

2.3.3 Troposphere, Stratosphere, and Mesosphere

Radiative transfer in the atmosphere tends to produce the highest temperatures at the lowest altitudes, as illustrated by the model calculation represented in Figure 2.5(a). At first sight, it might seem inevitable that convection would arise in this situation, since the hot, lighter, air initially lies under the cold, heavier air. In an atmosphere, the behaviour is somewhat more complex because gases are compressible, and pressures decrease with increasing altitude. A rising air parcel therefore expands, does work against the surrounding atmosphere, and is somewhat cooled. It can be that the temperature drop resulting from expansion would exceed the decrease in temperature of the surrounding atmosphere: in that case convection will not occur.

Simple thermodynamic arguments allow us to calculate the temperature decrease with altitude that is expected from the work of expansion. We imagine a packet of dry gas that is in pressure equilibrium with its surroundings, but thermally isolated from them. This packet of gas is allowed to move up or down in the atmosphere, and the rate of change of temperature with altitude (lapse

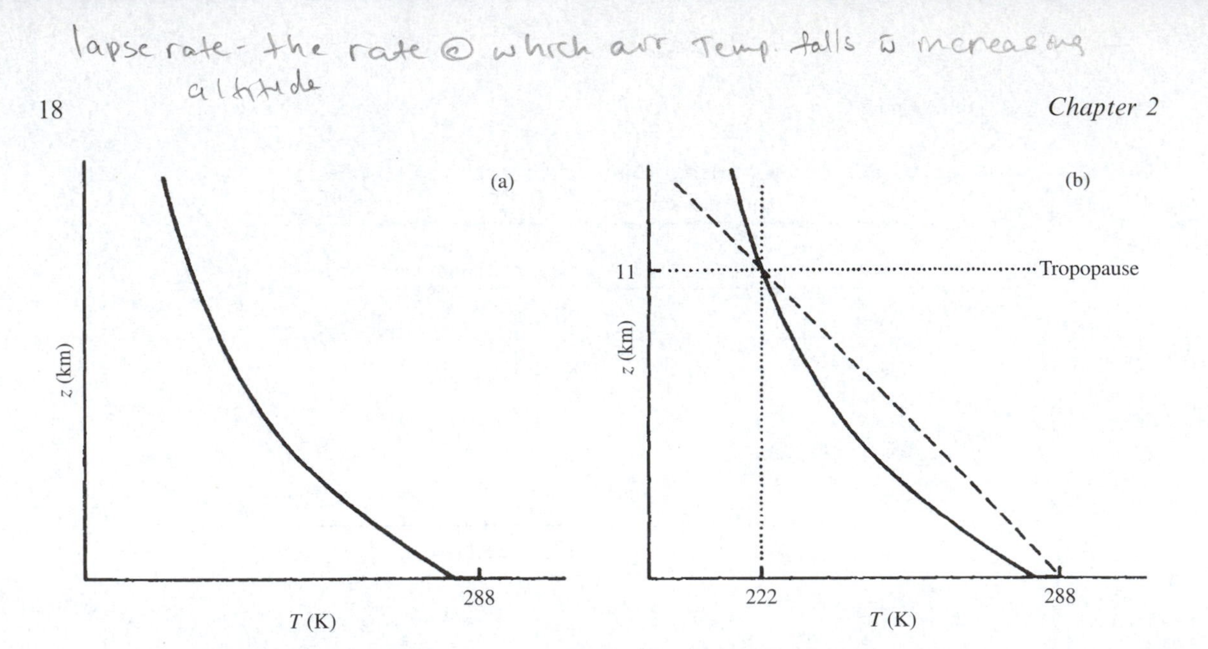

Figure 2.5 (a) Radiative equilibrium temperature T as a function of altitude, z. There is a temperature discontinuity at the lower boundary. The altitude scale is arbitrary at this stage. (b) Now a real altitude scale is provided. A dry adiabatic lapse line (~ 6 K km^{-1}) has been added to the radiative transfer model results of (a). The atmosphere is convectively unstable below ~ 11 km.

rate) is calculated. Since no heat flows (adiabatic conditions) and the gas is dry, the temperature profile, $-\mathrm{d}T/\mathrm{d}z$, calculated is the *dry adiabatic lapse rate*, which is given the symbol Γ_d. The result is

$$\Gamma_\mathrm{d} = \frac{g}{c_\mathrm{p}} = -\frac{\mathrm{d}T}{\mathrm{d}z} \tag{2.5}$$

where c_p is the heat capacity per unit mass of the gas at constant pressure. The dry adiabatic lapse rate, Γ_d, thus depends only on the acceleration due to gravity, g, on a planet, and the average heat capacity per unit mass of the atmospheric gases. For Earth, the calculated value of Γ_d is 9.8 K km^{-1}. This information now allows us to determine whether any particular dry atmosphere is stable or unstable with respect to convection. If the actual temperature gradient in the atmosphere, $-(\mathrm{d}T/\mathrm{d}z)_\mathrm{atm}$, is less than Γ_d, then any attempt of an air packet to rise is counteracted by cooling due to expansion that makes it colder and more dense than its surroundings. The atmosphere is stable. Conversely, if there were a tendency for $-(\mathrm{d}T/\mathrm{d}z)_\mathrm{atm}$ to be greater than Γ_d, convection would be set up. Such convection will restore the temperature gradient until the atmosphere is stable again, so that actual atmospheric lapse rates rarely exceed Γ_d by more than a very small amount.

The presence of condensable vapours in the atmospheric gases complicates matters. Condensation to liquid or solid releases latent heat to our hypothetical air parcel. For a saturated vapour, every decrease in temperature is accompanied by additional condensation. Qualitatively, it is obvious that the *saturated adiabatic lapse rate*, Γ_s, must be smaller than Γ_d. In the Earth's atmosphere, where the important condensable vapour is water, Γ_s varies from about 4 K km^{-1} near the ground at 25 °C to 6–7 K km^{-1} at around 6 km and -5 °C. The stability conditions in saturated atmospheres are the same as in unsaturated ones, but with Γ_s replacing Γ_d. Partial saturation is treated in two stages, with Γ_d being applicable until temperatures are low enough to cause condensation, and Γ_s thereafter. In these circumstances, a situation known as *conditional instability* can arise. If the atmospheric lapse rate is less than Γ_d but more than Γ_s, the atmosphere is stable unless some kind of forced lifting (*e.g.* by winds) raises gas to an altitude where condensation occurs. From this point on, the atmosphere is unstable.

We now have enough information to interpret, in broad terms, some further features of atmospheric structure. Figure 2.5(a) shows the temperatures in the Earth's atmosphere predicted by the very simple radiative transfer model. More sophisticated models show similar trends. At low altitudes, the negative slope of the temperature–altitude relationship greatly exceeds even the dry adiabatic lapse rate. Convective instability therefore mixes the atmosphere until the adiabatic lapse rate is nearly reached. In fact, the average lapse rate on Earth (resulting from an appropriate combination of Γ_d and Γ_s, as well as global circulation) is 6.5 K km^{-1}. If we draw a line of this slope together with the radiative transfer predictions on the same diagram, as in Figure 2.5(b), we see that the two lines intersect at $\sim$11 km. Below that altitude, convection keeps even the largest atmospheric lapse rate only marginally more than the adiabatic lapse rate. Above, however, the atmosphere is stable, and the atmospheric temperatures are, to a first approximation, determined by radiative transfer for that part of the atmosphere where trapping rather than escape of radiation occurs. The critical level corresponds to the tropopause, while below is the turning, turbulent troposphere and above the layered, stable stratosphere, as described without explanation in Chapter 1.

In reality, temperatures begin to rise again in the stratosphere as a consequence of the presence of ozone (Section 1.5), thus conferring additional stability. The temperature–altitude profile for Earth has already been shown pictorially in Figure 1.1. The troposphere and tropopause are evident enough, but beyond the tropopause the atmospheric temperatures only follow the radiative lapse rate in a very small region, and then they start to *increase* again in the stratosphere. Regions of negative lapse rate are said to constitute *inversions*, with hotter air on top of cooler. Obviously, inversion layers are peculiarly stable against vertical motions. Inversions can sometimes arise near the ground because of particular meteorological and geographical conditions. Such inversions can trap pollutants and prevent their dissipation, as we shall see in Chapter 11. The stratospheric inversion is a result of heating by absorption of solar ultraviolet and visible radiation in the ozone layer. Ozone is formed photochemically from O_2 and the 'layer' structure owes its existence to a peak in absorption and in reaction rates (Section 5.2.2). Too low in the atmosphere, there are insufficient short-wavelength photons left to dissociate much O_2, while too high, there are insufficient O_2 molecules to absorb much light and to associate with O atoms to make O_3. A series of chemical reactions concerned in the formation and destruction of ozone ultimately releases the chemical energy of O_2 dissociation, while the solar radiation absorbed by ozone itself is also liberated as heat. As a consequence, the heating in the stratosphere is related to the ozone concentration profile (although modified according to the exact mechanism of conversion of absorbed radiation to heat energy). It is noteworthy that the ozone layer absorbs wavelengths from the Sun that do not reach the Earth's surface, and that the heating is achieved *in situ* (except during the long polar nights) rather than by absorption of reradiated infrared. As well as its absorption bands in the ultraviolet region, ozone also absorbs in the visible region: although the absorption is weaker, the solar radiation is more intense, and the heating is comparable to that produced by the ultraviolet absorption. Figure 1.1 shows that at altitudes above about 50 km, the heating effect is too weak to compete with the cooling processes, and temperatures decrease again. Conventional nomenclature ascribes the name *mesosphere* to this next atmospheric region, and *stratopause* to the upper boundary of the stratosphere. The lapse rate is subadiabatic in the mesosphere, so that the mesosphere is stable with respect to convective motions. The lowest temperatures in the entire atmosphere of the Earth are found in the mesosphere, and are at a minimum over the summer polar regions (see Chapter 10). Finally, the temperature stops falling at the *mesopause*.

2.3.4 Beyond the Mesosphere

All atmospheres eventually become so thin that collisions between gaseous species are very infrequent. One consequence may be that energy is not equilibrated between the available degrees of freedom. Translational temperatures may be much higher than vibrational or rotational temperatures, and atomic species with high kinetic energy may not be sharing that energy with molecular species. Since the loss of energy to space from atmospheric species depends on radiation from infrared-active molecular vibrations, it follows that inefficient re-equilibration between energy modes will lead to a bottleneck for energy loss and excess translational temperatures. The region of a planetary atmosphere that shows, from these causes, increasing temperature with altitude is called the *thermosphere*. For Earth, the residual cooling mechanisms are probably infrared emission from transitions between the angular momentum levels of atomic oxygen, together with some emission from nitric oxide. The cooling efficiency is low, and thermospheric temperatures are correspondingly high. It is worth emphasizing that these high temperatures do not reflect a large energy source, but rather the extreme thinness of the atmosphere and its inefficiency at disposing of energy by radiative transfer. Thermal inertia is very small, so that there is a big diurnal temperature variation. At $z = 250$ km, translational temperatures are typically 850 K at night and 1100 K during the day; but at these altitudes the energetic particles that constitute the solar wind interact with the atmosphere, so that temperatures also depend on whether the Sun is 'quiet' or 'active'. We return in Section 3.3.4 to the question of *escape* of species from the uppermost parts of the atmosphere.

2.4 TRANSPORT IN THE ATMOSPHERE

Movement of the air leads to the mixing and redistribution of the chemical components of the atmosphere up and down, east and west, and north and south. Atmospheric motions in the vertical, longitudinal and latitudinal axes thus have a direct influence on atmospheric chemistry.

2.4.1 Vertical Transport

Vertical mixing and redistribution of atmospheric constituents is assured in the troposphere because of the turbulence that characterizes the region. Above the tropopause, however, where lapse rates are subadiabatic (or even negative, as in some regions of the Earth's atmosphere), how, and how fast, are chemical species transported in the vertical plane?

In Section 3.3.4, we consider the escape of hydrogen from our atmosphere. This escape is kinetically limited by the rate of diffusive motion through the upper atmosphere. Diffusion in one form or another does indeed provide a mechanism for transport through a non-turbulent atmosphere. The type of diffusion most familiar to chemists is *molecular diffusion*, in which molecules (or atoms) move in response to a concentration gradient, and in such a direction as to try to remove the gradient. At the level of individual particles, random molecular motion is destroying a special arrangement (higher concentration, and thus more molecules, in one volume element than in another). Thermodynamically, the system is behaving so as to maximize the entropy in accordance with the Second Law. Similar remarks apply to the velocities of particles as to concentrations. A system in which molecules with high velocity are separated from those with low velocity possesses a temperature gradient, and the statistical re-randomization corresponds to a heat flow, or thermal conduction, in response to that gradient. If the 'hot' molecules are also *chemically* different from the surroundings, as they might well be in an atmosphere, then thermal conduction also corresponds to an identifiable redistribution of matter, and the process is called *thermal diffusion*.

Molecular diffusion, and where appropriate thermal diffusion, are able to account for the rates of vertical transport in the upper mesosphere and thermosphere of the Earth's atmosphere. A number of indications suggest, however, that vertical mixing in the stratosphere is much more rapid than can be explained by simple diffusion mechanisms, even though the rates are also much less than those operating in the troposphere. Methods for investigating vertical transport in the stratosphere include measurements from satellites or balloons of the vertical concentration profiles of various minor constituents (*e.g.* CH_4 or N_2O). Observations of tracer species injected by volcanic eruptions or by nuclear-weapon testing have also been used.

Full three-dimensional models (see Section 4.4) of atmospheric behaviour include advection of chemical species within the atmosphere. Before such models were available, the concept of *eddy diffusion* was introduced. Although rather discredited as an artificial method of treating the mixing, it is instructive to examine the idea in the context of ordinary diffusion. Instead of individual molecules moving independently, small packets of gas are envisaged as executing eddy motions that it is assumed can still be treated by the methods of the kinetic theory of gases. The diffusion rates can be treated with exactly the same mathematical formulation as molecular diffusion, except that an *eddy diffusion coefficient*, K_z, with a value some 10^4–10^5 times larger than molecular diffusion coefficients, is adopted. Values of K_z are based on observations of a minor constituent whose chemistry is presumed to be well known. Chemical tracers (*e.g.* CH_4, N_2O) and radionucleides from past atmospheric nuclear tests (^{14}C, ^{90}Sr, ^{95}Zr, ^{185}W) have been used to provide source data. Simple calculations suggest that a given 'tagged' molecule might then be expected to move 1 km vertically in a period measured in hours for the eddy mechanism, but in years for the molecular mechanism.

Transfer of chemical constituents in both directions across the tropopause is clearly of great importance. Many of the minor reactants in the stratosphere are initially released, naturally or anthropogenically, to the troposphere. Downward transport may be an important stratospheric loss mechanism for those species or their products, as well as a source of ozone to drive tropospheric chemistry. For about fifty years, it has been supposed that the stratosphere contains very little water vapour (a few parts per million) because gases entered the stratosphere through a region cold enough to trap out the water. The requisite cold-trap temperature is in the region of 190 K, and the proposal is that troposphere-to-stratosphere exchange takes place in the tropics, which is where the lowest tropopause temperatures are experienced. These tropical regions are, as we shall see in Section 2.4.2, associated with rapidly rising moist tropospheric air masses. Mixing ratios of H_2O in the tropical stratosphere vary seasonally as the tropopause temperature passes through its annual cycle. Layers of air that have risen from the troposphere possess a water-vapour content determined by the tropopause temperature. Horizontal transport out of the tropical stratosphere is rather slow, so that this signature is retained for months or even years. Alternating layers of high and low water content reflect the annually varying saturated vapour pressure at the tropical tropopause, and these layers slowly move upwards within the stratosphere at velocities ranging from 0.2 to 0.4 mm s^{-1}. The system has been likened to an 'atmospheric tape recorder'; the air is 'marked' as it enters the stratosphere like a signal recorded on an upwards moving magnetic tape.

One or more of several mechanisms may be responsible for the transfer. Trace constituents, and water vapour in particular, could be transferred either by a steady rising motion or in violent tropical cumulonimbus cloud convection. There is also known to be an exchange of stratospheric and tropospheric air at middle and high latitudes as a result of *tropopause folding*. Thin ($\sim$1 km) laminar intrusions of stratospheric air enter the troposphere for perhaps 1000 km parallel to the tropospheric jet stream (Section 2.4.2), and then become mixed with the turbulent tropospheric air. Perhaps more generally important than these mechanisms is *wave-driven pumping* (see Section 2.4.2) within the stratosphere at mid-latitudes. This extratropical pumping leads to steady large-scale ascent of air in the tropics, and large-scale descent near the poles. The memory effect referred to in the previous paragraph may explain observations that the lowest water-vapour concentrations are found at altitudes several kilometres above the tropopause, since minima from the seasonal H_2O cycle could be propagated vertically. Another explanation that has been advanced is that cumulonimbus convection sometimes overshoots the tropopause by several kilometres.

Figure 2.6 is an attempt to represent one view of transport characteristics in the lower stratosphere. Above an altitude of about 16 km, the inner vortex region in both hemispheres is isolated from mid-latitudes. There is greater exchange between the *edge* of the vortex region and mid-latitudes, and also at lower altitudes. In the 'surf zone', air is drawn out of the polar vortex or the

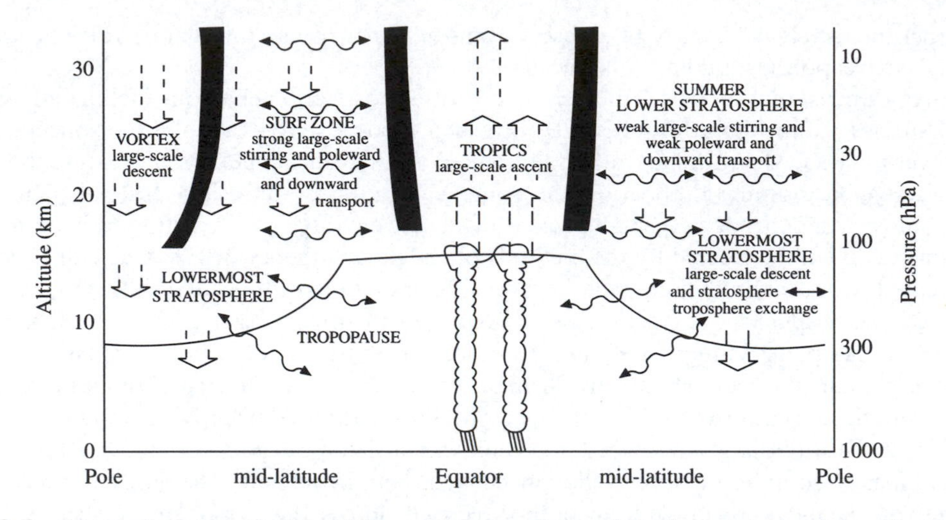

Figure 2.6 Schematic representation of the principal regions of the lower stratosphere, illustrating the several different transport characteristics. The diabatic circulation is represented by broad arrows, while the wavy arrows are indicative of a motion akin to stirring (along surfaces of constant entropy). The thick solid line is the tropopause. From *Scientific Assessment of Stratospheric Ozone: 1998,* World Meteorological Organization, Geneva, 1999.

tropics, and is stirred by the large-scale flow. This behaviour leads to the formation of sloping sheet-like structures that are observed in measurements of the distributions of chemical tracers. Horizontal cross sections through such sheets show filamentary structures, while vertical profiles demonstrate the presence of laminae. The filaments of polar or tropical air appear to survive for 20–25 days before they become mixed with their surroundings.

Large-amplitude planetary-scale waves in the stratosphere can make a large contribution to the transport and variability of inert and photochemically active species. A dynamical phenomenon of considerable interest in the Earth's upper atmosphere is the *sudden stratospheric warming* (SSW) that occurs about once every one to three years in the Northern Hemisphere winter. A large growth in wave amplitude occurs over a two-week period, and the normal cyclonic polar vortex then becomes highly distorted. There are strong poleward fluxes of heat, and temperatures at 40 km altitude can increase by as much as 80 K at high latitudes. This heating may be sufficient to reverse the usual latitudinal temperature gradients and winds, the normal westerlies temporarily changing to easterlies. Sudden stratospheric warmings play an important role in the budgets of trace species, as well as of heat, momentum, and energy in the stratosphere.

2.4.2 Winds

Horizontal motions of the atmosphere are central to the interests of meteorology. From the viewpoint of atmospheric chemistry, winds are responsible for the transport and mixing of chemical constituents. We shall examine in this section the barest outline of wind systems.

At and near ground level, winds are very variable and turbulent, gusting and changing direction frequently. Away from surface features and friction, the motions are more regular, and averaged over many days or weeks a clear and reproducible pattern emerges. The term *general circulation* is used to describe the global average winds that we need to discuss. Regular wind regimes have been recognized since the early days of sailing ships, with the tropical 'trade' winds in the Atlantic and Pacific (north-east in the Northern Hemisphere, south-east in the Southern) being very important. Even today, aircraft flying the transatlantic routes at mid-latitudes take longer in the east–west direction than on the return journey because of the prevailing high-speed westerly (from the west) *jet stream* winds at aircraft altitudes near the top of the troposphere.

Winds may be regarded as the air flow that is a response to pressure differentials between different locations on a planet. Temperature differences may be the cause of the pressure variations, as we shall see shortly. Thus, the winds, by transporting heat (in both sensible and latent forms), tend to remove those temperature differences, in accordance with thermodynamic expectations. Equator-to-pole temperature differences on Earth are certainly smaller than would exist in the absence of atmospheric motions, although the oceans also play a large role.

For Earth, we first consider the basic observations of circulation patterns, and turn later to possible explanations of them. On a global scale, hot air rises near the equator and sinks at higher latitudes. Each hemisphere has its own circulation cell—named, after its discoverer, a *Hadley cell*—and they converge near the equator in the *intertropical convergence zone* (ITCZ). Until recently, it was thought that the tropical trade winds were separated by a region of calm winds called the 'doldrums', but it now seems that the transition occurs a few degrees away from the equator. The ITCZ migrates with the Sun, being north of the equator in the northern summer, and south in the southern summer, its exact position being modified by the distribution of sea and land masses. The very narrow ITCZ belt is characterized by very strong upward motion and heavy rainfall. The returning surface air becomes saturated with water as it passes over the oceans. As it rises in the ITCZ, water condenses and heavy precipitation occurs. Release of latent heat by condensation increases the convective instability, reduces the lapse rate, and hence increases the driving pressure differential.

Hadley circulation might well consist of two cells each encompassing a hemisphere from equator to pole, and with the flows directly north–south, were it not for the rotation of the planet. For Earth, the real Hadley circulation only extends within the tropics (up to the 'horse latitudes'), and it has a strong westerly component aloft, and a corresponding easterly component on the return surface flow. Both the westerly directional component and the limited span of the Hadley cell are a consequence of planetary rotation. Atmospheric gases possess mass, and if they rotate more or less with the planet they therefore possess angular momentum. North–south motions imply a change in radius of rotation, decreasing to near zero at the poles. Yet angular momentum must be conserved, and the atmosphere achieves this conservation by developing zonal motion (that is, in the direction of rotation of the Earth). For example, an air parcel at rest with respect to the equator must develop a zonal (west to east) velocity by the time it reaches 30°N of $134\,\mathrm{m\,s^{-1}}$. The hypothetical force producing this motion perpendicular to the initial direction of transport is called the *Coriolis force*. The horizontal component of the Coriolis force is directed perpendicular to the horizontal velocity vector: to the right in the Northern Hemisphere and to the left in the Southern Hemisphere. The force is minimum at the equator and maximum at the poles. Trade winds on Earth clearly fit the requirements of angular-momentum conservation. The winds aloft, from equator to poles, gain a component from west to east in both hemispheres; the returning surface trade winds are north-easterly (from the north-east) and south-easterly, north and south of the equator. At mid and high latitudes, the Coriolis force increases and winds flow mostly in an east–west direction.

We noted in Section 2.4.1 that near the equator air lifts, and that it subsequently sinks at higher latitudes. What drives this process? The conventional explanation has been in terms of pressure differentials. Our experience on Earth tells us that surface pressures do not vary dramatically from one place to another. The same is not true aloft in the atmosphere, and the hydrostatic equation (2.3) indicates that pressures fall off more rapidly with altitude when the temperatures are low. If one location experiences a higher surface temperature than another, then, for similar lapse rates, the temperatures higher in the troposphere will bear the same relationship, and the pressure aloft over the warmer area will be higher. The consequent force will accelerate the air mass until kinetic energy losses (*e.g.* by friction) are matched. The concept then is that hot air rises near the equator and is forced towards the cooler poles by the higher pressures in the higher-temperature regions.

However, there is increasing evidence for the flows being generated by a 'pump' driven by atmospheric waves. Several kinds of *atmospheric oscillations* or wave motions have been identified. *Rossby waves*, or *planetary waves*, owe their existence to the variation of the Coriolis force with latitude. Such waves can cause air parcels to oscillate back and forth about their equilibrium latitude, and a pumping mechanism can be envisaged that will lead to mass transport of the kind seen in the Hadley cells.

Fluid-dynamic instabilities produce a different wave-like progression of low- and high-pressure regions (*cyclones* and *anticyclones*) around middle latitude areas of the Earth. Disturbances of this kind can be both reproduced in the laboratory in experiments with heated rotating fluids, and predicted by numerical models. The instabilities arise when latitudinal temperature differentials become too large; they are called *baroclinic instabilities*. Eddy motions (*baroclinic waves*), resulting from the instability, transport heat both poleward and vertically upward (thus reducing the tropospheric lapse rate). They also transport momentum to the upper troposphere and thus maintain the high-velocity jet stream. Four to six pairs of waves typically encircle the Earth at any one time.

Winds on the planet seem, then, to be explicable largely in terms of a thermally driven circulation modified by Coriolis forces and baroclinic instabilities. Additional features, such as stationary eddies (resulting from topographical or temperature contrast) and thermal tides also influence transport of heat and momentum, for example to the stratosphere, to produce the observed general circulation at all altitudes.

CHAPTER 3

Sources and Sinks of Atmospheric Species

3.1 THE LIFE CYCLE OF ATMOSPHERIC GASES

Our first two chapters considered the composition of the atmosphere and the way in which it behaves as a physical system. We now turn to the question of how the composition of the Earth's atmosphere is maintained. Almost all the gases and aerosols present in the atmosphere are removed from it by a variety of chemical and physical processes, which represent *sinks* for the species. Concentrations of some components of the atmosphere undergo changes between day and night, or between summer and winter, but averaged over a year the concentrations of most of the species do not vary wildly (but see Section 3.4). This behaviour provides evidence, if it is needed, that there must be *sources* of the individual species and that they more-or-less balance the sinks. In between birth at the source and death at the sink, the atom, molecule, particle or droplet is 'alive' in the atmosphere, and the duration for which it maintains its identity is the *lifetime* of the species, discussed in Section 3.5. As with living creatures, the atmospheric constituents do not necessarily die where they are born! In between release to and removal from the atmosphere, a species may be transported horizontally and vertically. The sequence is thus one of source–transport–sink, and the sink may be chemical or physical, or a combination of the two.

Figure 3.1 encapsulates these ideas, and indeed summarizes the essence of atmospheric chemistry. As we shall see in Section 3.2.2, ruminant animals such as cows really do emit large amounts of methane (from both ends!). The chemical processes in the atmosphere oxidize the CH_4 to CO_2 and H_2O. Several reaction intermediates are indicated in the atmosphere; those in red are free radicals and those in blue are 'ordinary' (closed-shell) molecules. Free radicals with unpaired electrons are exceptionally important in the chemical changes that occur in the atmosphere. Examples of such radicals that we shall meet time and time again are the hydroxyl radical (OH), the hydroperoxyl radical (HO_2) and the nitrate radical (NO_3). Atoms of elements that are diatomic in their natural form (*e.g.* H and O from H_2 and O_2) behave in a similar way. Radicals and atoms are generally highly reactive, which is why they are so important, but that also means that their concentrations are very small because they are removed so rapidly.

Photosynthesis in plants converts CO_2 and H_2O to carbohydrate and oxygen (Section 6.2), thus providing the food for the cow to eat. The concepts of sources, sinks and chemistry are neatly illustrated in this way, while at the same time a further concept, that of *cyclic* processes (Chapter 6), has appeared.

Atmospheric Chemistry
By Ann M. Holloway and Richard P. Wayne

Published by the Royal Society of Chemistry, www.rsc.org

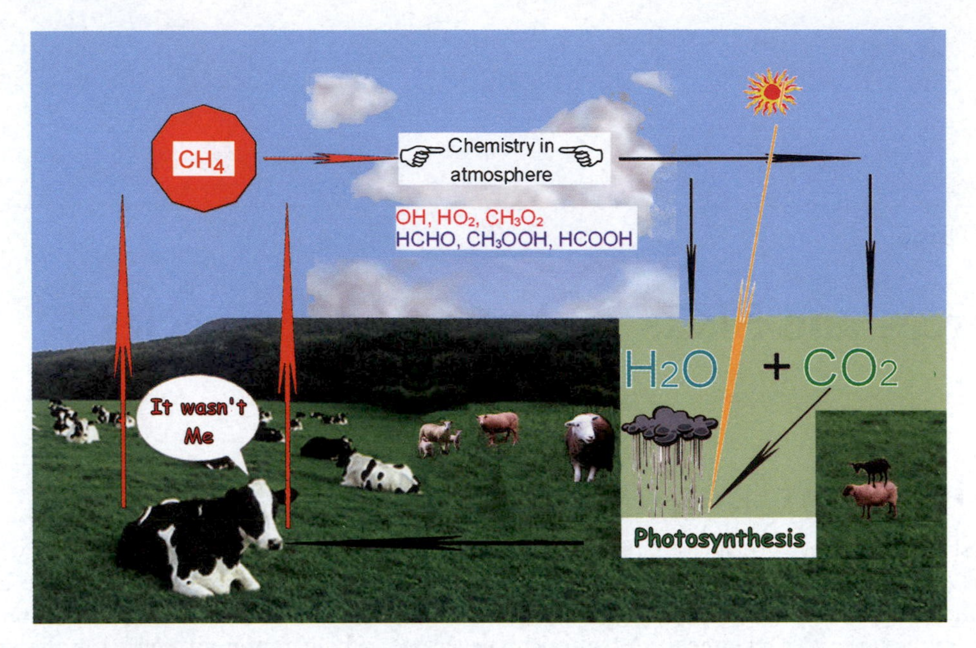

Figure 3.1 Atmospheric chemistry exemplified by the oxidation of methane from a ruminant animal. Despite the cow's protestations to the contrary, it has released significant quantities of methane to the atmosphere. The methane was generated by enteric fermentation of grass in the cow's stomachs. Once in the atmosphere, the methane is oxidized in a sequence of steps in which free radicals figure as important intermediates, and the end products are CO_2 and H_2O. The H_2O becomes rain, and it and the CO_2 are converted by photosynthesis in the grass to carbohydrate. The cow eats the grass, the carbohydrate ferments, and so the cycle begins again.

Physical removal processes include *dry deposition*, and *wet deposition*. Dry deposition refers to the removal on the Earth's surface of gases and particles by a direct transfer process, in which the species are deposited or absorbed irreversibly on soil, water, or plant surfaces and without the involvement of any precipitation. Wet deposition, on the other hand, is a term used for all deposition processes in which the gases or particles are carried to the surface dissolved or entrapped in water. The constituents are incorporated into precipitation elements (clouds, rain droplets, and aerosols). Fog, rain, hail, or snow may thus all be agents of wet precipitation. If there is no intervening chemistry, the source–transport–sink sequence defines a *physical lifetime* for the constituent. However, chemical changes may occur on a timescale comparable with, or smaller than, this lifetime. The *chemical lifetime* may then determine for how long the compound maintains its chemical identity after release, as discussed at greater length later (Sections 3.2.3, 3.3.1 and 3.5.1).

Degradation of one gas frequently generates another that survives long enough to have a separate existence. For example, released methane is not rapidly removed by physical processes, and is oxidized slowly (with a lifetime of several years). Formaldehyde is an intermediate in the oxidation, but is quickly photolysed (lifetime of a few hours). The products are hydrogen and carbon monoxide, which have lifetimes of several years and several months, respectively.

3.2 SOURCES OF ATMOSPHERIC CONSTITUENTS

Table 1.2 set out a list of the most abundant gaseous components of our atmosphere, with approximate values for the mixing ratios of each being provided. Table 3.1 focuses on the origins of the gases, rather than their abundances, and categorizes the trace species according to their

Table 3.1 Natural and Man-related sources of the gases of the atmosphere. (Updated from R.A. Cox and R.G. Derwent, *Specialist Periodical Reports Chem. Soc.* 1981, **4**, 189).

Compound	*Natural sources*	*Man-related sources*
Major Atmospheric Gases		
Nitrogen	Present in prebiological atmosphere Microbiological	
Oxygen	Photosynthesis	
Argon	Mainly radiogenic	
Carbon dioxide	Present in prebiological atmosphere Oxidation of natural CO; destruction of forests; respiration by plants	Combustion of oil, gas, coal, and wood; limestone burning
Water	Present in prebiological atmosphere Evaporation, especially from oceans	
Other relatively abundant gases not classified elsewhere		
Neon	Present in prebiological atmosphere	
Helium	Mainly radiogenic	
Hydrogen	Oceans; soils; oxidation of methane, isoprene, and terpenes *via* formaldehyde	Motor-vehicle exhaust; oxidation of methane *via* formaldehyde
Ozone	Stratosphere; natural NO/NO_2 conversion	Man-induced NO/NO_2 conversion
Carbon-containing gases		
Methane	Enteric fermentation in wild animals; emissions from swamps, bogs, *etc.*; natural wetland areas; oceans	Enteric fermentation in domesticated ruminants; emissions from paddies; natural-gas leakage; sewerage gas; colliery gas; combustion sources
Carbon monoxide	Oxidation of natural methane, natural C_5, C_{10} hydrocarbons; oceans; forest fires	Oxidation of Man-related hydrocarbons; incomplete combustion of wood, oil, gas, and coal, especially in motor vehicles; industrial processes; blast furnaces
Light alkanes, C_2–C_6	Aerobic biological source	Natural-gas leakage; motor vehicle evaporative emissions; refinery emissions
Alkenes, C_2–C_6		Motor-vehicle exhaust; diesel-engine exhaust
Aromatic hydrocarbons		Motor-vehicle exhaust; evaporative emissions; paints; petrol; solvents
Semiterpenes, C_5H_8 Terpenes, $C_{10}H_{16}$ Diterpenes, $C_{20}H_{32}$	Trees: broadleaves, and conifers; other plants	
Nitrogen-containing gases		
Nitric oxide, NO	Forest fires; anaerobic processes in soil; electric storms	High-temperature combustion (oil, gas, coal)
Nitrogen dioxide, NO_2	Forest fires; electric storms	High-temperature combustion (oil, gas, coal); atmospheric transformation of NO
Nitrous oxide, N_2O	Emissions from denitrifying bacteria in soil; oceans	Combustion of oil and coal
Peroxyacetyl nitrate, PAN	Degradation of isoprene	Degradation of hydrocarbons

chemical composition. The sources have themselves been divided into natural and Man-related. While not wishing to imply that Man is not part of nature, we shall see repeatedly that 'intelligence' allows Man to make disproportionate contributions to the atmosphere, even adding compounds unknown to Man-free nature. The table shows that the natural sources themselves are either geological or biological: the former include volcanic activity, outgassing, and *radiogenic* processes, and the *biogenic* sources include microbiological sources and photosynthesis. Man-related sources of atmospheric species are often referred to as *anthropogenic*, and are dominated by combustion products, accidental release in manufacturing processes, and the synthesis of chemicals. However, other activities of Man may lead indirectly to the release of gases to the atmosphere. Examples are the enhanced rates of release of N_2O resulting from the use of nitrate-based fertilizers, and the emission of methane by ruminant domestic animals (cows, sheep and horses). This is why we have been careful to say that the sources are Man-related, rather than stating that the gases are 'Man-made'. A brief note about the words *anthropogenic*, *biogenic* and *radiogenic* is also in order. The meanings of the adjectives as used by us are, in order, *Man-made* (or *Man-related*), *produced by living organisms* and *produced by radioactive decay*.

3.2.1 The Early Atmosphere

Hydrogen and helium were present in the universe almost immediately after the 'Big Bang', some 10–20 Gyr ago (1 Gyr = 1×10^9 yr; current best estimates for the age of the universe, based on Hubble space-telescope data, are 12 Gyr). Nuclear fusion transformed these elements into heavier ones such as carbon, nitrogen, oxygen, magnesium, silicon and iron. Elements of atomic number higher than that of iron are formed in supernova explosions that scatter material through the galaxies as tiny dust grains 1–1000 nm in diameter; the grains are probably composed mainly of graphite, H_2O-ice, and iron and magnesium silicates. Dust and gas in the universe are concentrated in the arms of spiral galaxies in which new stars, such as our Sun, are formed. Radiodating of meteorites and lunar samples provides ample evidence that the solar system was formed 4.6×10^9 years ago. In some regions, temperatures were low enough to permit crystallization of the metal silicates, but not of the ices of volatile elements such as H, C or N. Planets grow by agglomeration of rocky 'planetesimals', with diameters of up to a few kilometres, that form in the solar nebula. One pointer to the origins of the atmospheres of the inner planets comes from the noble gases present. Two types of noble gas can be distinguished: primordial and radiogenic. Primordial isotopes such as ^{20}Ne, ^{36}Ar, ^{38}Ar, ^{84}Kr, and ^{132}Xe were present in the solar system from the time of its creation. Radiogenic isotopes, however, have built up from the decay of radioactive nucleides: ^{40}Ar from the decay of ^{40}K, and ^{4}He from ^{232}Th, ^{235}U, and ^{238}U. The relatively large concentrations of Ar (Table 1.2) for the present-day atmosphere of Earth consist almost entirely of the ^{40}Ar, radiogenic, isotope (see Section 3.2.2).

In comparison with solar abundances, the Earth's atmosphere is depleted of ^{36}Ar by a factor of more than two million on a mass per unit mass basis, while for ^{20}Ne the depletion is 220 million. Evidence of this kind is taken as clear proof that Earth has lost almost all its primordial atmosphere, if such an atmosphere existed at all, and that the present atmosphere has been acquired later. The earliest precursor of the present-day atmosphere was formed by the outgassing from the interior of the planets of gases that were trapped inside as the solid materials coalesced. This outgassing would thus have released to Earth's atmosphere a mixture of gases similar in initial composition to those in the atmospheres of Mars and Venus. The gases were mainly CO_2, N_2 and H_2O, together with a little of the reduced gases CO and H_2. None of the original CO or H_2 has survived, of course, but the CO_2, N_2 and H_2O in the present-day atmosphere are mostly remnants of the early atmosphere (although they may have been recycled through the solids and liquids that lie beneath the atmosphere). The non-radiogenic noble gases must also be for the most part 'left-overs'. These are the materials that we have listed as present in the prebiological atmosphere in Table 3.1.

3.2.2 Sources Supplying the Contemporary Atmosphere

The great difference in composition between the atmosphere we find above us today and the early atmosphere just described has largely been brought about by living organisms. Chapter 1 provided a first indication of how the biota have transformed our atmosphere, and more detailed parts of this theme are taken up at various points of the book, most particularly in Chapters 6 and 7. The centrepiece of this biological activity is probably the conversion of CO_2 to O_2 by photosynthesis (Sections 6.2 and 6.3), with the biologically mediated removal of atmospheric CO_2 by formation of solid carbonates a close runner up. Plants are not only the source of atmospheric O_2, but also possess the metabolic potential to generate and emit a large variety of hydrocarbons and other volatile organic compounds (VOCs).

It is not appropriate in the present chapter to look in detail at even a modest selection of the sources listed in Table 3.1. Rather, we propose to illustrate here the general principles by considering just two examples. First, we extend somewhat our discussion of the non-biogenic noble gases, and then we examine the sources of the biogenic hydrocarbon methane, the simplest and most abundant organic compound in the atmosphere.

The noble gases provide valuable pointers to the processes that created our atmosphere and that, within limits, maintain its present composition. Of the gases in our atmosphere, only these elements have neither a biological nor an atmospheric source. Chemical inertness prevents loss of the noble gases (with the possible exception of xenon) to surface rocks. Except for the lightest noble gas, helium, they cannot readily escape to space (see Section 3.3.4) from any of the inner planets, Venus, Earth, or Mars. Abundances of the gases thus correspond to the cumulative quantity of gas present in and released to the atmosphere over the entire history of the planet.

Budgets of the inert gases may therefore allow us to probe how far the lithosphere–hydrosphere–atmosphere system is truly closed. Table 1.1 shows that argon makes up nearly one per cent of the contemporary atmosphere, but most of this argon is the radiogenic isotope ^{40}Ar, formed by decay of ^{40}K (^{36}Ar with some ^{38}Ar are the natural isotopes). See also Section 4.4. According to different estimates of the terrestrial potassium abundance, the present argon load (6.8×10^{16} kg) in the atmosphere corresponds to something between one-half and all of the ^{40}Ar generated within the Earth. No chemical sinks exist, and the oceans can dissolve only one per cent of the atmospheric argon. The mean annual input of ^{40}Ar to the atmosphere has thus been the present load divided by the lifetime of the Earth ($\sim 4.5 \times 10^9$ yr), and is 1.5×10^7 kg yr^{-1} globally. We shall use this result in Section 3.4. Helium is much less abundant than argon in the atmosphere: the ^4He isotope has a mixing ratio of 5.24 ppm, and ^3He is nearly a million times yet less abundant. Radioactive decay, this time of the ^{238}U, ^{235}U, and ^{232}Th series, is again the source of the noble gas. The relative atomic ratio of ^4He to ^{40}Ar in many samples of natural gas lies between 0.2 and 5, so that the order of magnitude of $^4He/^{40}Ar$ in gases entering the atmosphere is unity. Why, then, is present-day ^4He so much less abundant—by a factor of 1782—than ^{40}Ar? The answer must be that some helium has escaped from the Earth altogether: from the top of the atmosphere to interplanetary space. Possible mechanisms for such escape will be discussed in Section 3.3.4; other things being equal, a relatively light atom such as He will find it much easier to escape than a heavier one.

We turn our attention now to the sources of methane. Atmospheric CH_4 originates from both biogenic and non-biogenic sources; the biogenic sources may themselves be 'natural' or associated with Man's activities (such as rice agriculture, livestock, landfills, waste treatment and biomass burning). Non-biogenic CH_4 includes emissions from fossil-fuel mining and burning (natural gas, petroleum and coal), biomass burning, waste treatment and geological sources (fossil CH_4 from

Table 3.2 Sources of atmospheric methane: source strengths in units of 10^9 kg yr^{-1}.

Natural		*Anthropogenic (biospheric)*		*Anthropogenic (fossil fuels)*	
Wetlands	166	Enteric fermentation	81	Natural gas	40
Termites	22	Rice paddies	60	Coal mining	21
Oceans	10	Biomass burning	50	Coal burning	25
Hydrates	5	Wastes, landfill, sewage	61	Petroleum industry	20
Total	203	Total	252	Total	106

Based on table 7.6 of *Climate Change 2007 – The Physical Science Basis*, Contribution of Working Group I to the Fourth Assessment Report of the Intergovernmental Panel on Climate Change, Cambridge University Press, 2007. The report gives an estimate of emissions from all sources of 582×10^9 kg yr^{-1}, after including a number of smaller sources not listed here.

natural-gas seepage in sedimentary basins and geothermal/volcanic CH_4). Bear in mind, though, that all this CH_4 was formed biogenically in the geological past.

Table 3.2 shows the strengths of the main sources of CH_4 based on estimates given by the Intergovernmental Panel on Climate Change (IPCC) in 2007. Natural biological sources release about 200×10^9 kg of CH_4 to the atmosphere annually. This (large) quantity of CH_4 is, however, dominated by nearly twice as much from the anthropogenic-related biogenic and non-biogenic sources. The distribution of the 'natural biogenic' 203×10^9 kg yr^{-1} is displayed as a pie chart in Chapter 7 (Figure 7.3). Emissions of CH_4 from most of these sources involve ecosystem processes that result from complex sequences of events beginning with primary fermentation of organic macromolecules to yield acetic acid (CH_3COOH), other carboxylic acids, alcohols, CO_2 and hydrogen (H_2), followed by secondary fermentation of the alcohols and carboxylic acids to generate acetate, H_2 and CO_2, which are finally converted to CH_4 by *methanogenic* (methane-producing) bacteria (the *methanogenic Archaea*). Natural *wetlands* are responsible for approximately 81 per cent of global methane emissions from natural sources. Wetlands provide a habitat conducive to the methanogenic bacteria, which require environments with no oxygen and abundant organic matter, both of which conditions are found in wetlands. Methane is also produced by *termites* as part of their normal digestive process, and the amount generated varies among different species. Ultimately, emissions from termites depend largely on the population of these insects, which can also vary significantly between different regions of the world. The source of methane from *oceans* is not entirely clear, but two identified sources include anaerobic digestion in marine zooplankton and fish, and also methanogenesis in sediments and drainage areas along coastal regions. Methane *hydrates* are solid deposits composed of cages of water molecules that contain molecules of methane. The solids can be found deep underground in polar regions and in ocean sediments of the outer continental margin throughout the world. Some of this methane is released continuously to the atmosphere under present-day conditions, but the rate is dependent on temperature, pressure, salt concentrations, and other factors. Overall, the amount of methane stored in these hydrates globally is estimated to be very large, with the potential for large releases of methane if there are significant breakdowns in the stability of the deposits. Because of this large potential for emissions, there is much ongoing scientific research related to analysing and predicting how changes in the ocean environment affect the stability of hydrates. We shall return to this topic later in the present chapter, and again in Chapter 11 (Section 11.4).

Table 3.2 does not include geological sources, although there is evidence that significant amounts of CH_4 are released into the atmosphere through faults and fractured rocks, mud volcanoes on land and the seafloor, submarine gas seepage, microseepage over dry lands, and geothermal seeps. This CH_4 is produced within the Earth's crust (mainly by bacterial and thermogenic processes), and emissions from these sources are estimated to be as large as $40–60 \times 10^9$ kg yr^{-1}, thus adding very considerably to the natural budget.

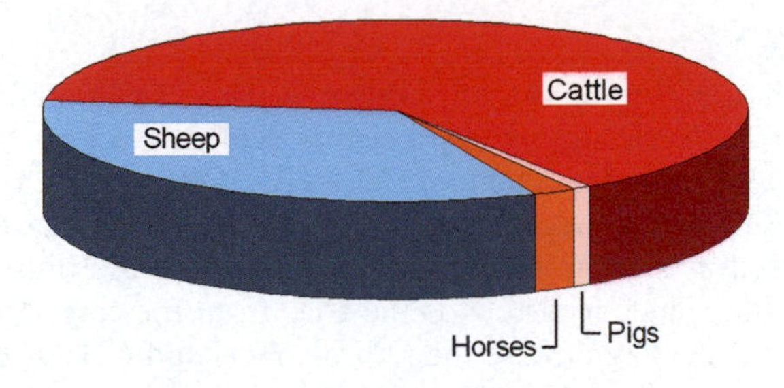

Figure 3.2 Methane emissions from domestic livestock in the UK. Figure drawn from population data drawn up by DEFRA for 2007. The total emissions of CH_4 for the UK that year are estimated to be 0.84×10^9 kg yr^{-1}.

Of the anthropogenic sources of methane, domesticated livestock seem at present to be the strongest source globally. *Ruminant animals* (cattle, buffalo, sheep, goats, and camels) produce significant amounts of methane as part of their normal digestive processes. In the rumen, or large fore-stomach, of these animals, microbial fermentation converts feed into products that can be digested and utilized by the animal. This microbial *enteric fermentation* produces methane as a by-product, which can enter the atmosphere from either end of the animal's digestive system (as hinted at with delicacy by Figure 3.1). Methane is also produced in smaller quantities by the digestive processes of other animals. Rough emission rates in g day^{-1} are cows: 150; horses: 49; sheep: 22; pigs: 4, although diet and husbandry practices lead to very wide variations. Figure 3.2 shows an approximate distribution between the different categories of domestic livestock for the UK, based on DEFRA estimates for the livestock populations in 2007. The grand total emission from the UK of *ca.* 0.86×10^9 kg yr^{-1} is just over one per cent of the worldwide total. Humans are definitely *not* ruminants, but even they are a source of methane, albeit a small one at about 0.1 g day^{-1} per head. With a population in the UK of about 61 million, that translates to more than 2000 tonnes a year, which would make 3×10^9 l of CH_4 at STP! However, our efforts with just one stomach luckily pale into insignificance compared with those of cows with four. Liquid manure management systems, such as lagoons and holding tanks, can cause significant methane production and these systems are commonly used at larger swine and dairy operations.

Table 3.2 shows that *rice cultivation* is another major anthropogenic source of methane. Flooded soils used for such cultivation are ideal environments for methane production because of their high levels of organic substrates, oxygen-depleted conditions, and moisture. The level of emissions varies with soil conditions and production practices as well as with climate.

Landfills, waste and sewage are major sources of methane worldwide, and in some countries may be the largest human-related source (accounting for 34 per cent of all methane emissions in the USA). Methane is generated in landfills and open dumps as waste decomposes under anaerobic conditions. The amount of methane created depends on the quantity and moisture content of the waste and the design and management practices at the site. Treatment processes for wastewater can produce methane emissions if organic constituents are treated anaerobically and if the methane produced is released to the atmosphere. In addition, the sludge produced from some treatment processes may be further biodegraded under anaerobic conditions, resulting in additional methane emissions.

Methane is the primary component of natural gas. Methane losses occur during the production, processing, storage, transmission, and distribution of natural gas. Because gas is often found in conjunction with oil, the production, refinement, transportation, and storage of crude oil is also a source of methane emissions. Methane trapped in coal deposits and in the surrounding strata is released during normal mining operations in both underground and surface mines, and handling of the coal after mining can liberate more methane.

3.2.3 Chemistry

Any chemical change that occurs in the atmosphere is, almost by definition, a source of the products of reaction. At the same time, the process is evidently a sink for the reactant species. Figure 3.1 offers some easy examples. The molecules HCHO (formaldehyde), CH_3OOH (methyl hydroperoxide) and HCOOH (formic acid) are intermediates in the oxidation of methane (the chemistry itself is discussed in Section 8.2.2). The chemical steps are thus sources of atmospheric HCHO, and so on. The immediate feedstock is the CH_4 from the cow. Along with the molecular products printed in blue in the diagram are the radicals HO_2 and CH_3O_2 shown in red, for which the reactions are likewise a source (OH is a special case since it drives the reaction, but is regenerated: see Section 8.2.2). All these species are *reaction intermediates* on the way to forming the end products CO_2 and H_2O, for which the next steps in the oxidation chain are a source. Viewed in less chemical detail, then, it might be argued (correctly) that the atmospheric oxidation of CH_4 is one source of CO_2.

One particularly interesting aspect of chemistry as a source of new atmospheric species is that the reaction product may condense into particles that are finely divided enough to remain suspended as *secondary aerosols*. This *gas-to-particle conversion* is a major contributor to the total atmospheric aerosol. For aerosols derived from naturally occurring precursors, the quantities produced by gas-to-particle conversion are comparable with those produced by direct processes (such as wind-raising of dusts and seasalt, or volcanic eruptions). In the case of anthropogenic aerosol, the secondary sources are stronger than the primary ones.

The oxidation of natural VOCs may generate secondary aerosol, as explained in Sections 7.2 and 8.2.7, in cases where the vapour pressures of the products are lower than those of the reactants. Inorganic compounds also contribute, especially to anthropogenic aerosol. Oxidation of biogenic dimethyl sulfide (DMS), and anthropogenic SO_2 and NO_x leads to the formation of aerosol sulfates and nitrates, NH_4HSO_4 and $(NH_4)_2SO_4$ being typical constituents of the sulfate aerosol. More detail about the oxidation steps and the formation of aerosol is provided in Sections 7.2.4 and 8.4.2. Secondary formation of biogenic iodine-containing aerosol is of potential significance in the coastal environment. Intense new aerosol formation has been shown to be linked to an iodine-compound precursor released from abundant shore-line beds of macroalgae. Bursts of particle formation occur overwhelmingly during daytime low tides, and emissions of molecular iodine (as well as of iodocarbons) have been suggested as precursors of IO and OIO radicals that go on to produce higher oxides I_xO_y that ultimately cluster and form particles. Section 8.5 describes this chemistry somewhat more extensively.

3.3 SINKS

3.3.1 Chemistry

A certain symmetry in our presentation can be achieved by balancing our discussion of the chemical sources of atmospheric constituents in the preceding section with a consideration of chemical sinks in the present section. We pointed out at the beginning of Section 3.2.3 that any chemical change represents at once a source of the products and a sink for the reactants. Figure 3.1 yet again illustrates the point. The moment that the cow's CH_4 is attacked chemically (by OH, in this case: see Sections 8.2.1 and 8.2.2), it has reached its sink. Of course, the carbon atom itself remains alive and well, and we *may* want to follow through to the point when that carbon has been removed from the atmosphere altogether, which the figure shows to occur when the CO_2 is taken up by the surface vegetation in photosynthesis. Even then, the disappearance is only temporary, because the cow will eat the grass and transfer carbon to the atmosphere in the form of CH_4 ... Nevertheless, the original CH_4 molecules have been lost from the atmosphere when the OH reacts with them, and it is only because we are assuming that the Sun will continue to shine and the cows continue to breed that we can make assumptions about the subsequent recycling.

The example of the atmospheric sinks for CH_4 turns out to be quite appropriate. Physical removal of CH_4 in the deposition processes to be described in the next sections is rather slow, and almost all the loss of this simple hydrocarbon is, indeed, chemical. Chemical conversions often play an essential part in the overall scavenging process for gases by generating some new compound that can be removed physically, for example by the wet deposition mechanism described in Section 3.3.3, which requires the target compound to be dissolved in cloud droplets or falling rainwater. A tendency exists for natural trace gases of biogenic origin to be reduced (CH_4, terpenes, H_2S, *etc.*) or only partially oxidized (CO, N_2O). These species are also of only modest solubility in water. By way of contrast, Man-related species, which frequently involve combustion sources, are often more highly oxidized (CO_2, NO_2, SO_2) and somewhat more soluble, or at least more readily hydrolysed. Chemical transformations may be precursors of deposition even for the latter species. For example, washout removal times for sulfur dioxide in moderate rainfall are estimated as several hours. Most of the sulfate found in precipitation, however, is a result of rainout of hygroscopic cloud-condensation nuclei (see next paragraph and Sections 1.3 and 7.2.4) involving SO_3 or H_2SO_4: that is, the SO_2 has already been oxidized before nucleation occurs. Acids such as HCl, HF, and HNO_3 are readily soluble, as are NH_3, SO_2, and NO_x after conversion to aerosol species: wet deposition is possible.

It is very difficult for water droplets to condense out of the vapour phase without the presence of nuclei on which to grow. The problem concerns the vapour pressure produced over a curved surface. The equilibrium vapour pressure at any temperature is larger for a drop than the saturated vapour pressure over a plane surface, and the excess vapour pressure increases as the radius of the drop becomes smaller. The atmosphere must therefore be supersaturated with respect to the thermodynamic (plane-surface) vapour pressure if droplets are to condense. One way out of the apparent dilemma is provided by condensation on nuclei. It is certainly suggestive that rainfall is greater over industrial areas where particle counts are highest, and that more rain falls on weekdays than at weekends. Thermodynamic arguments show that, once a critical droplet size has been reached, further increases in radius are favoured. Suppose water is condensed on a completely wettable nucleus of diameter 0.2 μm. The water film would be in unstable equilibrium with air supersaturated by as little as 0.6 per cent, a very reasonable atmospheric value. Slight increases in drop size would now lead to additional growth of the droplet. A most interesting situation arises if a condensation nucleus is soluble in water. Saturated vapour pressures are lower over solutions than over pure solvent, as a result of the lowered chemical potential of the solution. Solution droplets can therefore exist in equilibrium with their surroundings even when the relative humidity is less than 100 per cent (unsaturated air). This is, of course, exactly the situation with H_2SO_4 aerosol.

3.3.2 Dry Deposition

Dry deposition acts efficiently only where a specific chemical or biological interaction is available, and even then only when the trace gases are close to the planet's surface. Of the species listed in Table 3.1, only a handful undergo direct deposition. Dry deposition mechanisms exist for SO_2, O_3, CO_2, and SO_3, while microbiological sinks are known for soil removal of CO and H_2.

Three separate steps can be envisaged in the dry-deposition process. First, the species must be transported through the atmosphere to some region in close proximity to the surface, then they must cross to the surface itself (by diffusive or Brownian mechanisms), and finally there must be uptake on the surface. Turbulent atmospheric motions near the Earth's surface bring about the first step, whose efficiency is thus determined by the degree of turbulence. Uptake of the gas or particle on the surface is governed by a variety of factors, including shape and smoothness of the surface, the amount of moisture on it, and, of course, the solubility of the depositing species and any specific chemical or biological interactions. Moderately soluble compounds such as SO_2 and O_3 are reversibly absorbed, so that the dampness of the surface is critical, while highly soluble species such as HNO_3 absorb rapidly and irreversibly on virtually any surface.

Rates of dry deposition are often described in terms of a semiempirical equation

$$F = -u_d C \tag{3.1}$$

where F is the flux of dry deposition of a species whose concentration is C at some specified height above the surface. The equation then defines the *deposition velocity*, u_d. An assumption is that F remains constant up to the specified reference height; in reality, u_d and F are likely to depend on height above the surface. The equation is often used as a convenient boundary condition for the diffusion equation in atmospheric models in order to incorporate dry deposition. Its convenience should not, however, obscure the way in which the parameter u_d lumps together all of the factors in the three steps leading to deposition.

Experimental measurement of dry-deposition velocities has generally proved quite difficult, although it is evident that natural surfaces, such as plant leaves, act as relatively efficient sinks. For example, chamber experiments give velocities of deposition of ethene to spinach, grass, bare soil, and water as 74, 150, 27, and 1.6 $\mu m\, s^{-1}$, respectively, results that show clearly the much greater efficiency of grass compared with sea-water as a surface for dry deposition. The different surfaces afford a variation in the type of plant canopy and the leaf area, and also allow estimates to be made of the relative importance of deposition to the plant canopy and the soil. Different hydrocarbons behave in somewhat different ways; for example, deposition velocities for an alkane (*n*-hexane) were 43, 49, 22, and 3 $\mu m\, s^{-1}$ for the four surfaces in the same order, while for an alkyne (acetylene) the values were 2, 65, 16, and 1.6 $\mu m\, s^{-1}$. Some incompatibility between acetylene and spinach evidently exists! Regardless of the exact numerical values, all these deposition velocities are extremely small, and the experiments confirm the expectation that dry deposition of hydrocarbons is not important in the atmosphere.

3.3.3 Wet Deposition

Chemical species may either enter the cloud or fog that is the precursor of the precipitation, or they may enter the aqueous phase as the precipitation is falling. Wet deposition by incorporation into cloud droplets is termed *rainout*; removal by uptake in falling precipitation is *washout*. The processes are significant only for those gas-phase species that are water soluble. Particles, on the other hand, can serve as condensation nuclei and be incorporated into a droplet, or they can collide with a pre-existing droplet. A gas or particle taken up by a droplet may, of course, not reach the ground if the droplet evaporates during its descent.

Direct interception of a cloud (or fog) by the Earth is sometimes recognized as a modification to the wet-deposition process. Wet deposition can be thought of as occurring in several discrete steps as illustrated in Figure 3.3. First, the atmospheric condensed water ('*hydrometeor*') and the species to be removed must encounter one another. Secondly, the species must be taken up by the hydrometeor. Lastly, the hydrometeor must reach the Earth's surface. Virtually all the processes are reversible, and, at each step, a species may undergo chemical reaction, thus further adding to the complexity of the system.

Quantitative mathematical description of wet deposition is obviously a difficult task in the circumstances described. One of the problems is that the several contributing steps occur on vastly different scales of distance, ranging from molecular mean free paths of 10^{-10} m, through droplet physical processes occurring over 10^{-6} m, to atmospheric motions over distances of up to 10^3 km.

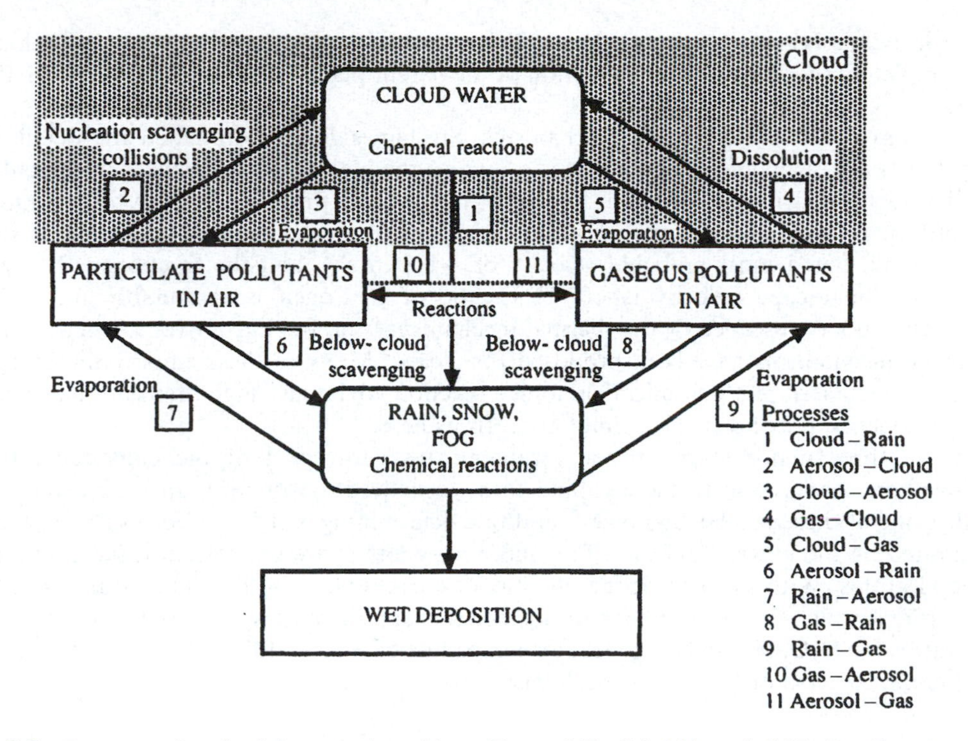

Figure 3.3 Processes involved in wet deposition. From J.H. Seinfeld and S.N. Pandis, *Atmospheric Chemistry and Physics*, John Wiley, 1998.

Formulation of the mathematics is made yet more daunting by the need to treat several physical phases (solids, liquids, and gases).

Although substantial progress has been made in the mathematical treatments, the complexity has recommended an altogether simpler approach to many atmospheric scientists. A semiempirical equation equivalent to equation (3.1) is written for the rate of transfer of the chemical species into the hydrometeor, with the deposition velocity being replaced by an *uptake* (or *scavenging*) coefficient, γ, which is the fraction of gas-phase molecules colliding with the surface that are actually lost. Rates of transfer of the chemical species from the cloud to rain together with the rate of below-cloud scavenging go to make up the overall flux of wet deposition. Several assumptions are implicit in this approach, two of the most serious being that the simple equation is applicable only if uptake is irreversible and if γ is independent of concentration of the species already contained in the condensed phase. Even within these limitations, γ remains a quantity that depends on the detailed meteorology and cloud physics pertaining to a particular place and time.

3.3.4 Escape

Species that move sufficiently fast may escape from the gravitational influence of the planet altogether, or perhaps go into orbit around it. Two conditions must be met for escape to occur. First, the escaping constituent must not suffer any collisions as it leaves the planet. A *critical level*, z_c, can be defined for which a proportion e^{-1} of particles capable of escape experience no collisions as they leave the atmosphere. Intuitively, it would seem that this altitude is the one at which the mean free path, λ_m (Section 2.2, equation (2.4)), becomes equal to the scale height for the particle in question, and, indeed, this condition can easily be shown to fit the definition of z_c. For the Earth's atmosphere, z_c is 400–500 km; the *exosphere* lies beyond that altitude. The second condition to be fulfilled for escape is that the particle has sufficient translational energy (*i.e.* starting velocity) to escape completely the gravitational attraction of the parent planet. The *escape velocity* for Earth is $11.2\,km\,s^{-1}$.

We now have to consider escape mechanisms. Sir James Jeans formulated the details of the process that bears his name. Escape is seen to involve the high-velocity particles in the tail of the Maxwell velocity distribution. Atomic hydrogen has a most probable velocity of $\sim 3\,km\,s^{-1}$ at 600 K, and the fraction of atoms with $\upsilon > 11.2\,km\,s^{-1}$ is just greater than 10^{-6}. Atomic oxygen, on the other hand, has a most probable velocity of $\sim 0.8\,km\,s^{-1}$, and the fraction with a velocity greater than the escape velocity is 10^{-84}! Escape of hydrogen is reasonably probable, but little oxygen can have escaped by the thermal mechanism from the Earth over its lifetime. Several non-thermal mechanisms have been proposed for escape. Many of these employ the energy of a chemical (particularly photochemical or ionic) reaction to impart high translational energy to product fragments, but the details cannot concern us here.

There are three major stages in the escape process: transport of the constituent through the atmosphere, conversion to the escaping form (usually atom or ion), and the actual escape. Normally, one of these will be the slowest, and rate determining. On Earth, for example, hydrogen loss is limited by the upward diffusion flux and not by any process involved in the conversion to escaping H atoms. With this knowledge, the flux of escaping hydrogen can be estimated from the diffusion rate through the stratosphere (and hence through the mesosphere and to the exosphere). Mixing ratios for H_2O, H_2, and CH_4 yield an escape flux of $\sim 2.7 \times 10^8$ H atoms $cm^{-2}\,s^{-1}$ from the atmosphere, *regardless* of the escape mechanism.

3.4 A BALANCE OF SOURCES AND SINKS?

Production and loss of atmospheric constituents have to be balanced if concentrations are not to vary. Over the lifetime of the Earth, the atmospheric composition has almost certainly undergone considerable modification (Chapter 7), and the balance is not perfect. On short timescales, however, a steady state obviously more or less holds for components such as nitrogen or oxygen.

In Section 3.2.2, we showed that the mean global rate of ^{40}Ar release as a result of radioactive decay is about $1.5 \times 10^7\,kg\,yr^{-1}$ and that roughly as many ^4He atoms are emitted as are ^{40}Ar, so that the emission rate of ^4He is *ca.* $1.5 \times (4/40) \times 10^7\,kg\,yr^{-1}$. The residence time of He within the atmosphere can be calculated as the atmospheric load divided by the rate of release, just as for any other atmospheric lifetime (Section 3.5.1). The mixing ratio of ^4He is 5.2 ppm (Table 1.2), which corresponds to a total mass of $3.6 \times 10^{12}\,kg$ in the atmosphere, giving a residence time of $(3.6 \times 10^{12}/1.5 \times 10^6) \leq 2.5\,Myr$. This lifetime is three orders of magnitude less than the Earth's age, so that most of the helium ever released from the crust and mantle has escaped. Of course, those 2 Myr are quite long compared with a human lifespan, and 99.996 per cent of the particular helium atoms in the atmosphere when a person is born are still there when he or she dies, which puts a rather different emphasis on the numbers!

The examination of noble-gas behaviour has highlighted two new features of the fluxes of elements into and out of atmospheres. First, material can be released, for example by outgassing, from the solid body of the planet. Gases can be formed, as in the cases of ^4He and ^{40}Ar, radiogenically, or they may have been trapped as the planet was created. Secondly, the extreme limit of the atmosphere forms an interface for the exchange of matter with 'space'. Our example was of escape, but material can also enter the Earth system at this interface. The Sun emits a continual stream of electrically charged particles, the *solar wind*, which impinges on the outer layers of a planetary atmosphere, or on the surface of a body without an atmosphere. Then, meteorites, asteroids, and even comets occasionally enter our gravitational field and burn up (*ablate*) or evaporate in the atmosphere. The atmosphere is thus slightly leaky. We stated in the first paragraph of this chapter that the yearly average concentrations of atmospheric constituents do not vary 'wildly', so that the leaks in and out must be nearly balanced. Of course, all the evidence is that over geological timescales quite considerable changes of composition have occurred. Even over the periods that Man has been able to make sufficiently accurate measurements, concentrations of gases such as CO_2, CH_4, N_2O and tropospheric O_3 have been showing a slow drift (generally upwards) over the course of tens of years. Nevertheless, the drifts are tiny compared with the amounts of these gases emitted to and removed from the atmosphere over the same time periods. Chemical kineticists would recognize this situation as being one in which it is legitimate to apply *stationary-state* methods, and that is what we shall do in the next section, where we will assume that source and sink rates are identical for any atmospheric constituent.

3.5 LIFETIMES AND TRANSPORT

3.5.1 Lifetimes

The concepts of physical *residence times* and chemical *lifetimes* were introduced in Section 3.1, and have been used several times already in this book. How long does a particular species remain in the atmosphere between the time that it is delivered by a source and the time that it is removed by a sink? This atmospheric *lifetime* is a revealing quantity because it gives some idea of the impact that the species may have, and it also links together concepts of source and sink strengths and the atmospheric concentration. An explanation of what these lifetimes mean usefully starts with considering *physical lifetimes*, for example of water that evaporates from the ocean. It is fairly obvious that the lifetime must represent the period during which the water is in the vapour phase between evaporation and precipitation as rain or snow. However, since different individual molecules survive for differing periods, the lifetime that we quote must represent some kind of average. Similar considerations would apply to fine dust that is wafted aloft by wind in a desert. Observe that the source liberates water or dust, and the sink removes water or dust. Atmospheric chemists are naturally often interested in *chemical lifetimes*. The ideas are just a modification of what we have just set out, with the added realization that the sink is a chemical one: the moment a given chemical species undergoes reaction to become a new entity, its life has ended. An interesting point about lifetimes is that they are determined by the rate at which a species is lost to the sink, and not by the rate of supply (even though those two may be numerically virtually identical). On the other hand, the rate of supply determines the atmospheric concentration. In fact, with the proviso that the rates of supply and removal are identical, a most important pair of results emerges that we will state here without the use of algebra! The concentration of a species is equal to the rate of its loss multiplied by its lifetime, and, since rates of loss and production are equal, the concentration is also equal to the rate of production multiplied by the lifetime. Perhaps the reader will find this result not unexpected, and say that indeed the concentration of a species should be equal to the rate at which it is supplied multiplied by the effective length of time that it is allowed to accumulate in the atmosphere. But the true value of our comments is that if you know two out of the three parameters under discussion—lifetime, concentration and source (or loss) rate—you can calculate the third. For example, if you have a measurement of the relevant rate coefficient (and thus of atmospheric lifetime) obtained in laboratory experiments, and a measurement of the atmospheric concentration, then you can estimate the source strength of a particular compound, and so on. We shall show now that considerable subtlety attaches to what appear to be simple ideas.

A connection between lifetimes, concentrations and supply rates can be formulated in a straightforward way. Consider a simple situation in which a liquid flows through a pipe without becoming turbulent and mixed. With such *plug flow*, any volume element entering the tube 'lives' in it for the time it takes it to pass from one end to the other of the tube. The total amount of material in the tube is thus the rate at which the material is supplied multiplied by this lifetime; material comes out of the tube at the same rate that it goes in, and we have agreed in Section 3.4 that this equality can be taken to apply to the atmosphere as well. It is convenient to express both the burden and the rates in concentration units (molecules or moles per unit volume). We could then write these definitions in the form

$$[\mathrm{A}] = \tau_u \times S = \tau_u \times L, \tag{3.2}$$

where τ_u is the lifetime of a species A in the unmixed system, [A] its atmospheric concentration, and S and L the rates of supply and loss in terms of concentration per unit time. Alternatively, the rearranged equations can be used to define τ_u

$$\tau_u = [\mathrm{A}]/S = [\mathrm{A}]/L \tag{3.3}$$

In the atmospheric context, the supply is the source of material and the loss is made up of the sinks, such as chemical loss, and physical removal by transport and deposition.

In reality, it seems very *un*reasonable to assume that the atmosphere is not mixed at all. If the atmosphere were *completely* mixed, any group of molecules introduced would slowly decrease in number as some were lost as a result of the chemical and physical removal processes. Although there is then some kind of characteristic mean time for which the molecules are in the atmosphere, certain of the molecules are lost at much earlier times, and some at much later ones. How then shall the lifetime be defined?

To reach a solution, we turn back to the ideas of reaction kinetics, which deal with the reaction and loss of large numbers of molecules. Let us assume that a trace constituent A decays by a (pseudo-) first-order process. The differential and integrated forms of the rate equation are

$$-\mathrm{d}[\mathrm{A}]/\mathrm{d}t = k'[\mathrm{A}]; \quad [\mathrm{A}]_t = [\mathrm{A}]_0 \exp(-k't), \tag{3.4}$$

where $[\mathrm{A}]_t$ is the concentration of A at time t and $[\mathrm{A}]_0$ the initial concentration. In a time $\tau_m = 1/k'$, the concentration of A will have decreased to a fraction $1/e$ (0.37) of its initial value (and in a time $\ln 2/k'$ to one-half of the initial value: that time is the *half-life*) in this fully mixed system. But

$$\tau_m = \frac{1}{k'} = \frac{[\mathrm{A}]}{-\mathrm{d}[\mathrm{A}]/\mathrm{d}t} \tag{3.5}$$

and $-\mathrm{d}[\mathrm{A}]/\mathrm{d}t$ is the same as the loss rate L of equation (3.3). Comparison of equations (3.3) and (3.5) then shows that τ_u in the unmixed system is numerically identical to τ_m in the mixed one, although the definitions of the two lifetimes are quite different. We recall that τ_u is the time required for *all* the material to be removed from the system, while τ_m is the time taken for $(1-0.37) \times 100 = 63$ per cent of A to be removed.

The difference between unmixed and fully mixed systems has been emphasized in this treatment so that the situation with regard to the atmosphere may be properly understood. In a real atmosphere, a particular trace species may, of course, exhibit behaviour lying anywhere between the two extremes. An atmospheric lifetime can always be expressed as the concentration of some species divided by the rate of its loss, but what that lifetime really means depends on the mixing behaviour within the atmosphere. We shall return to this aspect shortly when we consider the relation between lifetime and transport.

Whatever the particular extent of mixing, the mean lifetime apparently provides a clear measure of how long it would take a small perturbation to atmospheric concentrations to disappear. The only question to be resolved is whether 'disappear' means 100 per cent of the perturbation, 63 per cent, or something in between. In this sense, the lifetimes we have described are *adjustment times*.

One interesting complication that can arise in the real atmosphere is seen if addition (or removal) of one trace gas has a profound effect on those parts of atmospheric chemistry that themselves dominate the removal steps. An illustration of this phenomenon can be seen in the lifetimes of CH_4. In the troposphere, CH_4 is lost largely by reaction with the OH radical (Sections 3.3.1 and 8.2.2). Furthermore, one of the oxidation products is CO, which itself reacts rapidly with OH. Thus, addition of CH_4 depletes OH, and the rate of loss of CH_4 is consequently reduced. The result is that the lifetime of CH_4 is lengthened by the very act of increasing its atmospheric concentration. For this particular example, the lifetime of CH_4 estimated from the expected chemical and photochemical loss rates for the unperturbed situation would be about 10 years. But global models including the feedback indicate that the true adjustment time is nearer to 14 years.

An extension to the concepts is needed if there is more than one loss process for a species. For example, the species may be lost by physical processes, such as transport and wet and dry deposition, as well as by several different chemical and photochemical reactions. The method is entirely straightforward. An effective pseudo-first-order rate coefficient is written for each of the individual loss processes, and an overall value of k' obtained as the sum of them all.

3.5.2 Transport

We have emphasized several times that gases and aerosols in the atmosphere can be transported by atmospheric motions, so that they may be deposited or consumed chemically in a location different from the one where they were emitted or formed. For very unreactive species, transport can redistribute them globally over their (long) lifetimes. In contrast, species that are highly reactive may be born and die in effectively the same place.

Perhaps the most interesting situation is that in which the chemical lifetimes are comparable with transport lifetimes. The particular point of interest is how far a given air parcel might move within the mean lifetime of trace-gas species within it. Vertical motions within the troposphere, within the stratosphere, and exchange between the two regions, as well as E–W and N–S and hemisphere–hemisphere motions all occur on very different timescales. If the lifetime of the species involved is very short compared with the rate at which transport occurs between one part of the atmosphere and another, then the chemical kinetic behaviour of that species can be assessed as though it was confined to a well-defined volume element, just as in a laboratory system in which the reactant gases are held within a reaction vessel. At the other extreme, a trace species introduced to, or created within, the atmosphere at some location that is defined in altitude, latitude, and longitude, may become fully mixed throughout the atmosphere if its lifetime is sufficient. Reaction kinetics and atmospheric motions may thus be decoupled at the extremes of very short and very long mean lifetimes. Examples are easily found. The hydroxyl radical that is the major sink of CH_4 (see Section 3.3.1) has a lifetime in the troposphere of just seconds. At the other extreme, a molecule such as N_2O may have a lifetime of nearly 100 years in the troposphere, so that it must become completely mixed and redistributed in terms of geographical location before an 'average' molecule is lost. It is the situations in between that offer the difficulties.

Figure 3.4 provides a simple way of visualizing the comparison between the mean residence times of a variety of species and the timescales of different atmospheric transport processes. The feature to note especially is the distance or geographical scale over which a trace species can redistribute itself during its lifetime, because that will give an indication of the horizontal and vertical distances over which averaging of concentrations might be acceptable. Radicals such as OH or NO_3 live and die where they are born; molecules such as CH_4 and N_2O survive long enough to visit all parts of the (lower) atmosphere.

The figure pretends to show a lifetime for each chemical species that is independent of where it is formed or to where it is transported. For some atmospheric constituents, this idea is already an

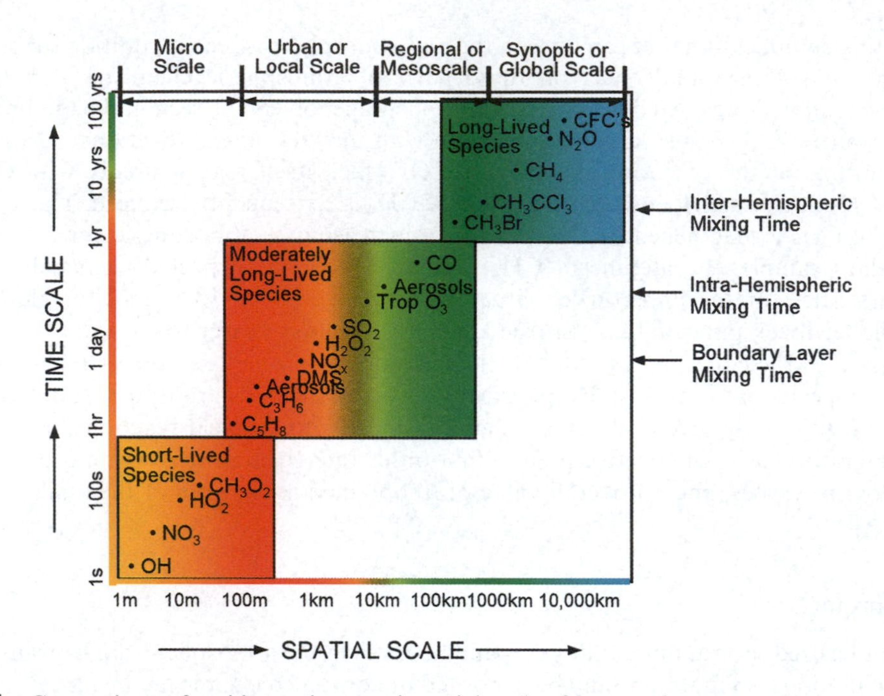

Figure 3.4 Comparison of residence times and spatial scales for some key atmospheric species. Derived from *The Atmospheric Sciences Entering the Twenty-First Century*, National Academy Press, Washington, DC, 1998.

oversimplification. A molecule such as CO, for example, has a very variable lifetime, which increases from a minimum of about a month in the tropics to approaching one year at high latitudes in the winter. One consequence is that CO released to the atmosphere in the tropics has a largely local effect, but that CO released during the winter to high latitudes can become widely mixed before removal, and the apparent lifetime is then *not* dependent on the exact location of the source.

The incomplete redistributions of the species of moderate lifetime, as well as this new complication of spatially variable lifetimes, make it evident that lifetimes are going to be given only very approximately by algebraic equations such as equations (3.3) or (3.5). Something much more sophisticated will be needed to work out how the chemical components should be distributed in the atmosphere. Numerical rather than algebraic techniques are called for, and the *models* that employ them are described in the next chapter (Section 4.4). Before that, however, we shall see how the components themselves are identified and their concentrations determined.

CHAPTER 4

Observations and Models

4.1 MEASUREMENTS OF ATMOSPHERIC COMPOSITION

A knowledge of the identity of the chemical species present in the atmosphere and of their concentrations is obviously at the very heart of the study of atmospheric chemistry. In the previous chapter, we saw the ways in which substances are released to the atmosphere and how they are removed from it. We also examined what chemical changes, if any, a particular species undergoes, and how long it lives until it is removed by physical and chemical processes. The substances initially released, the final products of the chemical transformations, and all the intermediates that participate in the steps of the chemistry will be present in the atmosphere, and may be detectable with suitable instrumentation. The discoveries of what species are present in the atmosphere have sometimes led to big surprises, because a substance found might be unexpected, and a new interpretation and deeper understanding often follows. Sometimes finding the expected adds to atmospheric scientists' confidence that they are on the right track. Observations may, for example, be able to provide the concentration of a compound hitherto unknown in the atmosphere. How and why is the compound there? Is it something that is being released from the surface, or is it the product or reaction intermediate of some chemical change that starts with a release? If the compound is identified as a primary release, then immediate tasks are to discover the sources and to employ what is known, if anything, about release rates, chemical kinetics and physical loss rates, and atmospheric concentrations, to come up with a coherent and consistent explanation of what is happening. We are beginning to stray here into what to do with the data, a topic more conveniently held over to Section 4.4. For the time being, our aim is to provide a brief survey of the ways in which the observations themselves can be made.

Measurements of trace species in the atmosphere can be categorized as (i) *in-situ* determinations of local concentrations, and (ii) remote sounding determinations, generally but not exclusively of non-local abundances. Table 4.1 gives some of the important characteristics of the two categories of measurement; there are advantages and disadvantages to each type, and compromises to be accepted. The altitude at which the information is sought obviously has a bearing on the choice of technique to be adopted. It may be considerably easier (and cheaper) to set up an *in-situ* experiment on the ground than to arrange for it to be transported on a balloon or a high-flying aircraft to examine the stratosphere. Remote sensing from a satellite can provide global spatial coverage in

Atmospheric Chemistry
By Ann M. Holloway and Richard P. Wayne

Published by the Royal Society of Chemistry, www.rsc.org

Table 4.1 Methods of measuring concentrations in the atmosphere.

	In-situ	*Remote*
Platforms	ground aircraft balloons	ground aircraft; balloons satellites; space shuttle
Main methods	grab sampling/later analysis spectroscopy: absorption or emission: microwave to UV resonance fluorescence chemiluminescence electrochemical	spectroscopy: absorption or emission: microwave to UV lidar
Advantages	high spatial and temporal coverage (especially with satellites)	measurements on regions distant from the instrument high spatial and temporal resolution
Disadvantages	measurements obtained only in vicinity of instrument: limited coverage in altitude, latitude, longitude and time	poor spatial resolution

latitude and longitude, yet it may be hard to achieve any altitude selectivity (although there have been great successes recently using sophisticated data-processing methods).

Figure 4.1 illustrates schematically some of the methods available. At the bottom right is an *in-situ* arrangement for looking at the composition of the boundary layer or lower troposphere, while at the top left is a concept for *in-situ* probing of the upper atmosphere. The tethered balloon shown may provide measurements for the mid-troposphere, but untethered balloons can reach very great heights and have been used for *in-situ* sounding of the upper stratosphere. The satellite in the middle of the top row is looking down into the atmosphere, and is thus a remote-sensing device, as is the *lidar* instrument (see later) of the middle of the bottom row, which is looking up from the ground.

Let us illustrate the different advantages and drawbacks of ground-based and satellite-platform techniques using measurements of stratospheric ozone as an example. Ground-based instruments can be calibrated periodically. However, the distribution of stations is not adequate to give a truly global average. Satellite observations can overcome the averaging problem, but so far have been available only for relatively short periods. The backscatter UV (BUV) instrument on the Nimbus 4 satellite was in operation for seven years from April 1970, but suffered from an apparent drift in sensitivity. A series of newer satellite instruments (see Section 4.3) subsequently took over the task.

Tropospheric measurements, especially of boundary-layer components, have historically relied rather heavily on localized *in-situ* methods. The very first experiments in atmospheric chemistry naturally examined the composition of air that could be sampled in the immediate vicinity. Stratospheric chemistry, in contrast, was initially reliant on remote sounding, because suitable platforms did not exist. These distinctions are slowly disappearing, as remote-sensing techniques are developed for tropospheric use, and as measuring devices can be carried into the stratosphere more easily. Satellite methods, in particular, are now being adapted widely to make tropospheric determinations of composition and concentration.

4.2 *IN-SITU* METHODS

Tropospheric *in-situ* methods are themselves sometimes perceived as falling into the two categories of 'spectroscopic' and 'chemical', although the borderline between the two is often blurred.

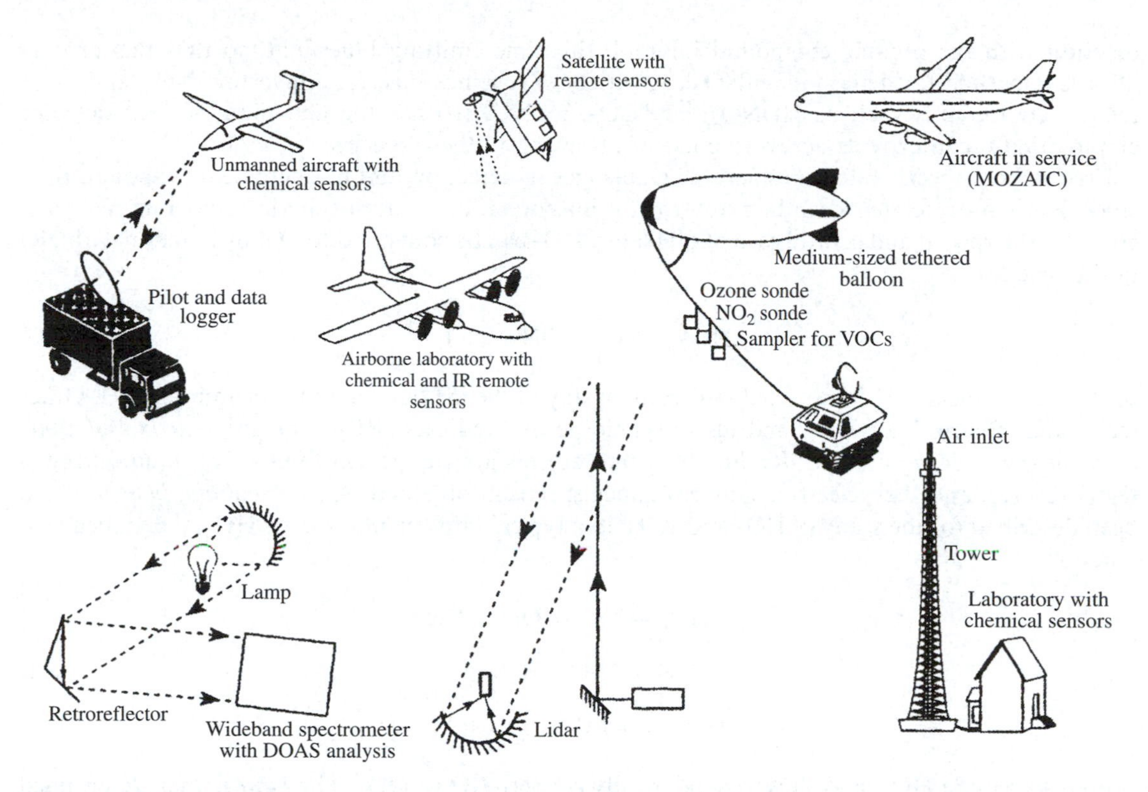

Figure 4.1 Some platforms used for measurements of tropospheric trace gases. After H.K. Roscoe and K.C. Clemitshaw, *Science*, 1997, **276**, 1065.

Christian Friedrich Schönbein is often regarded as the 'father of air chemistry' because, soon after his discovery of ozone in the laboratory in 1839, he demonstrated the presence of ozone in the atmosphere of several European cities. For this purpose, he employed test papers coated with potassium iodide. As Schönbein had found out already, ozone liberates iodine from the iodide ions, and the iodine can interact with the starch in the paper (or deliberately added) to form a dark-blue complex. By comparing the depth of colour with tints on a calibration chart after the test papers had been exposed to air for a fixed period, he was able to reach an estimate of the atmospheric ozone concentration. Many problems beset the use of 'Schönbein papers', but variants of the method of liberating iodine from iodide long remained one of the staples of atmospheric ozone measurements. Schönbein's experiments thus seem to have been the first application of chemical methods to the identification and determination of concentrations of a trace component in the atmosphere.

Present-day chemical sensors include gas chromatographs, mass spectrometers, and all the armoury of modern analytical chemistry. One method now of wide application is that of chemiluminescence. Certain chemical reactions can release part of their energy in the form of light. Two of these involve atmospheric species. Nitric oxide, NO, reacts with ozone

$$NO + O_3 \rightarrow NO_2^* + O_2, \tag{4.1}$$

where NO_2^* is electronically and vibrationally excited NO_2 that emits red light. Photoelectric measurement of the intensity emitted in the presence of a known amount of O_3 provides a quick, convenient, and specific way of determining [NO]. The system can be run in reverse, with NO added to the air sample, in order to measure [O_3]. Nitrogen dioxide, NO_2, undergoes a chemiluminescent reaction with the organic compound luminol, this time emitting blue light, so that this process affords a method for determining $[NO_2]$. A variety of schemes exist for converting NO_2, and other related compounds such as HONO and N_2O_5, to NO, so that the chemiluminescent detector employing O_3 can provide access to concentrations of all these species.

Free-radical species offer considerable challenges in detection and measurement, although their study is often especially rewarding in terms of interpretation of atmospheric behaviour. We shall consider the special and central case of OH later. HO_2 can be converted to OH by reaction with NO in the reaction

$$HO_2 + NO \rightarrow OH + NO_2, \tag{4.2}$$

so that this radical may be detected indirectly by any of the methods used for examining OH. Other techniques for probing HO_2, and also organic peroxy radicals (RO_2), include *matrix isolation–electron-spin resonance* (MIESR), in which the radicals are trapped in D_2O-ice at liquid-nitrogen temperatures, and their electron-spin resonance spectrum obtained. A *chemical amplifier* has also been described for the study of HO_2 and RO_2. In a typical implementation for HO_2, the sequence of reactions

$$HO_2 + NO \rightarrow OH + NO_2, \tag{4.2}$$

$$OH + CO \rightarrow H + CO_2, \tag{4.3}$$

$$H + O_2 + M \rightarrow HO_2 + M \tag{4.4}$$

is used to recycle HO_2, and thus to catalytically convert NO to NO_2. The NO_2 is then determined by, for example, an NO_2-specific chemiluminescence device.

Spectroscopy is very well suited to sensitive and specific *in-situ* applications. Ultraviolet, visible, and infrared regions are all widely used. *Diode-array detectors* are often used for wavelengths shorter than about 1100 nm, since they permit simultaneous examination of the whole of the spectrum of interest, in contrast to the point-by-point data gathering of older scanning spectrometers. At longer wavelengths, *Fourier transform infrared* (FTIR) spectroscopy serves a similar purpose.

Samples may be collected for off-line analysis in the usual way, but a more direct method is to measure the optical absorption in the atmosphere itself. For this purpose, observations may be made over straight horizontal paths of up to tens of kilometres in length, or the absorption beam can be reflected back to the location of the source, as illustrated schematically in the instrument on the bottom left of Figure 4.1.

One problem that besets optical measurements in the ultraviolet and visible regions is the scattering of the source radiation by both aerosol particles and by molecules, especially when long optical paths are employed. A great advance has been achieved in atmospheric studies by the adoption of the technique of *differential optical absorption spectroscopy* (DOAS). The spectrum is obtained in the ordinary way, but it then undergoes additional computer processing. Many molecules present in the atmosphere exhibit an absorption with sharp and discrete spectral features that result from the electronic, vibrational, and rotational quantum states that are populated. These absorptions are superposed on the spectral characteristics of the source radiation, and altered by atmospheric scattering. The DOAS procedure is to generate a second spectrum by computer that is deliberately degraded; the discrete features are smoothed out, but the general underlying shape is left behind. The difference between the original spectrum and this new one is then the true absorption of the gas-phase atmospheric molecules, which is relatively immune to the existence and characteristics of the interfering scattering. In the ideal situation, where the source output itself contains no sharp features, the procedure also takes into account the spectral distribution of the source, and even the spectral response of the spectrometer. DOAS is capable of very considerable sensitivity, and several species can be detected simultaneously with a modern wide-band instrument. Not only can relatively long-lived molecules such as O_3, HCHO, SO_2, HONO, and NO_2 be examined at concentrations down to a few ppb, but radical species such as OH, NO_3, and BrO can be detected at much lower concentrations (0.005 ppb for OH).

One technique for the *in-situ* detection of atomic and radical intermediates, which has proved particularly valuable in atmospheric studies, is that of *resonance fluorescence*. The basis of the technique is that electronic excitation of a species by absorption of light may be followed by fluorescent emission at the same wavelength: that is, the fluorescence is 'resonant' in energy with the exciting radiation. The wavelength(s) of absorption and emission are sharply defined for atoms and small radicals or molecules. For example, the ground-state hydrogen atom, $H(^2S)$, can absorb radiation at $\lambda = 121.6$ nm (the Lyman-α line of atomic hydrogen) to reach the first excited, ^2P, electronic state. One fate of $H(^2P)$ is radiation of the resonant Lyman-α line

$$H(^2S) + h\nu \rightarrow H^*(^2P) \qquad \text{absorption} \tag{4.5}$$

$$H^*(^2P) \rightarrow H(^2S) + h\nu \quad \text{resonance fluorescence.} \tag{4.6}$$

Since the radiation is isotropic, it can be detected off-axis from the exciting beam. Observation of Lyman-α fluorescence from a system illuminated by $\lambda = 121.6$ nm radiation (obtained from, say, a hydrogen-discharge lamp) thus demonstrates the presence of H atoms. The intensity of fluorescence is proportional to the concentration of hydrogen under suitable conditions. In the atmospheric experiments, a lamp and a shaped tube that is to act as a fluorescence cell are borne aloft on a balloon (or rocket) and parachuted down. Air passes through the tube, and the intensity of resonance fluorescence provides a measure of the concentration of the species for which the lamp provides specific excitation. A clever trick enables detection of radicals such as HO_2 and ClO, which do not fluoresce. By injection of NO, carried with the payload, into the test cell, OH or Cl can be produced stoicheiometrically. Both OH and Cl are fluorescent. The conversion involves the two reactions with NO that are also so important in the chemistry of the atmosphere itself, reaction (4.2) and its analogue

$$ClO + NO \rightarrow Cl + NO_2. \tag{4.7}$$

A modification to the balloon experiments allows the fluorescence probe to be lowered, like a yo-yo, more than ten kilometres from the balloon by a winch. The tethered instrument can thus make repetitive measurements of altitude profiles by reeling down or up, and the technique obviously represents a substantial advance over a conventional 'one-off' balloon experiment.

Aircraft campaigns have assumed an increasingly significant place in conducting stratospheric measurements. One aircraft that has been used is the ER-2 (it was formerly designated U-2, in an earlier incarnation when it was a spy plane). This aircraft can fly at such great heights that it can enter the stratosphere, and it has been equipped with a variety of instruments, including those that determine key free-radical species such as OH, HO_2, ClO, and BrO, as well as the oxides of nitrogen and ozone itself. A more recent stratospheric aircraft is the Russian Geophysica (Myasishchev M55), which can operate for up to five hours at an altitude of 21 km, well into the stratosphere. The level of instrumentation possible is illustrated by a campaign based on Ushuaia, Tierra del Fuego, in September–October 1999. The Geophysica carried three experiments designed for remote sensing of chemistry, and six for *in-situ* chemistry, as well as another five experiments for remote-sensing and *in-situ* physics.

Measurement of OH concentrations in the atmosphere is particularly important, and we now discuss briefly how this particular radical can be examined. Three methods currently exist for the time-resolved measurement of local hydroxyl radical concentrations. *Laser-induced fluorescence* (LIF) is a modification of the resonance-fluorescence method just described, in which lasers are used as the source of radiation resonant in wavelength with the $OH(A^2\Sigma^+ \leftarrow X^2\Pi)$ transition at $\lambda \approx 308$ nm. With continuous-wave (CW) lasers, molecular and aerosol scattering limit the sensitivity, because the scattered light is also at the same wavelength, and tends to mask the true fluorescence. One way round the problem is to use a pulsed laser. The scattered light follows the decay of the laser pulse, while the fluorescence decays with a longer lifetime that, in a vacuum, would be the radiative lifetime of the $A^2\Sigma^+ \leftarrow X^2\Pi$ transition. However, in air at atmospheric pressure, the fluorescence lifetime is much less than the radiative lifetime because of deactivating collisions, and the emission decays almost as fast as any available laser pulse. This difficulty has been overcome by expanding the sampled gas through a nozzle from 1 atmosphere (*ca.* 1 bar) to, say, 1 mbar, thus increasing the fluorescence lifetime, and permitting time discrimination from the exciting pulse. The technique is *fluorescence analysis with gas expansion* and is given the acronym FAGE.

Long-path absorption spectroscopy of OH has also been remarkably successful, especially when DOAS methods are used. Some results using DOAS will be shown later. In a recent high-resolution instrument, several absorption features of OH in the ultraviolet region are monitored simultaneously in order to overcome interference problems. The method gives absolute [OH] without the need for calibration.

4.3 REMOTE SENSING

Spectroscopy, in one guise or another, is at the heart of all remote-sensing experiments. We illustrate the available methods using stratospheric ozone as the example. Figure 4.2 shows some of the instruments that have been employed. Only one (the balloon-borne ozone sonde) of the nine represented is an *in-situ* device. The other eight all depend on remote sensing, and all use the absorption, emission or scattering of electromagnetic radiation. Here, five of the remote-sensing instruments are shown as ground-based instruments, while there are three on satellite platforms.

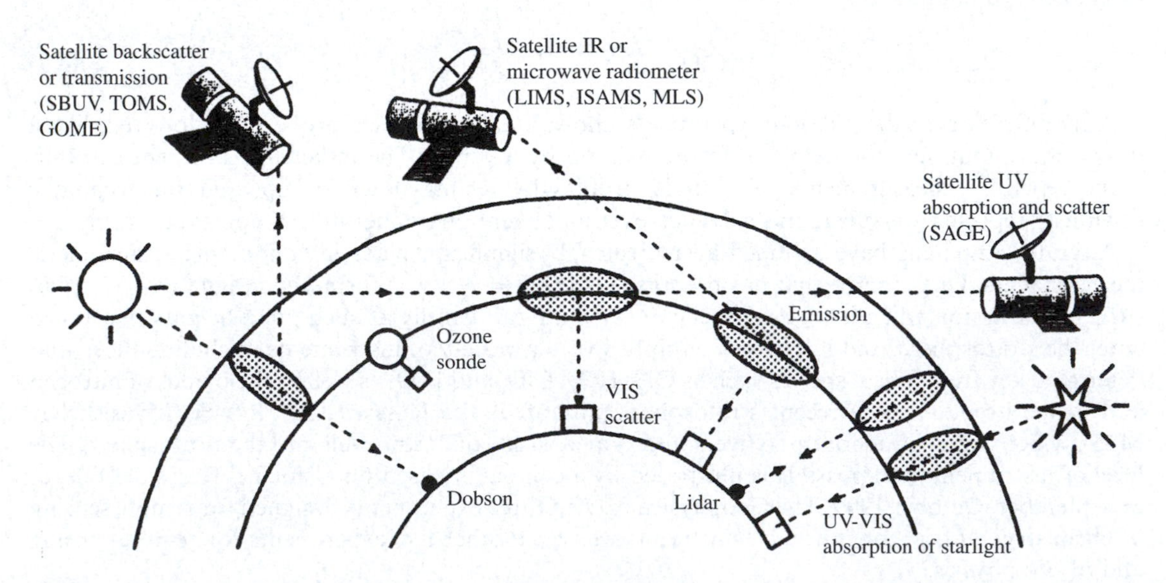

Figure 4.2 A selection of techniques that can be used for measuring stratospheric ozone concentrations from the ground, from airborne platforms, and from satellites.

The most famous of the instruments is probably the Dobson spectrophotometer. Developed in 1924, it is the earliest instrument used to measure ozone, and modern versions continue to provide data. They are the only long-term source of ozone data, with one station in Arosa, Switzerland, having provided continuous measurements since the 1920s. Dobson spectrophotometers can be used to measure both total column ozone and profiles of ozone in the atmosphere. Total ozone measurements are made by comparing a wavelength of the ultraviolet spectrum that is strongly absorbed by ozone with one that is not. Measurements can be based on light from the Sun, Moon, or stars. The vertical distribution of ozone is derived using the *Umkehr method* in which the intensities of reflected, rather than direct, UV light are measured. The ozone distribution is derived from the change in the ratio of two UV wavelengths with time as the Sun sets. An Umkehr measurement takes about three hours, and provides data up to an altitude of 48 km, with the most accurate information for altitudes above 30 km. Another, much more recent, application of optical spectroscopy is *lidar* (*li*ght *d*etection *a*nd *r*anging) which is the optical analogue of radar; it is represented in Figure 4.1 by the device in the middle of the bottom row, and in Figure 4.2 as the device next-to-last on the right. In lidar, absorption is observed in a laser beam that has been scattered downward by the atmosphere. A telescope is used to collect ultraviolet light that is scattered from two laser beams, one of which is absorbed by ozone ($\lambda = 308$ nm) and the other is not ($\lambda = 351$ nm). A short pulse of laser radiation is used, and the scattered light is monitored as a function of time after the pulse. Since the time is proportional to distance travelled, the technique allows vertical concentration profiles to be determined. By comparing the intensity of light scattered from each laser, a profile of ozone concentration as a function of altitude can be measured from 10 km to 50 km.

An important advance of the past decades has been the harnessing of satellite and space-shuttle platforms to provide long-term and global measurements of trace gases in the atmosphere. Several satellites have carried or currently carry the Total Ozone Mapping Spectrometer (TOMS) that provides crucially important daily global maps of stratospheric ozone concentration. At the time of writing, the TOMS record is obtained by an instrument mounted on the AURA satellite, which achieved its five-year mission requirement in July 2009. These craft sense ozone concentration by looking down at backscattered ultraviolet light. Another satellite, the European ERS-2, was launched in April 1995, carrying additional equipment called Global Ozone Monitoring Experiment (GOME). GOME is a nadir-looking instrument that measures solar radiation transmitted through or scattered by the Earth's atmosphere. In March 2002, the European Space Agency (ESA) launched ENVISAT. This large satellite carries many different instruments, of which those of major importance in studies of atmospheric chemistry are GOMOS (Global Ozone Monitoring by Occultation of Stars), MIPAS (Michelson Interferometer for Passive Atmospheric Sounding), and SCIAMACHY (Scanning Imaging Absorption Spectrometer for Atmospheric Cartography).

4.4 MODELS OF ATMOSPHERIC CHEMISTRY: DIAGNOSIS AND PROGNOSIS

Determination of the identities and concentrations of the chemical species in the atmosphere undoubtedly has an intrinsic interest, but the information really only becomes of true value when it can be interpreted in terms of chemistry and physics. What has become known as a *model* provides the necessary link. Models can range from extremely simple mental pictures of the processes occurring to highly complex implementations that can be tackled only by the fastest computers available. A rough calculation, perhaps requiring just mental arithmetic, of source strength of some compound from a concentration measured in the atmosphere and a rate coefficient determined in the laboratory is in truth a simple model. However, we know already (see Section 3.5.2) that a species may have a lifetime that is short compared with the time taken for atmospheric motions to transport it a significant distance; or that the lifetime may be so great that the species is essentially evenly mixed throughout the global atmosphere; or that the situation lies somewhere between these extremes. A species that has lived long enough to travel a significant distance from its place of birth may find itself in a chemical and physical environment very distinct from that of its origins. Temperatures, pressures, solar intensities, and the other chemical species present may be quite different. It may therefore be necessary to make progressive assessments of the chemistry occurring as the species travels vertically, latitudinally (north–south), and longitudinally (east–west). In a real atmosphere, there may be hundreds of chemical species participating in hundreds of processes. Some of the species may affect the temperature, globally or locally, thus leading to a change of reaction rates. The Sun certainly goes up and down every day (in most places!), and the seasons change. What we see emerging is a system involving a multitude of reactants and reactions, and in which there are interactions and feedbacks between the different parts of that system. We have now reached something beyond the capabilities of the most skilled mental arithmetic!

Let us return to the mental arithmetic that allows us to calculate the source strength for our hypothetical compound. If the rate of release of the compound is actually known from measurement, or some other *independent* observation, the match (or lack of it) between the related quantities becomes a *diagnostic* indicator of how well we have really understood the system. The concentrations of trace gases and particles in general give better diagnostic information than the concentrations of the bulk gases of the atmosphere, such as N_2 and O_2. The reason lies in part in the atmospheric lifetimes of the trace components; for those of relatively short lifetime, the variability with time since release, and time of day and season on the one hand, and with altitude, latitude and longitude on the other, provide a rich basis for interpretation. Richer yet may be the information yielded by the concentrations of the intermediates, including atoms and free radicals, of the chemical transformations. Their lifetimes are likely to be considerably shorter than those of the parent compounds, and their concentrations more sensitive to the details of the chemistry and interactions occurring. Unfortunately, a short chemical lifetime, which implies a high reactivity, will mean that atmospheric concentrations of such reaction intermediates will be small, and obtaining accurate measurements will be correspondingly taxing.

Once a model has proved itself diagnostically, and confidence has been reached about its validity, it may be employed to predict the atmospheric response to situations that have not yet arisen. *Prognostic* models are of interest in evaluating the future behaviour and evolution of an atmosphere subject to changes in inputs of trace species produced, for example, by the biosphere, by volcanic eruption, or by Man, and to changes in natural parameters such as solar intensity. The predictions of reliable prognostic models can (and should) be used to formulate a variety of policy decisions.

One of the reasons for emphasizing the complexity of the atmospheric systems with regard to lifetimes and transport is to provide an explanation for the key role that atmospheric models play in the way that we currently think about atmospheric chemistry. Even the chemistry on its own is not susceptible to straightforward algebraic treatment, except in the very simplest circumstances. Add the other factors of the interactions and feedbacks between the chemistry and temperatures, pressures, light intensities, and so on, and it is evident that recourse must be had to numerical models that can be solved by computer methods (although the most complete treatments offer almost insurmountable obstacles even to the highest capacity of present-day computing power).

Numerical models are currently used very widely in the sciences in order to describe complex, interacting, and non-linear systems such as those encountered in atmospheric chemistry. In one sense, such models attempt to replicate by computer the behaviour of the natural system. The results of the computer simulation can then illuminate various aspects of the real-life atmosphere, such as the causes of certain observations or the probable effects of projected or actual perturbations.

The starting point in the simulations is often to construct diagnostic models, as just noted. These models are used to assess hypotheses about the physics and chemistry of atmospheres from a knowledge of present-day physical and chemical structure. Information appropriate to the type of model is fed into the calculations, which provide as output further information about the other parameters in the system. These parameters can then be compared with measurements in the atmosphere. The question of what reaction scheme should be used is highlighted by the discovery of the Antarctic ozone 'hole', which is discussed in detail in Sections 9.4 and 11.3. Models current at the time of the discovery showed no hint of the atmospheric behaviour actually observed. The reason was that their chemical reaction schemes were incomplete, and that they omitted the reactions on ice and other particles that have subsequently been recognized as playing a key role. In this instance, then, a lack of compatibility between the observations and model predictions required a revision of our interpretation of the fundamental physical chemistry. The new understanding has been employed to further refine models of atmospheric behaviour so that they can incorporate heterogeneous chemistry in their schemes. We note also that, since the chemical reaction schemes in the models perforce calculate concentrations of all the reaction intermediates, the predictions are often used to estimate atmospheric concentrations for species for which present-day experimental techniques cannot provide the true measurements.

It is now sensible to examine how typical models are constructed. One counsel of perfection would be to consider a volume element in an atmosphere small enough to be uniform with respect to all variables such as temperature, density, composition, and so on. If the rate of flux to and from that element of heat, radiation, matter, *etc*., from all other atmospheric elements is calculated, then the rate of change of the various physical parameters can be established. Chemical change within the volume element requires only slight modification for the local alterations in composition, and possibly in physical conditions.

Consider first what happens if more matter enters the volume element than leaves it. The density of the element evidently has to increase, and the local rate of change of density is equal to the net flow of mass into unit volume per unit time. This kind of balancing equation is referred to as a *continuity equation*. Solution of the continuity equations for every physical and chemical parameter of interest, and for every volume element in the atmosphere, should then lead to a self-consistent model of atmospheric behaviour that mimics in all respects the temporal and spatial changes in the real atmosphere. According to this view, with sufficient input information, and a fine enough grid size for the volume elements, all meteorological as well as chemical phenomena could be simulated by the model. Such three-dimensional models (3D) would be ideal for studying atmospheric chemistry. Computer-numerical solutions are naturally used in models, but the 3D models are extremely demanding of computer time and memory, and the chemical content may have to be limited. Simpler models may thus provide more useful and detailed chemical information than the 3D models. For this reason, we base the present brief overview on models employing the greatest approximations to begin with, and then build up to models of greater complexity.

The simplest of all possible models is the box model. The box model is a zero-dimensional model, and assumes uniform mixing of individual constituents of an atmosphere. Figure 4.3 shows the components of such a model. Chemical species are brought into the box horizontally and vertically. In the horizontal dimension, the flow is along the wind direction. Vertically, gases may enter or leave by the top and bottom of the box. If the box is situated at ground level, as is the case for the box shown in our illustration, then there may be surface sources and sinks of the various molecules.

In the calculation, each chemical species is handled separately. An equation is written for the rate of change of concentration as determined by physical input and output in horizontal and vertical dimensions and by chemical and photochemical change within the box.

The chemistry is treated exactly as in a well-mixed laboratory system. The time evolution of each species is controlled by the chemical interactions alone. Rates of reaction are calculated as described in the previous section, with the steady-state hypothesis applied if necessary and appropriate. For long-lived species, diurnal (but probably not seasonal) averaging may be

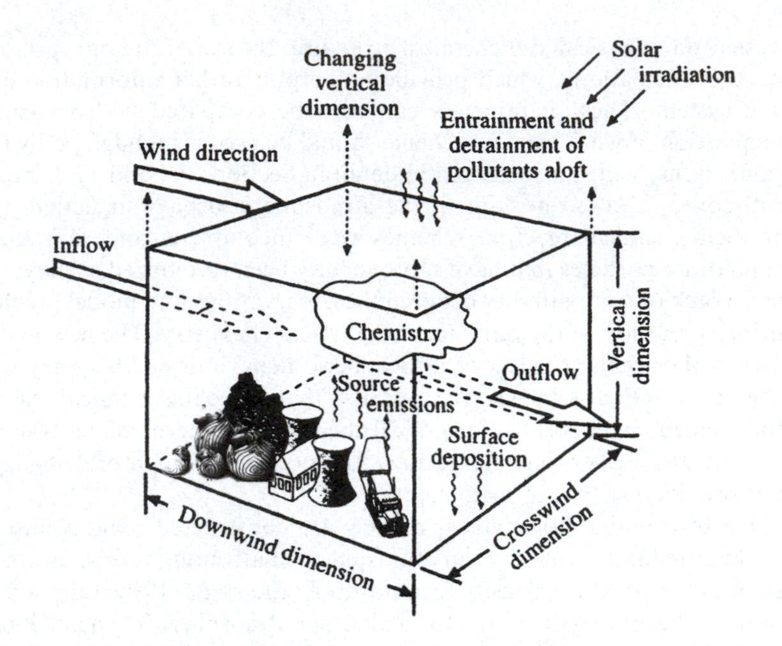

Figure 4.3 Components of a zero-dimensional box model. Adapted from T.E. Graedel and P.J. Crutzen, *Atmospheric Change*, W.H. Freeman, New York, 1993.

reasonable. Such models are most useful in the analysis of global budgets of long-lived species, and of local budgets of short-lived ones.

Despite the gross oversimplifications inherent in box models, they do allow a first appraisal of many atmospheric problems, and the relatively small demand on computing power makes them very attractive for preliminary investigations.

One further feature of atmospheric models should be introduced at this stage. Two distinct types of model have been developed for use in atmospheric chemistry, which will be illustrated here for the box model so far developed, although they can be extended to models of higher dimensionality. In the first approach, the box is located at a fixed geographical location, as implied by the description already given. Such a formulation is a *Eulerian model*. If atmospheric motions and the transport of chemical species are to be treated more or less realistically, that is achieved in Eulerian models of higher dimensionality by providing realistic terms for transfer in and out of the box from some or all of its faces. In a *Lagrangian model*, on the other hand, a completely different approach is taken. The hypothetical box containing the chemical components is now allowed to follow the atmospheric motions themselves, and the chemistry in this air parcel is allowed to occur as it moves around. A knowledge of the meteorology, obtained by observation or by forecasting, provides the information about the route taken by the box. Each type of model has its virtues and its disadvantages of course. One great advantage of the Lagrangian approach is that there are now no terms for input or output of chemical species from the faces of the box. On the other hand, factors such as the source strengths, temperatures, pressures, and solar irradiances may all vary as the box moves from place to place. Another problem is that it may be difficult to make real time-dependent atmospheric measurements, because the instruments would have to follow the air parcel around.

One-dimensional (1D) *models* are the next step up in the model hierarchy. They are designed to simulate the vertical distribution of atmospheric species, but not any horizontal variations. The models thus represent horizontal averages, but do include atmospheric transport, attenuation and scattering of solar radiation, and detailed chemistry. Figure 4.4 illustrates schematically how such a model is constructed by making a stack of the zero-dimensional boxes of Figure 4.3. Where vertical

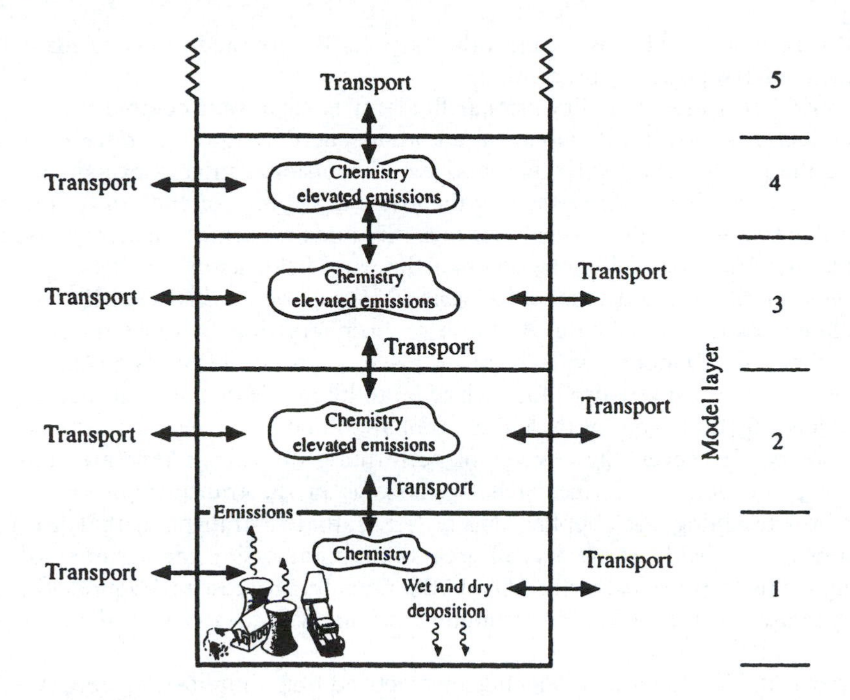

Figure 4.4 A one-dimensional box model: several of the boxes of Figure 4.3 are stacked on top of each other. Source as for Figure 4.3.

winds are weak or absent, as in the Earth's stratosphere, vertical transport is treated by adopting the *eddy diffusion* mechanism (Section 2.4.1).

Averaging of the 1D model results must be performed to achieve a link with physical reality. A sensible altitude spacing must be adopted; frequently, a greater number of thinner layers are set up at the lower altitudes, where changes of concentration with altitude are most rapid, and where the chemical reactions are fastest. The results must also be calculated over averaged time steps. Diurnal averaging procedures permit considerable savings of computer time when solution is required of time-dependent problems over long durations. Fully time-dependent solutions are more useful diagnostically and prognostically, and must be used for many short-lived species. Such solutions are, however, complicated, and expensive computationally.

Two-dimensional (2D) *models* recognize that atmospheric conditions are a strong function of latitude as well as altitude. They therefore provide spatial resolution in vertical and meridional directions (*i.e.* latitudinal, but not longitudinal, resolution). Horizontal as well as vertical motions must now be accommodated. Averaged meridional velocities can be obtained from direct observation for both troposphere and stratosphere. Some problems exist about interpretations of eddy motions, which include, in two dimensions, large-scale internal waves. Nevertheless, 2D models are far more realistic than 1D models. The motivation for constructing 2D models is that meteorological quantities on a rotating planet are not expected to show variations with longitude if averaged over several days. In practice, latitudinal gradients in the Earth's atmosphere are generally larger than those in the longitudinal direction. However, at low altitudes, where the sources and sinks of trace gases may be highly localized, ignoring the longitudinal variations may make the models unrealistic. Two-dimensional models are thus possibly most suited to the study of stratospheric problems. The use of zonal means is even then a limitation, since processes such as those involving heterogeneous chemistry can also be highly localized. Some attempts have been made to overcome this difficulty by using parametrizations that permit the incorporation of some effects of zonal asymmetries. Coupling of transport and heating effects is an important part of the models. Feedback between chemistry and other aspects of the model should ideally be accommodated because of the potential interactions.

Because atmospheres are three-dimensional fluids, it is clear that complete simulation of the radiative–chemical–dynamical behaviour of an atmosphere requires a *three-dimensional* (3D) *model*. One of the problems with the one- and two-dimensional models explained so far is that vertical motions and mixing are described by the artificial concept of eddy diffusion. Yet it is not generally possible to *measure* the vertical motions. 3D models, such as those now used in weather prediction, may yield the vertical components of the wind fields as one of the explicit or implicit results. *Regional*, *mesoscale*, and even global *general circulation models* (GCMs) have been developed to simulate weather and climate. Addition of chemistry thus becomes the next goal but, as pointed out earlier, full 3D models with chemistry place enormous demands on computer power. A set of kinetic equations must be solved for each of up to hundreds of chemical species in the model, and at each of possibly thousands of three-dimensional grid points for which the physical equations must also be solved. However, the rewards of performing the calculations are considerable. 3D models can, in principle, include feedback mechanisms in the transport process, and they are indispensable in examining the coupling effects (*e.g.* radiative transport) that are incompletely treated in simpler models. They are also of great value in assessing the accuracy of spatial- and time-averaging techniques needed in the 1D and 2D models. Progress in 3D modelling is currently rapid both because of increasing programming efficiency and the ever-increasing power of computers.

Early attempts at 3D chemical modelling involved adding simplified chemistry schemes that utilized just a small subset of the important reactions. Another way of dealing with the complexities inherent in 3D chemical models allows the chemistry to be run *'off-line'*, in *decoupled chemical transport models* (CTMs). In this approach, the model is first run without chemistry, in order to obtain the meteorological parameters such as winds and temperatures. This information is subsequently fed into another 3D *chemical* model, which then advects the trace species and updates the chemistry. Yet another approach uses analyses of meteorological observations to provide the wind and temperature fields. Large reductions in the complexity of the calculations can be obtained in this way, but the decoupling necessarily leads to a drift away from reality. In particular, the feedbacks between chemistry and meteorology are lost. Since the two-way impacts of chemistry, radiation, and dynamics on each other are evidently one of the proclaimed virtues of the full 3D models, the simplification is bought at considerable cost.

The general principles in constructing models of tropospheric chemistry follow those just set out. The models are not so very different in concept from those used for the stratosphere, but they may present much greater difficulty in implementation. Sophisticated modelling of transport is essential, and clouds and other particles need to be accommodated by the physics and the chemistry. Furthermore, the transport of trace constituents in the atmosphere is intimately related to the hydrological cycle, so that a high-quality simulation of the hydrological cycle is a necessary prerequisite for successful modelling of trace gases in the free troposphere.

Three-dimensional (3D) chemistry–transport models are being employed increasingly widely for the representation of the complex interactions between the main precursors of O_3, such as NO, NO_2 and CO, and the highly variable meteorological conditions encountered in the troposphere. A first requirement is that they properly reflect transport properties. One way of evaluating these properties is through tracer experiments. A tracer particularly well suited to this purpose is ^{222}Rn, a radioactive isotope of radon. The gas is emitted from soils, and since the half-life for radioactive decay is 5.5 days, ^{222}Rn distributions are quite sensitive to relatively small-scale meteorological processes. In general, the tracer tests show that many current models perform well in representing the effects of synoptic weather systems, but are much less good at simulating behaviour in the upper troposphere, suggesting that the effects of deep convection are not being correctly modelled. Since the oxides of nitrogen and other short-lived species in the free troposphere are influenced by vertical exchange processes, the poor performance of the models in this respect indicates one direction in which improvement must be made.

As computer technology advances, it seems certain that future fully coupled 3D models will be susceptible to calculations even when full chemical schemes are run 'on-line'. Such chemical models are already being run, and they can typically be integrated for simulations covering periods from a few months up to years. The task of reducing approximations, and thus increasing realism, looks like one that can be tackled with continuing hope.

4.5 'FAMILIES' OF REACTANTS AND INTERMEDIATES

Even for 1D and 2D approximations, the chemistry may impose an excessive computational load. The chemistry of the stratosphere is 'simple', yet involves about fifty chemical species in nearly 200 reactions. There are thus fifty (simultaneous) continuity equations to be solved at each grid point. Many more reactants and reactions may be required to describe the troposphere completely. Direct numerical solution of the differential equations may require inordinate computer memory and run times. The problem is obviously particularly pressing when a model is used to test prognostically the effects on an atmosphere of altering concentrations of trace species, because the calculation must be repeated many times. A particular difficulty in the computations is that the differential equations for the kinetics are often '*stiff*'. That means that the concentration of one species may change very rapidly, while the concentration of another may vary much more slowly. Although there are numerical methods designed to tackle stiff differential equations, there is always a price to be paid in choosing, for example, sensible time steps over which to perform the integrations. One approach used to reduce the computational needs consists of identifying closely coupled chemical species that can be summed into *families*. The idea is that the members of the family can be interconverted on a timescale short compared with that over which other concentrations vary. 'Odd oxygen' is one of the most important of such families in the Earth's atmosphere. The word 'odd' used in this context means 'a not even' number of oxygen atoms, and the members of this family are O, O_3 but not O_2. The point is that during daylight hours, the two reactions

$$O_3 + h\nu \rightarrow O(^1D) \xrightarrow{M} O(^3P) \tag{4.8}$$

$$O(^3P) + O_2 \xrightarrow{M} O_3 + M \tag{4.9}$$

rapidly interconvert O and O_3. The symbol M in these equations is a 'third body' that removes energy: in the Earth's atmosphere, M represents the bulk constituents, mainly O_2 and N_2. The ultraviolet photolysis of ozone in step (4.8) produces initially an excited oxygen atom, $O(^1D)$, but for altitudes of 30 km and lower it is quenched by M to the ground-state $O(^3P)$ with a lifetime of $<0.1\ \mu s$. The lifetime for regeneration of O_3 at these altitudes is somewhat longer (< 25 ms), but these numbers mean that the interconversions are so rapid that the concentrations of O and O_3 can readjust almost 'instantaneously' on the timescale of any transport or changes in solar intensity. Thus, in the lower stratosphere and below it is only necessary to include O_x ($x=1$, 3) in the modelling, not the individual members of the family. Other important families, as we shall see, include NO_x ($x = 1$, 2), HO_x ($x = 0$, 1, 2) and ClO_x ($x = 0$, 1, 2). Another important representation, NO_y is frequently used; it stands for 'odd nitrogen' and is defined as NO_x *plus* all oxidized nitrogen species that are sources or sinks of NO_x on short timescales. These additional species include N_2O_5, HNO_3, HNO_4, $ClONO_2$ and $CH_3CO(O_2)NO_2$. A reduced set of equations is solved for the families, and then each family is partitioned into its components. This latter stage requires some (inspired) guesswork, but 1D models using 'family' and direct solutions compare well.

4.6 COMPARISON OF MODEL PREDICTIONS WITH ATMOSPHERIC OBSERVATIONS

Atmospheric measurements provide the concentrations and distributions of trace species; models can predict the same information, and the quality of the match between the measurements and calculations indicates how well or badly we understand the aspect of atmospheric chemistry under consideration. Most of the comparisons between what we can crudely call experiment and theory obviously belong along with the discussion of the topics that will be presented in the chapters that follow in this book. However, to round off this chapter on observations and models, we give just two illustrative examples, one from the stratosphere and one from the troposphere. The down-side of providing examples at this point is that the reader hoping to find an exposition in strictly logical order will now discover measurements of species whose significance and value can only be appreciated after understanding the later chapters. To reduce the effects of this difficulty, we have chosen for both examples measurements of the HO_x species OH and HO_2, mentioned several times as being highly sensitive indicators of atmospheric behaviour.

Figure 4.5 shows stratospheric measurements of HO_x obtained by a balloon-borne far-infrared Fourier transform spectrometer (FIRS-2). The solid line is the altitude profile generated by a model in which O_3, H_2O and CH_4 were constrained by simultaneous measurements. These model calculations demonstrate that, by making certain adjustments to the kinetic parameters, existing chemical reaction schemes can reproduce the observations quite well. One particular conclusion from these results is that HO_x observations in the upper stratosphere appear to require a reduction in the rate coefficient for the reaction

$$HO_2 + O \rightarrow OH + O_2 \tag{4.10}$$

by about 25 per cent from the currently recommended value.

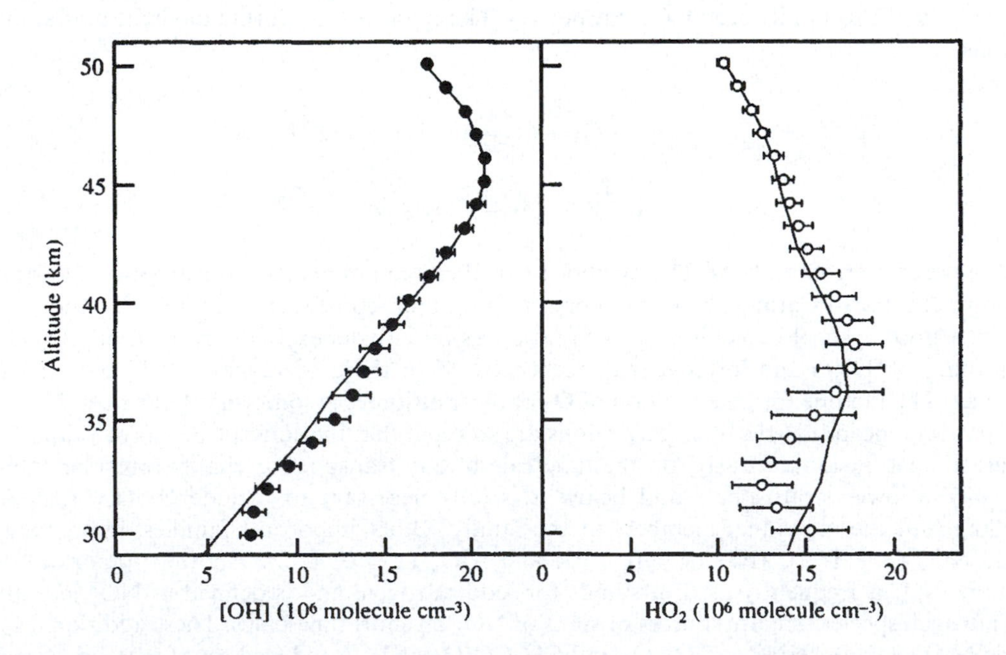

Figure 4.5 Concentration profiles of OH and HO_2 measured by a balloon-borne instrument at 65°N (spring, mid-morning). The solid lines are the results of calculations using selected kinetic parameters. From K.W. Jucks *et al.*, *Geophys. Res. Lett.*, 1998, **25**, 3935.

Despite the apparently good agreement between observation and theory that the results of Figure 4.5 imply, there remain uncertainties about how well HO_x concentrations are understood. Although balloon-borne observations of OH generally agree with modelled concentrations to within about 10 per cent, shuttle-borne observations are 30 to 40 per cent *lower* than the predicted values, while observations of total column concentrations are up to 30 per cent *higher* than the modelled ones. Different techniques, platforms, and instruments sometimes produce conflicting data about HO_x concentrations. At present, it seems that the most sensible interpretation is that the results of these extremely difficult experiments confirm expectations well enough to enhance confidence in the reaction schemes presented so far. Nevertheless, there is doubtless a place for further refinements in both observation and theory.

So far as the troposphere is concerned, it has long been evident that verification of the theories of OH chemistry would benefit from good measurements of the spatial and temporal distribution of concentrations of the radical. Validation of a time-dependent model requires simultaneous determination not only of [OH] itself, but also of all the parameters that control the local OH chemistry.

Since OH is very reactive, its chemical lifetime is so short that atmospheric transport plays only a minor role in determining concentrations and distributions of the radical. Comparison of model predictions with experimental data is thus possible with a simple zero-dimensional model that takes into account only the chemical part of the budget. Without agreement at this level, there is little point in evaluating models of higher dimensionality and complexity that incorporate the same chemistry. Figure 4.6 shows the results of one comparison between calculated and measured OH concentrations in the troposphere. The field observations were made at two rural sites in Germany, and the predictions come from a simple box model that should be perfectly adequate. According to the figure, there is evidently an excellent correlation between the observed and predicted concentrations over an [OH] range that spans a factor of five. In absolute terms, the model overpredicts concentrations by about 20 per cent, but that discrepancy can be readily accommodated by the *combined* uncertainties in all the rate constants that are needed for the reactions in the model.

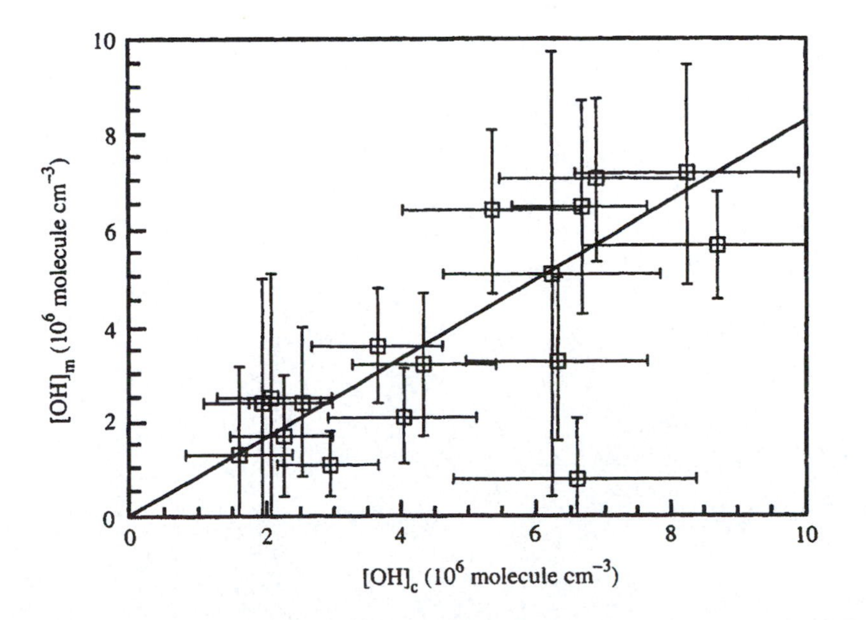

Figure 4.6 Correlation of measured and calculated OH concentrations at two rural sites in Germany. From D. Poppe, J. Zimmermann, and M. Kuhn, in *Tropospheric Modelling and Emission Estimation*, ed. A. Ebel, R. Friedrich, and H. Rodhe, Springer-Verlag, Berlin, 1997.

The qualitative and quantitative behaviour of the model thus gives considerable confidence in the basic chemical scheme that underlies our current understanding of tropospheric chemistry, and illustrates why measurements of OH are so important in achieving this kind of validation.

Daytime [OH] ranges from 1 to 10×10^6 molecule cm^{-3}, in line with the general predictions of realistic models of the troposphere, winter-time values of 1–5×10^6 molecule cm^{-3} being realistic, and 5–10×10^6 molecule cm^{-3} being generated during the summer. At night, concentrations drop to $< 5 \times 10^6$ molecule cm^{-3}. Recent *in-situ* observations show that, especially in the pollution-free atmosphere, OH concentrations are controlled by O_3 photolysis and the subsequent reaction of $O(^1D)$ with water vapour (see Section 5.1). Under more polluted conditions, the presence of NO_x allows the reaction

$$HO_2 + NO \rightarrow OH + NO_2 \tag{4.2}$$

to recycle HO_2; O_3 concentrations will also increase because NO_2 is involved in its tropospheric formation (see Section 5.2). In turn, OH concentrations then become enhanced.

Models predict that the highest OH concentrations of all will arise in the tropics, where large light intensities lead to rapid photolysis of O_3, and thus rapid production of $O(^1D)$, and high humidities favour its conversion to OH. Concentrations are also expected to be about 20 per cent higher in the Southern Hemisphere than in the Northern. This asymmetry is a consequence of the higher abundance in the Northern Hemisphere of anthropogenic carbon monoxide, CO, which removes OH in the rapid reaction

$$CO + OH \rightarrow CO_2 + H. \tag{4.11}$$

Largely because of the differences in NO_x concentrations and the occurrence of the key reaction (4.2), [OH] is about five times larger over continental areas than over the oceans.

CHAPTER 5

Ozone

5.1 OZONE: CHEMISTRY AND PHOTOCHEMISTRY

Ozone (trioxygen, O_3) was discovered by Christian Friedrich Schönbein in 1839 as a product of passing an electrical discharge through air. The substance possessed a strong odour, and Schönbein named it ozone from the Greek *ozein*, to smell. He thought that he had found a new halogen such as chlorine, bromine and iodine. Indeed, it liberates elemental I_2 from solutions of iodides, just as can Cl_2 and Br_2, although it subsequently became apparent that ozone contains oxygen atoms alone. Ozone is an endothermic substance (positive enthalpy of formation from the standard state of the element)

$$\tfrac{3}{2}\,O_2 \rightarrow O_3; \quad \Delta H^{\ominus}_{298} = 143\,\text{kJ mol}^{-1} \tag{5.1}$$

Put another way, it means that O_3 possesses stored internal energy; in a reasonably pure state it can explode violently to yield O_2 again.

For such a simple molecule, ozone has long held a fascination for laboratory scientists: its chemistry, spectroscopy and photochemistry are all enormously rich. Ozone also plays a peculiarly significant part in the chemistry of the Earth's atmosphere, even though it is a 'minor' species in terms of abundance. The aeronomy and meteorology of atmospheric ozone have been studied for about 170 years, since the suggestion by Schönbein in 1840 of an atmospheric constituent having a peculiar odour. Schönbein's detection of ozone in the atmospheres of several European cities (see Section 4.2) surely represents the first step in the modern study of air chemistry. Ozone's existence in the troposphere was established in 1858 by chemical means, and subsequent spectroscopic studies in the visible and ultraviolet regions by W.N. Hartley showed, as early as 1881, that ozone is present at a higher mixing ratio in the 'upper' atmosphere than near the ground. By the early part of the 20th century, quantitative analysis, particularly by Fabry and by Dobson, had shown that this upper-atmospheric ozone was to be found in an *ozone layer*.

We now know that ozone is found in trace amounts throughout the atmosphere, with the largest concentrations in a well-defined layer at altitudes between about 15 and 30 km. The mixing ratio of ozone with respect to the entire atmosphere is a few tenths of a part per million. If the ozone in a column of the atmosphere were collected and compressed to 1 atm pressure, it would occupy a column about 3 mm tall. Total amounts at a particular location of a trace gas in the atmosphere are

Atmospheric Chemistry
By Ann M. Holloway and Richard P. Wayne

Published by the Royal Society of Chemistry, www.rsc.org

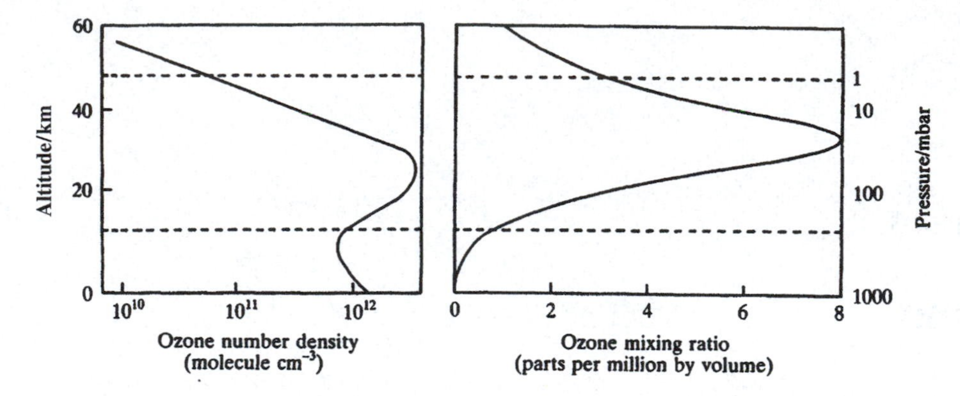

Figure 5.1 Variation of atmospheric ozone concentration with altitude, expressed as an absolute number density and as a mixing ratio. From *Stratospheric Ozone 1988*, UK Stratospheric Ozone Research Group, HMSO, London, 1988. Reproduced under the terms of the Click-Use Licence.

calculated as *column densities*, the number of molecules in an overhead column of unit area (normally in units of molecule cm^{-2}). The geophysical literature almost always uses *Dobson units* (DU) to describe atmospheric column ozone abundances in order to honour Dobson's pioneering work on stratospheric ozone (see also Section 4.3). The definition of 1 DU is that thickness in units of hundredths of a millimetre that the ozone column would occupy at standard temperature and pressure (STP: 273 K and one atmosphere). Thus, the compressed ozone column of 3 mm thickness corresponds to an atmospheric abundance of 300 DU. In this book, we generally keep to molecular units more familiar to chemists and physicists for concentrations, but use DU for column abundances. The conversion factor is $1\,DU = 2.69\times10^{16}$ molecule cm^{-2} for $T = 273$ K.

Figure 5.1 shows a typical concentration–altitude profile for ozone in two ways: as an absolute number density (concentration), and as the mixing ratio (fractional composition). The concentration peaks sharply at about 30 km, and most of the ozone molecules lie within a layer 20 km thick (remember that the number density scale for the left-hand panel is logarithmic). The dashed lines represent the approximate positions of the tropopause and stratopause, thus showing that the ozone layer lies within the stratosphere: indeed, as we shall see later, it is the presence of ozone that is responsible for the existence of the stratosphere. As stated in Section 1.5, peak fractional abundances in the ozone layer can approach 10^{-5}(10 ppm), and peak concentrations in the troposphere reach more than 10^{12} molecule cm^{-3}.

Several factors contribute to ozone's importance, and they will be a recurrent theme in our later discussions. Perhaps the outstanding feature is the relationship between the absorption spectrum of ozone and the protection of living systems from the full intensity of solar ultraviolet radiation (see Sections 5.5 and 7.2.2). In the present section, we note certain chemical features that contribute to ozone's atmospheric role.

Energy absorbed by ozone from the solar ultraviolet radiation is ultimately degraded to heat; so, indeed, is the solar energy originally used in the formation of the ozone. The net result is a heating of the Earth's atmosphere in the region of the ozone layer that has a profound influence on atmospheric temperature structure and vertical stability (Section 2.3).

Examination of the chemical routes that lead to the release of thermal energy is also instructive. Because ozone is an endothermic species, reactions involving it are often exothermic. Further, rates of these reactions tend to be relatively high, so that ozone is an important chemical reaction partner. In the particular case where ozone itself absorbs a quantum of light ($\lambda \leq 310$ nm), the energy of the system is yet further elevated. A manifestation of this energization is that the ozone *dissociates* to form an atomic and a molecular fragment (*i.e.* O and O_2), *both of which are electronically excited.*[i] The oxygen atom has an excitation energy of $190\,kJ\,mol^{-1}$, and this excess energy makes the atom highly reactive. Processes such as

$$O^* + H_2O \rightarrow OH + OH; \quad \Delta H^{\ominus}_{298} = -119\,kJ\,mol^{-1} \tag{5.2}$$

$$O^* + CH_4 \rightarrow OH + CH_3; \quad \Delta H^{\ominus}_{298} = -178\,kJ\,mol^{-1} \tag{5.3}$$

$$O^* + N_2O \rightarrow NO + NO; \quad \Delta H^{\ominus}_{298} = -340\,kJ\,mol^{-1} \tag{5.4}$$

(where O* represents an electronically excited oxygen atom), are all exothermic; we note that reactions (5.2) and (5.3) are endothermic with ground-state oxygen atoms for which enthalpies of reaction are $190\,kJ\,mol^{-1}$ less. All the reactions are fast because they have low activation energies. Water vapour, CH_4, and N_2O are important minor atmospheric constituents, while the radicals OH, CH_3, and NO are themselves highly reactive and are involved in atmospheric chemical changes of paramount significance. In each case, a driving force for radical production can be the absorption of solar radiation by ozone.

5.2 SOURCES OF OZONE IN STRATOSPHERE AND TROPOSPHERE

5.2.1 Photolysis Mechanisms

Almost all the ozone in the stratosphere and troposphere is formed by the addition of atomic oxygen to molecular oxygen

$$O + O_2 + M \rightarrow O_3 + M \tag{5.5}$$

In this equation, the species M is a *third body*: that is, a species that can carry off the excess energy of formation of O_3 and without which the newly formed O_3 would fall apart again. In the atmosphere, M is air, largely N_2 and O_2.

In the stratosphere, relatively short-wavelength ultraviolet radiation arrives from the Sun. Atomic oxygen is formed by the photolysis of O_2 at $\lambda < 242$ nm.

$$O_2 + h\nu \rightarrow O + O \ (\lambda < 242\,nm). \tag{5.6}$$

At $\lambda < 195$ nm, one of these atoms is born excited, as the O* we have mentioned earlier, but that is not important for the present purposes. Ultraviolet radiation is absorbed most strongly by O_2 at the shorter wavelengths, but far more radiation with $\lambda = 195$–242 nm reaches the stratosphere, and this spectral range is dominant in generating O atoms, at least in the mid and lower stratosphere.

By the time we reach the top of the troposphere, say at 10 km, there is no longer enough short-wavelength solar radiation to make reaction (5.6) a significant source of atomic O. There is, however, one minor constituent of the atmosphere at these altitudes that *can* be photolysed to yield

[i] Electronic excitation implies that the electrons in an atom or molecule are not in their lowest-energy ('ground state') arrangement. In the case of atomic oxygen, the lowest-energy *configuration* is $1s^22s^22p^4$, but this configuration gives rise to three electronic *states*. The outer-shell p^4 electrons can have their *spin* paired (*singlet states*) or unpaired (*triplet states*). More subtly, the unit of electronic angular momentum possessed by each p electron can be coupled to the momentum of other p electrons in three ways. Out of the six apparent possibilities, only three are permitted by the Pauli Exclusion Principle, and spectroscopists assign the term symbols ^3P, ^1D and ^1S to identify them. The ground state is the ^3P, and O* in the present discussion is the ^1D state.

O, and that is nitrogen dioxide, NO_2. The process

$$NO_2 + h\nu \rightarrow O + NO \ (\lambda < 400\,\text{nm}) \tag{5.7}$$

followed by reaction (5.5) is thus the main (perhaps the only) *in-situ* source of tropospheric ozone.

Competing with the photochemical formation of ozone are processes that remove ozone, so that the concentration does not rise without limit. Some of the loss is photochemical because the molecule absorbs solar radiation during the day, and some of the loss is a result of chemical reactions. The loss processes in the stratosphere and troposphere are different in nature, and the chemistry involved is one of the threads running through this book (see, for example, Sections 6.1, 7.3, 11.2 and 11.3, and the whole of Chapter 9). However, to illustrate the nature of the processes, it is convenient to introduce here one of the simplest of the loss steps, the reaction between atomic oxygen and ozone

$$O + O_3 \rightarrow O_2 + O_2. \tag{5.8}$$

5.2.2 Chapman Functions and the Ozone Layer

Regardless of the identity and rates of the loss processes it is evident that, other things being equal, the concentrations of ozone in the atmosphere will be greatest where the *production* rate is largest. For the stratosphere, this statement is equivalent to saying that the concentrations will be largest where the rate of O-atom production from O_2 (reaction 5.6) is maximum. It is this consideration that explains why ozone in the stratosphere is found in a layer, as first mentioned in Section 1.5. The existence of a layer-like structure is expected for any species such as O_3 whose concentration depends on the photochemical production rate in an atmosphere of varying optical density. At high altitudes the solar intensity is also high, but the concentration of O_2 is too low to give large rates of O atom formation by photolysis. At low altitudes, there is plenty of O_2, but little radiation that can dissociate it. Somewhere in between there is a compromise that maximizes the rate of O-atom production. Sydney Chapman first discussed the formation of layers in this way in the early 1930s, and the mathematical function that describes the shape of such a layer is called a *Chapman function*. Chapman developed the concept of stratospheric ozone production from a set of reactions involving an 'oxygen only' chemical scheme, of which the significant reactions are the production of odd oxygen (see Section 4.5), in reaction (5.6), conversion of O to O_3 in reaction (5.5), and the loss of odd oxygen in reaction (5.8). One further important oxygen-only step is the photochemical one that interconverts O_3 and O, introduced as reaction (4.8) in Section 4.5. The suite of four reactions in the (simplified) Chapman scheme can now usefully be set out next to each other

$$O_2 + h\nu \rightarrow O + O \quad (\lambda < 242\,\text{nm}) \tag{5.6}$$

$$O + O_2 + M \rightarrow O_3 + M \tag{5.5}$$

$$O_3 + h\nu \rightarrow O(^1D) + O_2 \xrightarrow{M} O(^3P) + O_2 \tag{5.9}$$

$$O + O_3 \rightarrow O_2 + O_2 \tag{5.8}$$

The rate of O_2 photolysis, and the altitude of the maximum rate, depend on how nearly overhead the Sun lies (a parameter quantified in atmospheric science by the *solar zenith angle*, θ, the angle between the Sun's rays and a line drawn perpendicular to the Earth's surface). A simplified form of the Chapman function can readily be derived.

Figure 5.2 is a sketch of the form of this simplified function for three zenith angles. A clear maximum exists in the rate of light absorption. The altitude of maximum light absorption depends

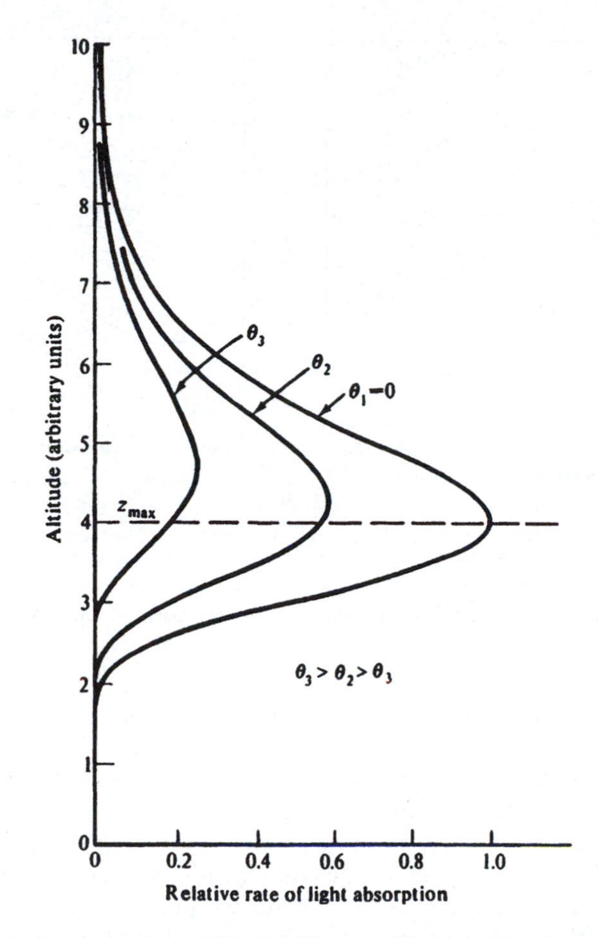

Figure 5.2 Photochemical energy deposition in Chapman layers. The figure shows the rate of light absorption as a function of altitude for three different zenith angles, θ, as calculated from a simplified Chapman function. For $\theta = 0$, z_{max} is the altitude at which the rate is a maximum.

only on the absorption cross section, σ_a, and zenith angle, θ. For absorption by molecular oxygen in the Earth's atmosphere at $\lambda \sim 220$ nm, that altitude is about 35 km with an overhead Sun ($\theta = 0$). The altitude of the maximum in the photolysis rate, P, is thus predicted by this theory to be centred on 35 km. In Chapman's oxygen-only scheme, stratospheric loss of O_3 involves reaction with O atoms, so that O both creates and destroys O_3. If there is a steady state for O_3 (*i.e.* production is balanced by loss in reaction (5.8)), it may be shown that the ozone concentration is proportional to $P^{1/2}$. The maximum ozone concentration is thus predicted also to be at about 35 km. Increasing zenith angles lead (Figure 5.2) to a decrease in the magnitude of the maximum and an increase in its altitude.

5.2.3 Comparison of Measurements with Predictions

The simplified treatment of oxygen photolysis, given in the last section, can be improved to allow for latitudinal variations of solar intensity and zenith angle, real atmospheric temperatures, curvature of the Earth, and so on. Each wavelength makes its own contribution to the photolysis rate, so that the ozone-production rate has to be summed over all photochemically active wavelengths. The mathematical demands of these calculations obviously make this improvement a

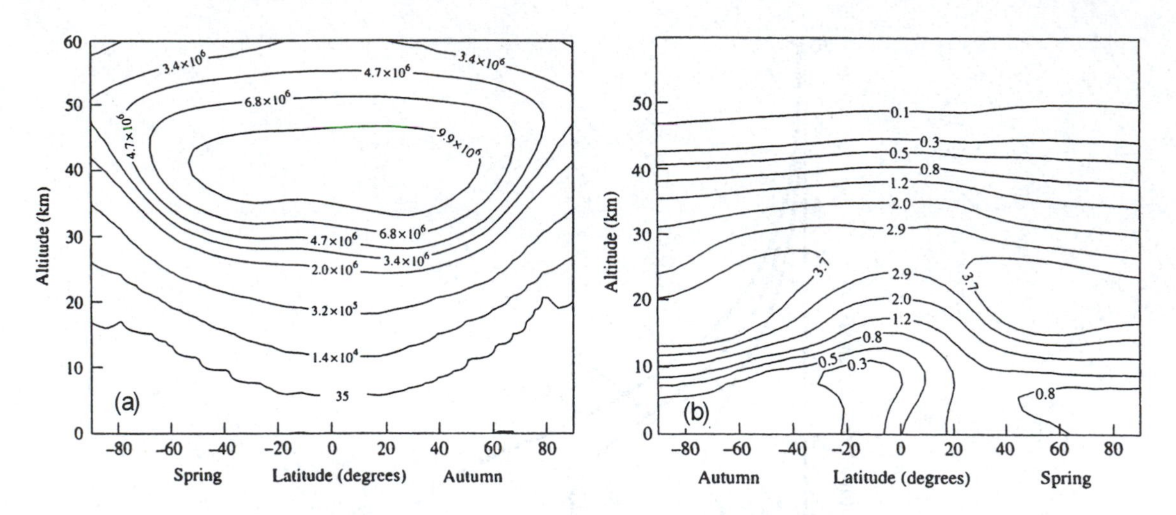

Figure 5.3 (Left) Rate of ozone formation from the photolysis of O_2 (zonally averaged, and in units of molecule cm^{-3} s^{-1}). (Right) Ozone concentrations as a function of altitude for spring (22 March; zonally averaged, and in units of molecule cm^{-3}). Drawn from data provided by D.E. Shallcross in June 2009, and derived from his calculations performed in 1999 using the Cambridge 2D model.

candidate for numerical modelling (Section 4.4). Figure 5.3(a) gives rates of *formation* of ozone *via* reactions (5.6) and (5.5) calculated in this way. The contour lines are zonal averages (over longitude at a given latitude) for the Northern Hemisphere spring equinox (22 March). Atmospheric ozone *concentrations* for the same conditions are shown in Figure 5.3(b).

Comparison of Figures 5.3(a) and 5.3(b) shows straight away that the highest oxygen photolysis rates do not correspond with the highest ozone concentrations, even though the balance between production and loss of odd oxygen suggests that concentrations and production rates ought to be directly linked through the $P^{1/2}$ relation just mentioned. Photolysis rates are highest near the equator, while ozone concentrations are at a maximum in northern regions. At the equator, the ozone layer is centred on about 25 km where the production rate is insignificant, while the production of atomic oxygen reaches a maximum at ~40 km. Near the poles, the maximum production rate is displaced to higher altitudes, but the largest ozone concentrations are found at lower altitudes, and there is a marked north–south asymmetry. The steady state oxygen-only photochemical model predicts that ozone concentrations should be proportional to $P^{1/2}$, so that the lack of correspondence of $[O_3]$ and P indicates an inadequacy of the model. Vertical and horizontal transport are clearly of great importance in redistributing stratospheric ozone. Indeed, the comparison of the two figures can be said to reveal the occurrence and direction of air motions in the stratosphere. Observations confirm this view. The ozone layer is a highly variable phenomenon: ozone column densities near the pole can increase by 50 per cent in ten days in spring, and daily variations of up to 25 per cent have been recorded. Concentrations at particular altitudes show even greater variability, eight-fold increases in a few days being not unusual in the lower stratosphere (say 13–19 km altitude). The rapid build-up of ozone at the spring pole is strongly correlated with northward and downward air transport that could bring ozone in from high-altitude equatorial regions where the production rate is highest.

We now have to ask whether or not the oxygen-only mechanism for the production and loss of ozone can account for the absolute ozone concentrations observed, if due allowance is made for horizontal and vertical transport. Answers to this question can be obtained in several ways. Models can be set up to predict vertical concentration–altitude profiles for comparison with experiments. Such models generally match the *shapes* of the observed profiles (such as those in Figure 5.1), with the altitude of ozone maximum being more or less correctly predicted. Invariably, however, the *absolute concentrations* calculated from the oxygen-only scheme are too high. Total overhead column abundances of ozone are also found experimentally to be lower than those calculated. Typical measured column abundances are 8.8×10^{18} molecule cm^{-2}, while 1D models including vertical transport give about twice this value with oxygen-only chemistry. Because the rate of reaction (5.8) is approximately proportional to $[O_3]^2$, the result suggests that the reaction accounts for only a quarter of the actual ozone loss. Similar results are obtained by examining the predicted global rates of production and destruction of ozone in order to remove errors arising from horizontal transport. For the spring equinox data, the integrated rate of ozone production is 4.86×10^{31} molecule s^{-1} of which 0.06×10^{31} molecule s^{-1} are transported to the troposphere (see next section). The calculated rate of loss of odd oxygen in reaction (5.8) is 0.89×10^{31} molecule s^{-1}, leaving 3.91×10^{31} molecule s^{-1} unaccounted for. If this figure really represented an unbalanced rate of ozone formation, the present global quantity of ozone would double in two weeks! Ozone is being produced five times faster than it is being destroyed by the Chapman mechanism. Since ozone concentrations are not rapidly increasing, we conclude that something other than the Chapman reactions is very important in destroying ozone in the natural stratosphere. This question receives some answers in Section 6.1, where it is seen that reaction (5.8) can have catalytic pathways that are faster than the direct mechanism that we have been discussing.

The catalysis of reaction (5.8) is a central point of our current understanding of ozone chemistry in the stratosphere, and the catalysts and reaction channels are explored more fully throughout Chapter 9. Before we come to this detail, however, we should take up the thread of the present section. We have established that transport processes have to be taken into account, and we know that the chemical schemes have to be made more extensive. With these adjustments made, how well are the observations reproduced by the best models? Figures 5.4(a) and 5.4(b) provide the basis for one comparison, that of total ozone columns as a function of latitude and month of the year. The first of these figures gives observational data obtained by two ESA satellites averaged over the four years 2004–2007, while the second shows the results of a sophisticated 3D model with full chemistry, run to obtain simulated mean column amounts averaged over the same years. The point of the averaging is to smooth out anomalies that might apply to any one year. The correlation between observed and calculated ozone columns is remarkable.[ii] The general contours are reproduced well in the simulations, and very low ozone columns during the Antarctic spring (September–October) are a notable feature that we shall have occasion to look at in greater depth later on (Sections 9.4 and 11.3).

5.3 TRANSPORT FROM STRATOSPHERE TO TROPOSPHERE

In Section 2.4.1, we discussed the *stratosphere–troposphere exchange* (STE) of chemical species. Recent observations, including those of hemispheric asymmetries in water-vapour mixing ratios, suggest that the stratosphere can be divided into three regions. First, there is the 'overworld', where transport is controlled by non-local dynamical processes, such as those involving the pumping mechanism. Next, there is a tropically controlled transition region, containing relatively young air that has passed through, and been dehydrated by, the cold tropical tropopause. Finally, there is the lowermost stratosphere, in which stratosphere–troposphere exchange occurs by processes such as tropopause folding. Figure 2.6 attempts to summarize the key information pictorially.

[ii] In the eyes of the authors! Those who provided the results specially for this book quibble, of course. Dr Douglass wrote "Ten years ago we would have done the scientist dance of joy for something that looked this good, but now we are still puzzling over some problems that don't go away". There are (small) differences between the two plots, perhaps a result of a bias in the measurements or the way in which the (Brewer–Dobson) circulation is handled in the model. What is clear is that identification of discrepancies like this exemplify our statements that differences are sometimes more interesting than agreements in stimulating further research. But we still dance for joy.

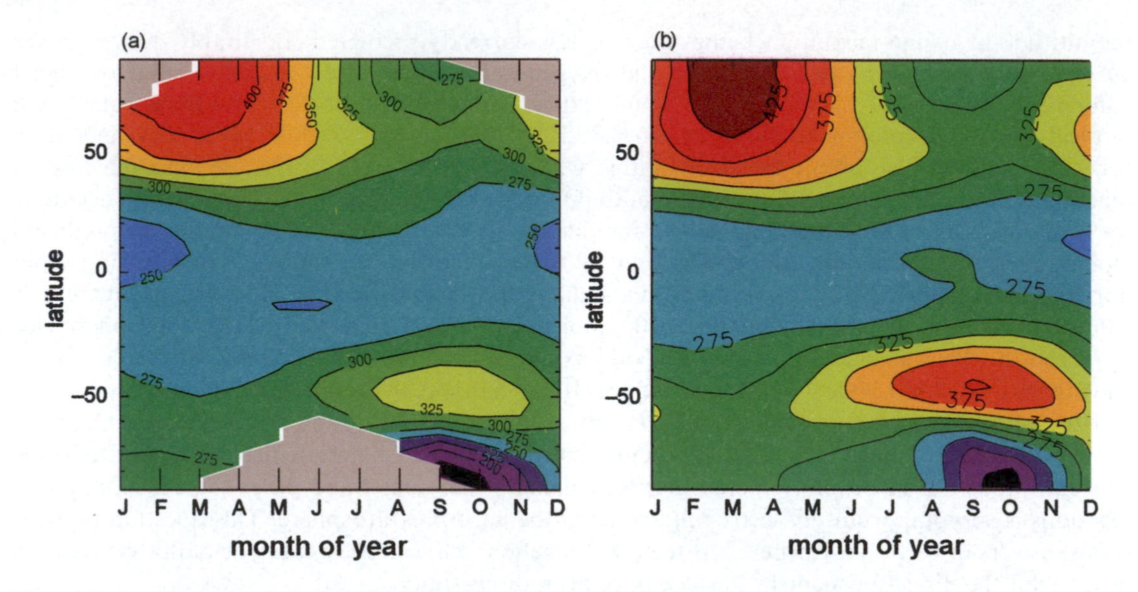

Figure 5.4 Atmospheric ozone as a function of latitude and time of year. The contour lines are total overhead column abundances in Dobson units, and are four-year averages for 2004–7 of monthly mean ozone columns. (a) Observations, retrieved from SCIAMACHY nadir observations. Plot kindly provided by J.P. Burrows and M. Weber, Institute for Environmental Physics and Remote Sensing, University of Bremen, June 2009. (b) Simulations, obtained using the NASA Global Modeling Initiative Chemistry and Transport Model. Meteorological fields are from the Goddard Earth Observing System Data Assimilation System. Figure kindly provided by Anne Douglass, Goddard Space Flight Center, June 2009.

Stratosphere–troposphere exchange of O_3 adds a very considerable complication in the interpretation of tropospheric ozone concentrations and distributions, because it provides an additional source term that may be difficult to quantify. A few sets of experimental evaluations of the contribution exist. One such study put the global mean flux of O_3 to the troposphere at $510 \times 10^9\ \mathrm{kg\,yr^{-1}}$. However, most estimates of the extent of STE depend not on atmospheric experiment, but on 3D chemistry–transport models. Most of these models indicate an STE flux of $> 400 \times 10^9\ \mathrm{kg\,yr^{-1}}$. The contribution of STE to O_3 in the lower troposphere remains the subject of debate. Some researchers suggest that STE events can sometimes penetrate towards the surface, but most believe that only elevated locations (including some of the high-mountain ozone-monitoring sites) experience the direct influence of STE. Nevertheless, after transport of O_3 to the upper troposphere by STE, further downward transport and mixing can bring the stratospheric O_3 to lower altitudes. Models suggest that, on a global scale, STE can account for about 40 per cent of the total ozone found in the troposphere. In the Southern Hemisphere, up to 45 per cent of the *surface* O_3 may have arrived by STE. STE-derived ozone makes the largest fractional contribution at middle and high latitudes during winter; conversely, it has only a very small influence in the tropical troposphere.

5.4 INITIATION OF ATMOSPHERIC OXIDATION BY OH, NO_3 AND O_3

Ozone stores solar energy as a result of the way it is formed from O atoms that are generated photochemically in reactions (5.6) or (5.7). The energy of the O–O_2 bond formed in the addition of O to O_2 in reaction (5.5) is available to the newly formed O_3 molecule. The energetics of the ozone molecule, together with the large atmospheric abundance of its precursor O_2, mean that ozone is singled out as one of the most important of all atmospheric reactants.

Subsequent chapters of this book explore in detail the chemistry of specific altitude regions of the atmosphere, but a very brief summary here usefully rounds off our preliminary survey of the chemical behaviour of ozone in the atmosphere. We have seen at the beginning of this chapter (Section 5.1) that ozone is photolysed by solar ultraviolet radiation, and that an *energy-rich* ('excited') oxygen atom, O*, is formed in the process. This O* is highly reactive, and can interact with atmospheric constituents such as H_2O, CH_4 and N_2O to generate radicals such as OH, CH_3 and NO. These radicals are themselves key reactive intermediates in various aspects of atmospheric oxidation and other chemical transformations. In the stratosphere, OH and NO are involved in catalytic removal of O_3 (see, for example, Sections 6.1, 7.2.2 and 9.3). Meanwhile in the troposphere, the OH radical initiates atmospheric oxidation by acting as the first agent of attack on many trace gases (see Section 8.1, 8.2, 8.4 and 8.5). Reaction of OH with methane

$$OH + CH_4 \rightarrow H_2O + CH_3 \tag{5.10}$$

is illustrative of a typical initiating step that, through subsequent reaction of the methyl radical (CH_3), ultimately results in the formation of CO_2 and H_2O, the fully oxidized 'combustion' products. Hydroxyl radicals also add to unsaturated molecules such as alkenes to form a radical adduct that reacts in ways similar to the CH_3 formed in reaction (5.10).

Thus far in this section, we have suggested that the significance of ozone is related to its photochemical behaviour in acting as a precursor of O*. But formation of O* can only occur during the daytime. Is there any complementary chemistry involving ozone that could occur during the hours of darkness? There is, indeed! Ozone reacts with nitrogen dioxide, NO_2, to form the nitrate radical,[iii] NO_3

$$O_3 + NO_2 \rightarrow NO_3 + O_2 \tag{5.11}$$

The radical is rapidly decomposed photochemically, so that its lifetime during the day is short. At night, however, its concentration can build up. Since NO_2 is the precursor of the tropospheric ozone that is formed *in-situ* (see Section 5.2 and equation (5.7)), generation of NO_3 in reaction (5.11) is possible wherever there is NO_2 in the atmosphere.[iv] Reactions of NO_3 with organic trace constituents are similar to the reactions of OH, and include both abstraction and addition processes that may be formulated as

$$NO_3 + RH \rightarrow HNO_3 + R \tag{5.12}$$

$$NO_3 + {>}C{=}C{<} \rightarrow {>}C(ONO_2)\dot{C}{<}. \tag{5.13}$$

NO_3 is generally less reactive than OH in terms of rate coefficients, but it can also be orders of magnitude more abundant at night than OH is during the day. The consequence is that night-time oxidation initiated by NO_3 can compete in rate with day-time oxidation initiated by OH. For certain species (notably some terpenes and some sulfur compounds) the night-time reactions can

[iii] Not only is NO_3 a radical, but so are both nitrogen monoxide (nitric oxide, NO) and nitrogen dioxide (NO_2) since they possess an electron with an unpaired spin. Unlike most free radicals, and unlike NO_3, NO and NO_2 are long lived and, indeed, can be purchased in cylinders. Another atmospherically important oxide of nitrogen is dinitrogen pentoxide, N_2O_5, which is a closed-shell species: it is the acid anhydride of nitric acid, HNO_3. The anhydride of nitrous acid, HNO_2 is dinitrogen trioxide, N_2O_3, while NO_2 behaves as a mixed anhydride since hydrolysis yields a mixture of HNO_2 and HNO_3.

[iv] Nitrogen dioxide is found in the atmosphere over land, and especially over urban areas: there are both natural and anthropogenic sources. (The oceans are not a significant source of NO_2, and any NO_2 in the atmosphere over them has probably originated over land.)

even dominate over the day-time ones. But the key point to note is that the energy that drives day and night processes is in each case solar; and, in each case, ozone acts as the energy store.

Emphasis has been laid so far on the initiation of the atmospheric oxidation of volatile organic compounds (VOCs) by hydroxyl radicals during the day, and by nitrate radicals at night. Both these radicals indirectly owe their existence to ozone in the troposphere. However, unsaturated VOCs such as alkenes and dienes may be attacked in the atmosphere not only by OH and NO_3, but by O_3 itself. Organic radicals, and even OH, may be formed as products, and a chain oxidation might follow the initial step. In circumstances when solar intensity is low (resulting in low [OH]) or NO_x levels are small (resulting in low $[NO_3]$), the direct reactions with O_3 can make a significant contribution to the loss of VOC. This chemistry will be explored much more fully in Section 8.3.2, and we shall confine ourselves here to illustrating the type of reaction that occurs with a VOC such as propene. One of several reaction pathways can be represented by the equation

$$O_3 + CH_2{=}CHCH_3 \rightarrow CH_3 + HCHO + CO + OH, \tag{5.14}$$

although the reaction involves more than a single step. For the present purpose, it is the formation of CH_3 that is of significance, since this radical leads into the oxidation steps. However, we cannot abandon equation (5.14) without remarking on the simultaneous generation of OH *at night*. We have apparently stumbled on a source of this most important of initiators of oxidation that does not need sunlight. But the reaction still depends on ozone that itself was born out of light.

5.5 ABSORPTION OF UV AND ITS RELEVANCE

The macromolecules, such as proteins and nucleic acids, that are characteristic of living cells are damaged by radiation of wavelength shorter than about 290 nm. Major components of the atmosphere, especially O_2, filter out solar ultraviolet with wavelengths <230 nm; at that wavelength, only about 1 part in 10^{16} of the intensity of an overhead Sun would be transmitted through the molecular oxygen. But at wavelengths longer than ~ 230 nm, the only species in the atmosphere capable of attenuating the Sun's radiation is ozone. Ozone has an unusually strong absorption just at the critical wavelengths (230–290 nm), so that it is an effective filter in spite of its relatively small concentration. For example, at $\lambda = 250$ nm less than 1 part in 10^{30} of the incident (overhead) solar radiation penetrates the ozone layer.

Ozone is formed in the atmosphere from molecular oxygen, the necessary energy being supplied by the absorption of solar ultraviolet radiation. We have already noted that the oxygen in the Earth's contemporary atmosphere is largely biological in origin. Now we see that ozone, needed as an ultraviolet filter to protect life, is itself dependent on the atmospheric oxygen. These links further emphasize the special nature of Earth's atmosphere. Actually, the interactions are even more subtle than we have suggested. Absolute concentrations, and, indeed, the height distribution of ozone depend on a competition between production and loss. Loss of ozone is, as we have already stated in Section 5.2.3 (and shall discuss in more detail in Sections 6.1, 7.2.2 and 9.3.1), regulated by chemistry involving some of the other trace gases of the atmosphere, such as the oxides of nitrogen, which are themselves at least partly of biological origin. Biological processes thus influence both the generation and destruction of ozone. We shall certainly have to examine if human activities could interfere with the delicate balance (Sections 11.2 and 11.3).

CHAPTER 6

Cyclic Processes

6.1 CHEMICAL AND PHYSICAL CYCLES

To a first approximation, the land, sea, air—lithosphere, hydrosphere, atmosphere—system is a 'closed' one: that is, material substance neither enters nor leaves it. In that case, the total quantity of each chemical element is fixed, although the distribution between the elemental and combined forms can alter. This conservation of quantity means that if a species appears in the atmosphere (at a rate equalled by its disappearance), then the elements involved must be passing through a series of *cyclic* chemical and/or physical transformations.

In Section 5.2 of the previous chapter, we discussed the 'oxygen-only' chemistry of stratospheric ozone. Written in simplified form, the key reactions for the formation and loss of ozone are

$$O_2 + h\nu \rightarrow O + O \tag{6.1}$$

$$O + O_2 \xrightarrow{M} O_3 \tag{6.2}$$

$$O_3 + h\nu \xrightarrow{M} O + O_2 \tag{6.3}$$

$$O + O_3 \rightarrow O_2 + O_2 \tag{6.4}$$

This scheme of reactions now serves to introduce the concept of cyclic processes in atmospheric chemistry. Destruction of O_3 in the atmosphere ultimately yields O_2, so that the number of elemental O atoms is conserved in the O_2–O_3 system by a cyclic process occurring solely within the atmosphere (largely within the ozone layer). Several oxygen-containing species are involved in the cycle: O, O_2, O_3. In the four reactions (6.1)–(6.4), the starting material is O_2 and the end product is also O_2. Solar ultraviolet radiation is absorbed in reactions (6.1) and (6.3), with shorter wavelengths being needed for O_2 photolysis than for O_3 photolysis. Energy is liberated in the reactions: some of the energy appears as internal excitation of the O_2 and O_3 molecules formed, some as translational energy of the O_2 molecules from reaction (6.4), and some is carried off by the third-body M. The outcome is the degradation of energy in the form of ultraviolet photons into 'heat'. On the other hand, there is apparently no overall chemical change: the cycle has led back to the starting point. However, much has happened chemically along the way, including the formation of the reactive intermediates O and O_3.

Atmospheric Chemistry
By Ann M. Holloway and Richard P. Wayne

Published by the Royal Society of Chemistry, www.rsc.org

The O_2 cycle just described involves changes occurring within a specific altitude region of the atmospheric system. Other chemical steps link land, water and air in cyclic processes. Meanwhile, physical cycles link the atmosphere with solid land, the oceans, lakes and rivers through processes such as evaporation, transport, condensation, and precipitation of water. The land mass (the lithosphere) itself exchanges gases with the atmosphere either directly or indirectly *via* the hydrosphere. Let us use carbon dioxide as our example. On Earth, carbonate rocks such as limestone ($CaCO_3$) account for 100 000 times as much CO_2 as there is in the atmosphere. Thus the lithosphere is a potentially greater reservoir of CO_2 than are the oceans, although the rates of release may be much smaller. One indirect release process on Earth involves weathering of limestone by CO_2 dissolved in water; the process returns bicarbonate ions, HCO_3^-, to the oceans

$$CaCO_3 + H_2O + CO_2 \rightleftharpoons Ca^{2+} + 2HCO_3^- \tag{6.5}$$

HCO_3^- is in equilibrium with the carbonate ion, CO_3^{2-},

$$HCO_3^- \rightleftharpoons CO_3^{2-} + H^+ \tag{6.6}$$

and both the bicarbonate and carbonate ions can be used by living creatures to build their shells of calcite (calcium carbonate, $CaCO_3$). After the organisms die, their shells sediment out to end up as limestone. The processes therefore transfer CO_2 from the hydrosphere and the atmosphere to the lithosphere.

Inorganic weathering that converts silicate rocks such as diopside ($CaMgSi_2O_6$) to carbonate

$$CaMgSi_2O_6 + CO_2 \rightleftharpoons MgSiO_3 + CaCO_3 + SiO_2 \tag{6.7}$$

is another major mechanism for the transfer of atmospheric CO_2 to the lithosphere. O_2 is involved in a further key cycle, with CO_2 being converted to O_2 by photosynthesis, and oxidation of organic species regenerating CO_2 (see Sections 6.2 and 6.3).

One of the most significant aspects of Earth's atmosphere is the way in which life modifies its composition. With the exception of the noble gases and water vapour, *all* the gases listed in Table 1.2 for Earth have a biological or microbiological source, which for some species (such as oxygen and ammonia) may be the *only* significant one in the contemporary atmosphere. Biospheric sinks exist for many of the atmospheric gases at the land/sea interface with the atmosphere. Respiration and other oxidative processes remove O_2, photosynthetic organisms (Sections 6.2 and 7.2.2) remove CO_2, and certain species of plant–micro-organism systems fix nitrogen, for example. This influence of life on the atmosphere is so important, and makes Earth's atmosphere so different from that of any other planet in the solar system, that a separate chapter (Chapter 7) is devoted to the subject.

Even apparently abiological changes such as the weathering of silicate rocks by CO_2 can be modulated by biological influences. Partial pressures of CO_2 in the soil where weathering occurs are 10–40 times higher than the atmospheric pressure, and these high partial pressures are maintained by soil bacteria. The cycles are thus *biogeochemical* in nature, and the term *biosphere* is used (by analogy with lithosphere, *etc.*) to represent the biological component of the Earth's surroundings. The source regions of the biosphere are divided into oxic (containing free oxygen) and anoxic. Upper layers of oceans and soils are oxic, and produce fully oxidized species (*e.g.* CO_2) as well as partially reduced species (*e.g.* N_2O from bacterial processes and NH_3 from decay of animal excreta). Reduced species (*e.g.* H_2S, CH_4) are produced in anoxic environments such as lower soil regions or the interiors of animals. Transport and change in the biosphere itself may influence the nature and amount of gas reaching the atmosphere. Thus, bacterial reduction of continental shelf sediments probably produces much H_2S, but less than one per cent reaches the atmosphere, the remainder being taken up by bacterial oxidation. Similarly, the decay of organic materials under anaerobic conditions gives rise to H_2 as the major primary product. Several groups of micro-organisms generate the intermediate (*e.g.* H_2S) or released (*e.g.* CH_4, N_2O, N_2) product.

6.1.1 Cycles within Cycles

The oxygen-only stratospheric cycle used to introduce this section is a component cycle within a much larger oxygen cycle that encompasses the entire lithosphere–hydrosphere–atmosphere system. A more detailed examination of the oxygen cycles will be the subject of Section 6.3, but at this point we note the widespread occurrence of cycles within cycles. Indeed, even the apparently simple stratospheric O_2–O_3 cycle contains within it additional cyclic complexity. Reaction (6.4), which is the sink of 'odd oxygen' ($O + O_3$), proceeds in the stratosphere by the direct route implied by the equation to only a minor extent. Substantially more odd oxygen is lost by indirect routes involving other species that act as catalysts. Since one of the defining features of a catalyst is that it is neither created nor destroyed, it follows that the reactions involved must themselves be cyclic. Catalytic ozone loss is the subject of its own sections (particularly Sections 9.2 and 9.3), but we must give a brief explanation of the concept here. The cycles are made up of steps that may be generalized in the pair of processes

$$X + O_3 \rightarrow XO + O_2 \tag{6.8}$$

$$XO + O \rightarrow X + O_2, \tag{6.9}$$

where X is the catalytic species. It is immediately seen that the overall chemical effect of the two reactions is the same as that of reaction (6.4), and that X is not lost. The pair of processes can dominate over the direct step because they proceed more rapidly at low stratospheric temperatures than does the rather temperature-sensitive step (6.4). A real and important example for the identity of X would be NO, in which case XO is NO_2

$$NO + O_3 \rightarrow NO_2 + O_2 \tag{6.10}$$

$$NO_2 + O \rightarrow NO + O_2. \tag{6.11}$$

For our immediate concern, the point to be emphasized is that within the stratospheric O_2 cycle, itself a component of larger cycles, there exist several more 'internal' cycles.

Having used stratospheric ozone to exemplify cyclic behaviour, we might ask ourselves now if there are chemical cycles concerned with tropospheric oxygen and ozone. As explained in Chapter 5, the ultimate driving force for almost all tropospheric chemistry is solar energy, but ozone is the initial vehicle for carrying the energy into chemical reactions. Radicals such as OH and NO_3 owe their origins to O_3, and O_3 itself can react directly with some trace components. Recall that the ozone that is formed *in-situ* in the troposphere is generated by the photolysis of nitrogen dioxide

$$NO_2 + h\nu \rightarrow NO + O \tag{6.12}$$

followed by the addition of the O atom to O_2 in reaction (6.2). It is easy enough to find cyclic pathways that lead back from O_3 to O_2, but it is more instructive in this case to look at the NO_x (NO and NO_2) species to provide a new example. Since NO_2 is a minor species, it is particularly important to determine if it can be recycled to provide a continuing source of O_3 during daylight hours. The three-body reaction

$$NO + NO + O_2 \rightarrow 2NO_2 + O_2 \tag{6.13}$$

is far too slow to make any significant contribution to recycling at atmospheric concentrations of NO. Reaction (6.10) between NO and O_3, on the other hand, is of some importance, especially in polluted urban environments where $[O_3]$ is high. The reaction not only recycles NO to NO_2, but also converts O_3 back to O_2. However, chemistry involving oxygenated organic radicals is the dominant mechanism by which NO is oxidized to NO_2 in much of the troposphere. The processes are usually described in terms of the oxidation of organic molecules, and our main presentation is therefore found in Chapter 8. One of the central groups of species in the oxidation mechanisms are peroxy radicals, such as methylperoxy, CH_3O_2, encountered in the oxidation of methane. Such radicals bring about the conversion of NO to NO_2

$$CH_3O_2 + NO \rightarrow CH_3O + NO_2, \tag{6.14}$$

as well as producing the methoxy radical, CH_3O. Subsequent reactions of CH_3O lead to the formation of another peroxy species, this time the hydroperoxy radical, which can also oxidize NO

$$HO_2 + NO \rightarrow OH + NO_2. \tag{6.15}$$

Each of the three hydrogen atoms in CH_3O can potentially yield an HO_2 radical, so that the participation of reactions (6.14) and (6.15) can convert up to four molecules of NO to NO_2. Although these processes are more than adequate to achieve the recycling of the NO_2 that was lost in the first step of ozone generation, reaction (6.12), the situation has become a little more subtle than just a simple NO_2–NO cycle. The initial step in the oxidation of methane and the formation of the organic radicals is the attack of OH on the CH_4. Reaction (6.15) regenerates OH, so that there is clearly an OH cycle linked to the NO_x one. But the OH itself has a starting point in the photolysis of ozone (Section 5.4). Thus, O_3 depends on NO_2, OH depends on O_3, CH_3O_2 depends on OH, and NO_2 depends on CH_3O_2, which after a few more steps is a source of OH. We set out at the beginning of this paragraph to examine an NO_x cycle to avoid the monotony of using O_3 cycles for every example, and we have ended up after all with an O_3 cycle as well as NO_x and OH cycles! If all this seems rather daunting and complex, Chapter 8 should bring clarity.

Figure 6.1 summarizes the complex interactions that go to make up the biogeochemical cycles. The figure shows the many ways in which volatile materials can be delivered to the atmosphere or be removed from it. Solid–gas cycling occurs on a timescale of hundreds of millions of years. Volcanic gases, primarily CO_2 and H_2O, are released to the atmosphere from the Earth's crust. Carbon dioxide participates in weathering reactions that result in the deposition of carbonate sediments. Sea-floor *plates* are finally *subducted* into the mantle, volatiles trapped are released at high temperature and pressure, and recycled into the atmosphere through volcanoes along the plate interface. On a shorter timescale, the hydrologic cycle exchanges water between the atmosphere and the condensed phases on the surface. Water vapour precipitates into the oceans, and on average the CO_2 and H_2O cycles are balanced so that hydrosphere and atmosphere are maintained at roughly constant volume. Interactions with the biosphere determine the detailed composition of the atmosphere, both with regard to major components such as O_2, and in respect of almost all the trace gases. The figure also shows that the cycles are not completely closed, as already anticipated in Section 3.4. Some material, *juvenile* in the sense that it has not hitherto been cycled, is released from the Earth's interior; the solar wind and debris left by stray bodies entering the atmosphere also contribute to the inward flux. Escape from the Earth's gravitational field to the interplanetary medium constitutes an outward flux.

This chapter continues with a discussion of some of the major cycles occurring in the Earth's troposphere. Our concern will be to see what reservoirs there are for the species, and what determines the rate of physical and chemical interconversions. In other words, we wish to study the *budgets* involved in the cycles and the *lifetimes* of the chemical species. Quite apart from the

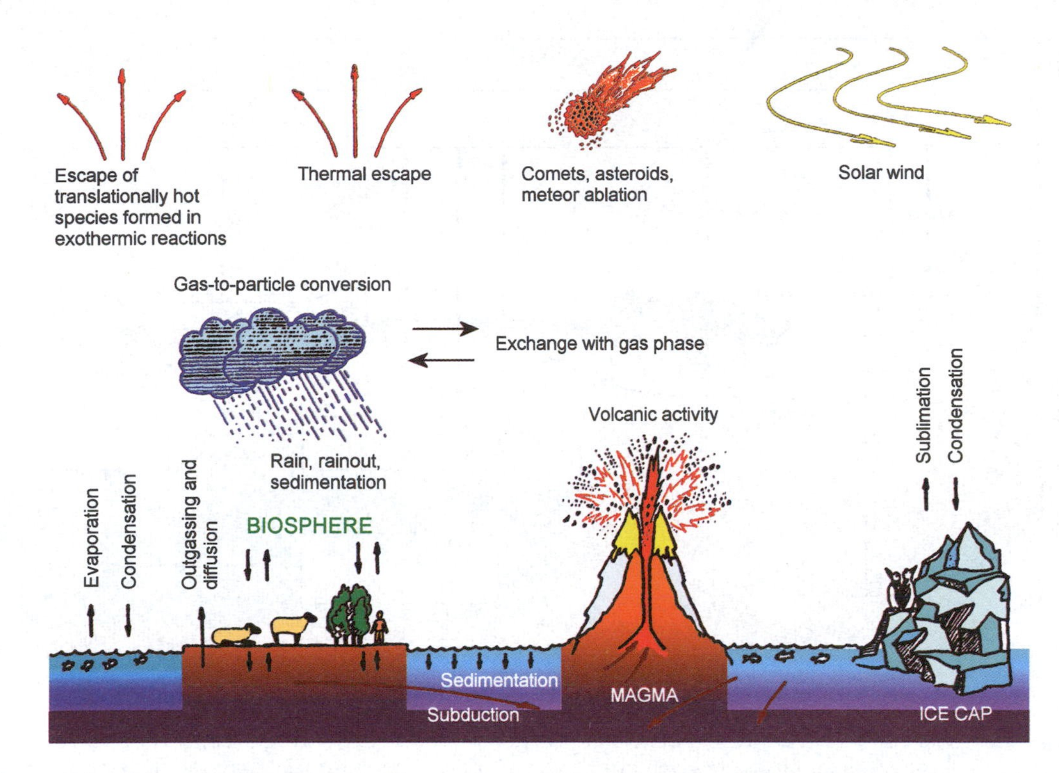

Figure 6.1 Representation of the cyclic processes of biogeochemistry that exchange constituents between air, land, and sea. At the top of the picture, solar particles and extraplanetary objects bring matter into the atmosphere, while a certain amount escapes. Most material, however, is recycled. Even solids deposited on land and on the ocean beds can eventually be subducted to become molten, and components returned to the atmosphere through volcanic activity. Drawn by Sophie McLaughlin, with biospheric species from Carol Wayne and David Koslow.

fundamental issue of how the Earth's atmospheric disequilibrium is sustained, an understanding of the budgets will also allow us to assess whether *anthropogenic* (man-made) sources of various species could be comparable with or larger than natural ones, and thus pose a threat of local or even global pollution (Chapter 11). It should be emphasized that the actual numbers given here are representative only, and many are very uncertain. Some difficulty also arises over the concept of the 'natural' troposphere. It is certain that interactions with the biosphere largely determine atmospheric composition. Human life can potentially alter the composition *out of proportion* to human life's biological importance. Man's activities undoubtedly contribute to present-day budgets and reservoirs in carbon, nitrogen, and sulfur cycles. Our 'natural' troposphere ignores these contributions, and is to that extent artificial. It will, however, allow us to show up the anthropogenic sources more clearly as perturbations.

6.2 CARBON CYCLE

The major carbon species in the troposphere are carbon dioxide ($\sim$385 ppm), methane (1.7–1.8 ppm) and carbon monoxide (0.04–0.20 ppm). Partial pressures of CO_2 today and in the past are of prime geochemical importance since they may influence global temperatures, the composition of marine sediments, rates of photosynthesis, and ultimately the oxidation state of the atmosphere and oceans. A pictorial representation of the complex coupling of inorganic and organic chemistry through CO_2 is given in Figure 6.2. The boxes represent components of the atmosphere, biosphere, hydrosphere,

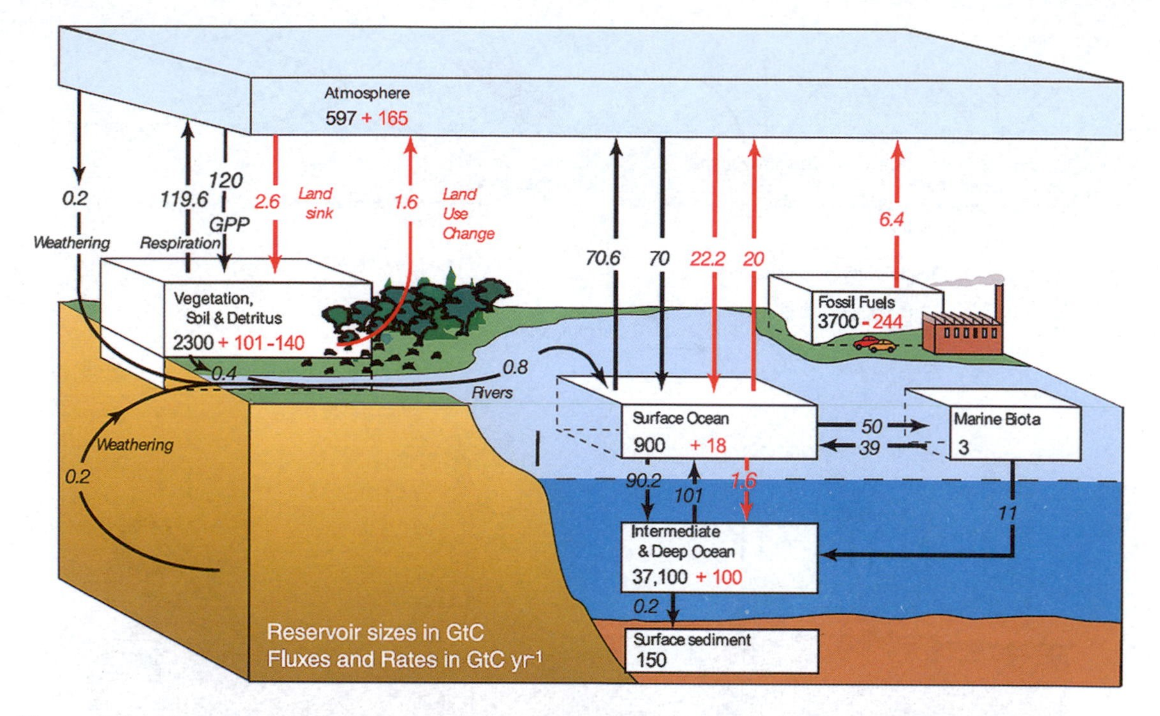

Figure 6.2 The carbon cycle. Representative global values for the 1990s are indicated for reservoirs and fluxes. Annual fluxes are given in 10^{12} kg carbon yr^{-1}: preindustrial 'natural' fluxes are in black and 'anthropogenic' fluxes in red. From *Climate Change 2007 – The Physical Science Basis*, Contribution of Working Group I to the Fourth Assessment Report of the Intergovernmental Panel on Climate Change, Cambridge University Press, 2007.

and lithosphere. Arrows show routes for conversion between one component and another; several closed cycles can be identified on the diagram, and some cycles constitute smaller loops within larger ones. Estimates of transfer rates and reservoir capacities are given in the figure. Atmosphere–biosphere interactions completely dominate over the geochemical parts of the cycle. About 217×10^{12} kg of carbon are transferred each way each year, so that about 25 per cent of the CO_2 content of the atmosphere is converted annually. Photosynthesis by plants and micro-organisms is responsible for the intake of CO_2, while respiration and decay account for the reverse process. From the point of view of generation of organic material, the overall photosynthetic process consists of the formation of carbohydrates by the reduction of carbon dioxide,

$$nCO_2 + nH_2O \xrightarrow{\text{light}} (CH_2O)_n + nO_2, \tag{6.16}$$

where $(CH_2O)_n$ is a shorthand for any carbohydrate. Glucose, for instance, is one of the carbohydrates for which $n = 6$. The essence of the process is the use of photochemical energy to split water and, as a result, to reduce CO_2. Molecular oxygen is liberated in the reaction, although it appears at an earlier stage in the sequence of steps than the reduction of CO_2. Light absorption is achieved by pigment systems involving various chlorophyll–cell structures. Chlorophylls are peculiarly suited to this purpose since their optical absorption is in the visible region, just where the photochemically active part of the solar radiation is most intense at ground level. Furthermore, the structures of the chlorophylls make them particularly efficient photosensitizers. Isotope experiments show that photosynthetically produced O_2 is derived exclusively from the H_2O and not the CO_2. The biochemical reactions and cycles involved in the photosynthetic process are fascinating, but since they are not directly relevant to our present interest we must abandon the topic here.

So far as the atmospheric cycle is concerned, the photosynthetic liberation of O_2 can be balanced by the return of CO_2 to the atmosphere. Respiration of living organisms is the reverse of reaction (6.16),

$$(CH_2O)_n + nO_2 \rightarrow nCO_2 + nH_2O + \text{heat} \tag{6.17}$$

while oxidation or decay of carbohydrates likewise returns both CO_2 and H_2O from the biosphere.

The biosphere–atmosphere cycles in Figure 6.2 are very nearly closed, and can be treated in isolation from the rest of the carbon cycle over short times. Balances between the rate of photosynthesis and the rates of respiration and decay obviously alter diurnally and seasonally. As an example of diurnal change, it has been shown that CO_2 concentrations in a forest can rise to 400 ppm at night, and drop to 305 ppm at noon when photosynthetic activity is highest. Figure 6.3 (red points) shows how the CO_2 concentrations at a site in Hawaii have oscillated with season over many years: high CO_2 is associated with winter and spring, when there is least photosynthesis. Similar observations made near the South Pole (blue points) show very much smaller fluctuations, because there is little local photosynthetic activity, and the CO_2 remains close to the global average. We note in the figure an overall upward trend in CO_2 concentrations over the years. This increase is almost certainly a result of the burning of fossil fuels, and will be investigated further in Chapter 11.

Simple models for the distribution of carbon between reservoirs in the closed biological cycle lead to a surprising result. The way in which the rate of photosynthesis depends on CO_2 concentration is

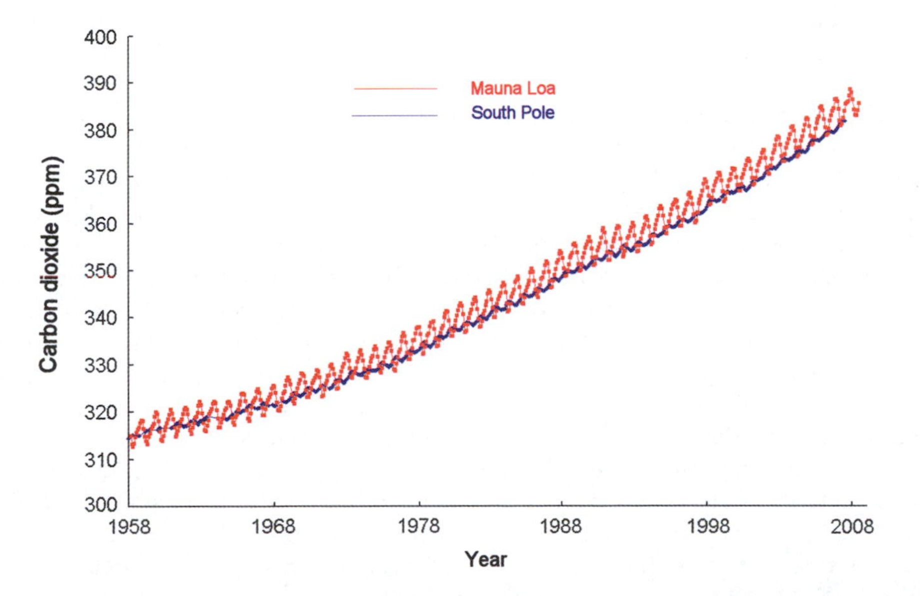

Figure 6.3 Atmospheric carbon dioxide concentrations over more than forty years, (a) in Hawaii, (b) near the South Pole. An increasing trend is present in the data from both stations, and is a consequence largely of Man's combustion of fossil fuels. Superposed on the trend are annual oscillations caused by seasonal changes in photosynthetic activity, which consumes carbon dioxide: the amplitude of oscillation at the South Pole is smaller than that in Hawaii because there is no local vegetation, and CO_2 levels closely reflect values averaged for the globe. The dots indicate monthly average concentrations plotted from data available at http://cdiac.oml.gov/trends/co2/sio-keel.html, and are from a series of measurements started in 1958 by Charles D. Keeling of the Scripps Institute of Oceanography (SID). The results were obtained in a cooperative programme of the US National Oceanic and Atmospheric Administration (NOAA) and SID.

known; it is likely that the rates of transfer from the living biosphere to dead matter and of return back to the atmosphere are linear functions of the mass in the reservoir. It can then be shown that the carbon content of each reservoir is determined only by the total carbon content of all of them. That is, the carbon content of the atmosphere is fixed, and the rate of photosynthesis has no direct influence. In reality, of course, there are slow leaks to the hydrosphere and lithosphere: some dead matter (mainly marine, with a smaller terrestrial contribution) does not return to the atmosphere, and about 0.2×10^{12} kg yr^{-1} becomes buried. Losses from the rapidly cycling biosphere–atmosphere system are what determine atmospheric CO_2 pressure and the carbon content of the biosphere.

Chemical weathering, included in Figure 6.2, both adds and removes CO_2 from the atmosphere. Oxidation of elemental carbon (C^0) and organic compounds in rocks adds CO_2; decomposition of calcium and magnesium silicates (reaction (6.7)) and solution of carbonate minerals (reaction (6.5)) removes it. The estimates of flux rates for all the processes suggest that carbon leaks from the atmosphere–hydrosphere–upper-lithosphere system at a rate of 0.08×10^{12} kg yr^{-1}, or roughly 0.1 per cent of the turnover by photosynthesis. This rate of removal, if continued, would deplete the atmosphere in (762/0.08) $\sim$ 9500 yr, and the oceans in (38 118/0.08) $\sim$480 000 yr. No geological evidence exists for *large* fluctuations of CO_2 over such periods, and it is likely that the shortfall of 0.08×10^{12} kg yr^{-1} is made up by degassing and volcanic release from the interior of the Earth of juvenile carbon or subducted carbonates. Whatever the explanation, it is clear from the calculations that the balance of CO_2 in atmospheres and oceans is a delicate one, and that a serious alteration in the leak rates could change CO_2 concentrations quite rapidly.

Methane and carbon monoxide participate in cycles that are linked with each other and the main carbon cycle. Concentrations of methane are at present about 1.75 ppm. They are essentially independent of altitude within the troposphere but show a slight latitudinal variation: mean concentrations are about six per cent higher in the Northern Hemisphere than in the Southern Hemisphere, largely because the sources are situated on land. Primary ‘natural’ CH_4 production mechanisms are enteric fermentation in animals, mostly cattle, and microbiological anaerobic decomposition of organic matter in wetlands, swamps, and paddy fields (see Section 3.2.2). In radiocarbon (^{14}C) dating, it is assumed that living organisms possess the same $^{14}C/^{12}C$ content as the atmosphere, in which the ^{14}C is continually replenished by cosmic-ray bombardment of ^{14}N. Dead or fossilized material no longer incorporates new ^{14}C, and radioactive decay reduces the ^{14}C content. The half-life of ^{14}C is only about 5600 years, so that the radiocarbon has almost completely disappeared in fossil gases, oils, and solids. The ^{14}C content of atmospheric methane is about 80 per cent that of modern wood less than 20 per cent of the total CH_4 supply comes from natural gas leakage. Current estimates suggest that roughly one per cent of all photosynthetically produced organic matter decays to produce atmospheric methane. Given the atmospheric load of 4.8×10^{12} kg, the methane could be supplied in 7 to 12 years. In fact, there is some evidence that production and loss are not quite balanced, and that the concentration of CH_4 has been increasing by about 0.01 ppm yr^{-1}, starting about 150 years ago. Long-term changes could alter both the chemistry and the climate of the atmosphere, and will be discussed in Chapter 11. Loss processes for CH_4 include bacterial consumption in the soil, but by far the most important loss mechanism is the inorganic oxidation chain (Section 8.2) initiated by hydroxyl radicals

$$OH + CH_4 \rightarrow CH_3 + H_2O. \tag{6.18}$$

One net destruction cycle can be written in the simplified form

$$CH_4 + O_2 \xrightarrow[\text{light}]{\text{OH}} H_2O + CO + H_2 \tag{6.19}$$

to show that CO may be a product. A few per cent of the emitted methane molecules cross the tropopause to enter the stratosphere. Above the tropopause, CH_4 concentrations drop quite rapidly by a factor of about four between 15 and 40 km, suggesting that the CH_4 is rapidly destroyed. Stratospheric loss is not, in itself, important for tropospheric methane budgets, but the process does represent a chemical transport of hydrogen-containing species from troposphere to stratosphere.

Oxidation of methane is a major natural source of carbon monoxide. Non-methane organic compounds such as terpenes may also be oxidized, and contribute up to 50 per cent of the atmospheric carbon monoxide. Smaller amounts of CO are emitted by plants and micro-organisms. It used to be thought that most of the atmospheric CO was anthropogenic in origin, and that the difference in concentrations for Northern (0.07–0.20 ppm) and Southern (0.04–0.06 ppm) Hemispheres reflected the differences in location of industrial sources. However, the asymmetry in the natural oxidation sources could also match the concentration asymmetry. According to recent estimates, the global source of CO is about $2.2\times10^{12}\,kg\,yr^{-1}$, of which 60 per cent is natural. For steady-state CO concentrations globally averaged at 0.12 ppm, the corresponding residence time is about two months; without the natural sources, the residence time would have to be up to years. Strong hemispheric asymmetries are obviously more easily maintained if lifetimes are short. Loss of CO is, as with methane, largely ($\sim$90 per cent) dependent on reaction with OH in the very fast process

$$CO + OH \rightarrow CO_2 + H. \qquad (6.20)$$

The hydrogen atom can reform OH radicals in a number of ways (Section 8.2), so that the oxidation of CO to CO_2 may be regarded as catalysed by hydroxyl radicals. A microbiological sink for CO at the surface of the soil accounts for about 10 per cent of the total loss.

6.3 OXYGEN CYCLE

Short-term oxygen fluxes are dominated by the photosynthetic cycle. Transfers of CO_2 between atmosphere and biosphere will correspond stoicheiometrically to oxygen transfers in the opposite direction: thus atmospheric oxygen is released by photosynthesis and consumed by respiration and decay. Figure 6.4 illustrates these aspects of the oxygen cycle. Photosynthesis accounts for the annual release of $\sim 507\times10^{12}$ kg of oxygen. This value can be calculated by scaling the 'natural' carbon fluxes of Figure 6.2 by the relative mass of O_2 (32) compared with that of C (12). The atmosphere contains 1.2×10^{18} kg of O_2, so that oxygen cycles through the biosphere in $(1.2\times10^{18}/507\times10^{12})$ $\sim$2400 yr, a period much longer than the equivalent one of a few years for carbon because of the much greater atmospheric oxygen reservoir. Seasonal fluctuations in concentration, seen for CO_2 (Figure 6.3), are therefore damped out in the case of oxygen.

As we discussed in Section 6.2, there is a small leak of carbon out of the atmosphere–biosphere system, which has a strong influence on atmospheric CO_2 concentrations. This same leak is also important in the geochemistry of oxygen. Marine organic sediment deposition (Figure 6.2) buries $\sim 0.2\times10^{12}\,kg\,yr^{-1}$ of C without decay, thus releasing $\sim 0.5\times10^{12}\,kg\,yr^{-1}$ of O_2. That is, the leak could cause atmospheric O_2 to double in concentration in $(1.2\times10^{18}/0.5\times10^{12})$ $\sim 2.4\times10^6$ yr. However, in the contemporary atmosphere oxygen is consumed in oxidation of rocks. Weathering rates of elemental carbon to carbon dioxide, sulfide rocks to sulfate, and iron (II) rocks to iron (III) roughly match the leak rate. Oxidation of reduced volcanic gases (*e.g.* H_2 or CO) is a smaller, but not negligible, additional balancing process. Because O_2 is less soluble in water than CO_2, the oceans provide far less of a stabilizing reservoir for O_2, and marked variations in the atmospheric O_2 level may have occurred over geological time.

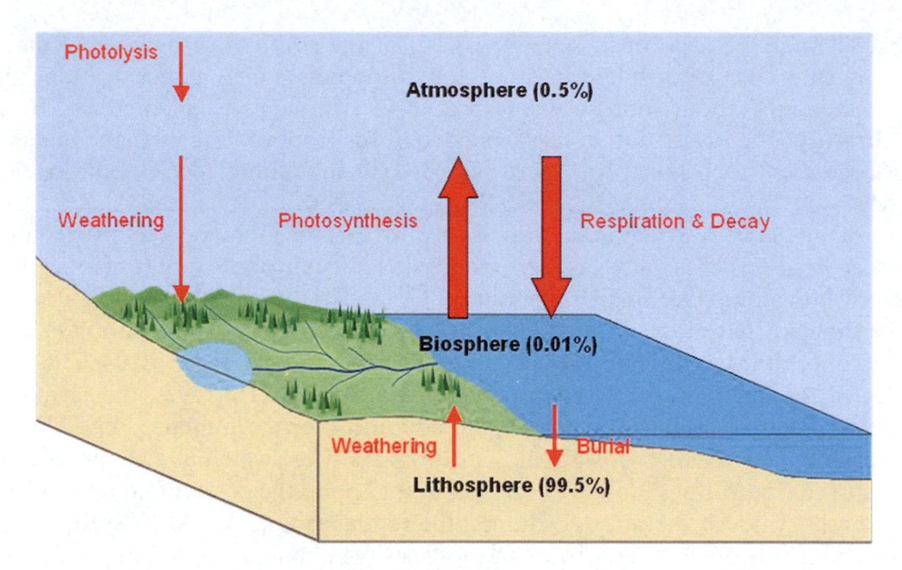

Figure 6.4 The oxygen cycle. The cycle consists of the movement of oxygen within and between its main reservoirs: the atmosphere, the biosphere, the lithosphere and the hydrosphere. Most of Earth's oxygen is in the lithosphere, but the greatest fluxes are to and from the biosphere. The main driver of the oxygen cycle is photosynthesis. Figure from: http://en.wikipedia.org/wiki/Oxygen_cycle

6.4 NITROGEN CYCLE

Figure 6.5 shows the key components of the nitrogen cycle. Molecular nitrogen is chemically rather inert, partly because the large N–N bond energy of 945 kJ mol^{-1} makes most reactions endothermic, or at least kinetically limited because of a large activation energy. Natural processes can *fix* nitrogen (bring it into combination). Lightning within the atmosphere can produce the higher oxides of nitrogen (NO, NO_2, *etc.*), which are converted to acids (HNO_2, HNO_3) and rained out. Biological fixation is, however, of even greater importance, at least over land. Independent microorganisms can fix nitrogen into soils, but, for the planet as a whole, the greatest sources of naturally fixed nitrogen are symbiotic organisms found in the root nodules of the pulses or leguminous plants (peas, beans, *etc.*). Assimilation of nitrogen is catalysed by the enzyme complex *nitrogenase* that brings about the reduction of N_2 to NH_4^+ ions. Other links in the soil microbiological chain involve *nitrifying* bacteria that oxidize NH_4^+ to NO_3^-, *denitrifying* bacteria that reduce NO_3^- to N_2, and *ammonifying* bacteria in which wastes and remains of animals and dead plants are reconverted to ammonia. The microbiological chains maintain the enormous disequilibrium between atmospheric N_2 and O_2 concentrations and those of NO_3^- ions in sea-water, while the greater solubility of common nitrate minerals compared with common carbonates is a partial cause of the dominance of N_2 over CO_2 in the Earth's atmosphere.

Estimates of global budgets and reservoirs are even more imprecise for N_2 than those for carbon, but they still indicate the major aspects of the nitrogen cycle. Weight ratios for C/N are about 7.9 for terrestrial plants and 5.7 for marine plants, so that the biospheric reservoirs and transfer rates in Figure 6.2 may be converted to provide rough values for the nitrogen cycle. In both terrestrial and marine organic systems, most of the nitrogen is recycled within the system, although a smaller fraction is transferred between biosphere and atmosphere. For the hydrosphere, $\sim 130 \times 10^9$ kg yr^{-1} of nitrogen are transferred from the atmosphere, as against 4.0×10^{12} kg yr^{-1} cycling in the marine biosphere. Without this recycling, the lifetime of atmospheric N_2 would be about 6×10^7 yr; but with it the calculated lifetime is nearer 1.6×10^6 yr. About 40×10^9 kg yr^{-1} of N_2 are buried as dead

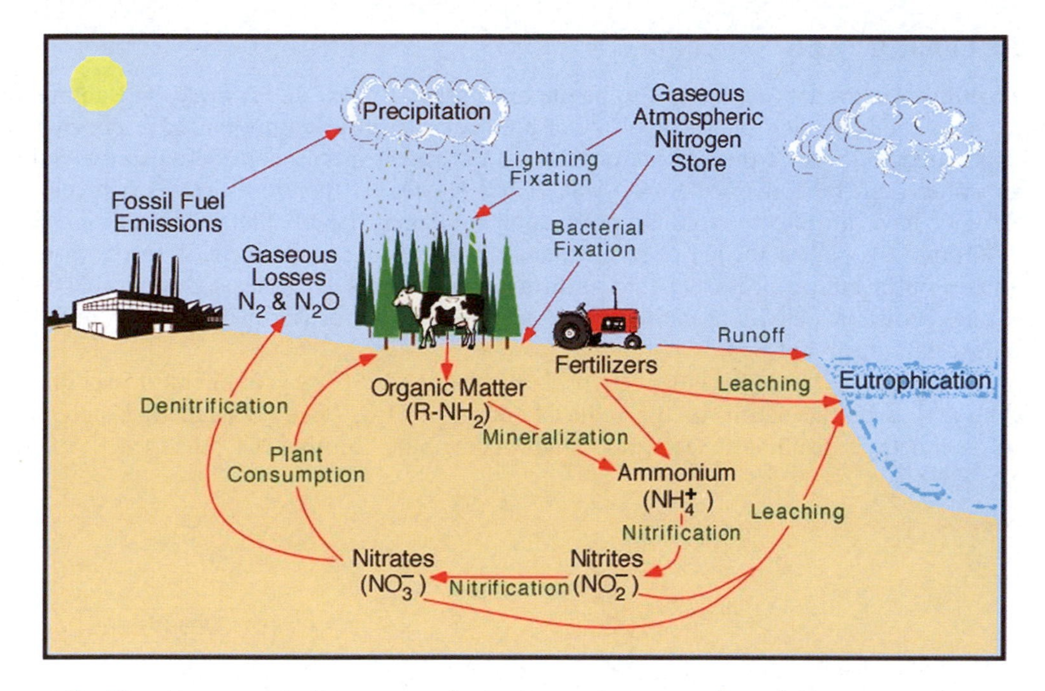

Figure 6.5 The nitrogen cycle. Some atmospheric nitrogen is converted to soluble compounds through the effects of lightning, but most transferred from the atmosphere is biochemically fixed within the soil by specialized micro-organisms, some of which are hosted symbiotically on the root nodules of leguminous plants. Biological activity fixes globally $\sim 1.40 \times 10^{11}$ kg yr^{-1} N. Nitrogen stored in organic matter is returned to the atmosphere in a series of bacterial oxidation steps that form first ammonium salts, then nitrites and nitrates, and finally the gases N_2 and N_2O. Source: M. Pidwirny, *The Nitrogen Cycle* in *Fundamentals of Physical Geography*, 2nd edn, 2006. [eBook at http://www.physicalgeography.net/fundamentals/9s.html]

organic matter, mostly from the marine biosphere. Exposure, weathering, and conversion of organic and inorganic deposits, and release of juvenile nitrogen, can balance the burial rate. However, even without the restoring sources, it would take $\sim 2 \times 10^8$ yr to consume all atmospheric nitrogen. That is, the imbalance is small even on geologically long timescales because of the large nitrogen content of the atmosphere.

Nitrous oxide (dinitrogen oxide, N_2O) is liberated from soils as a result of incomplete microbiological nitrification or denitrification. Indications are that the biological source is dominated by nitrification. The yield of N_2O depends in a non-linear way on oxygen concentrations, ranging from 0.03 per cent (referred to overall NH_4^+ oxidation) at high oxygen levels to more than 10 per cent at low oxygen levels. Global mixing ratios seem to have increased from 0.292 ppm in 1961 to 0.321 ppm by 2008. Nitrous oxide is chemically significant in the atmosphere largely as a precursor of the higher oxides of nitrogen. Reaction with excited oxygen from ozone photolysis

$$O^* + N_2O \rightarrow NO + NO \tag{6.21}$$

produces nitrogen monoxide (nitric oxide, NO), which may be further oxidized to nitrogen dioxide (NO_2). Both NO and NO_2 are involved in much of the chemistry of the troposphere and stratosphere (Chapters 8 and 9). Transport upwards across the tropopause provides a supply of N_2O from the troposphere, and hence of NO and NO_2 to the stratosphere.

6.5 SULFUR CYCLE

Figure 6.6 illustrates what are thought to be the main aspects of the sulfur cycle. The abundances, sources, sinks, budgets, and photochemistry of atmospheric sulfur compounds are poorly understood compared to those of the carbon, oxygen, and nitrogen species considered so far. Sulfur in the atmosphere can be converted to SO_2, SO_3, and H_2SO_4. It therefore acts as a precursor of aerosol, may have an effect on cloud production, and may be of climatological significance. The H_2SO_4 may also lower the pH of rain-water with deleterious consequences. In the case of the atmospheric sulfur budget, it is clear that anthropogenic sources are at least comparable with the natural ones. Volcanic activity probably produces $7\times10^9\,kg\,yr^{-1}$ of sulfur in the form of SO_2 (but see the next paragraph but one).

Decay of organic matter in the biosphere probably yields $58\times10^9\,kg\,yr^{-1}$ of sulfur over land, and $48\times10^9\,kg\,yr^{-1}$ over the ocean, in the reduced forms of H_2S, $(CH_3)_2S$ (dimethyl sulfide), and $(CH_3)_2S_2$ (dimethyl disulfide). Sea-spray might contribute another $44\times10^9\,kg\,yr^{-1}$ of sulfur

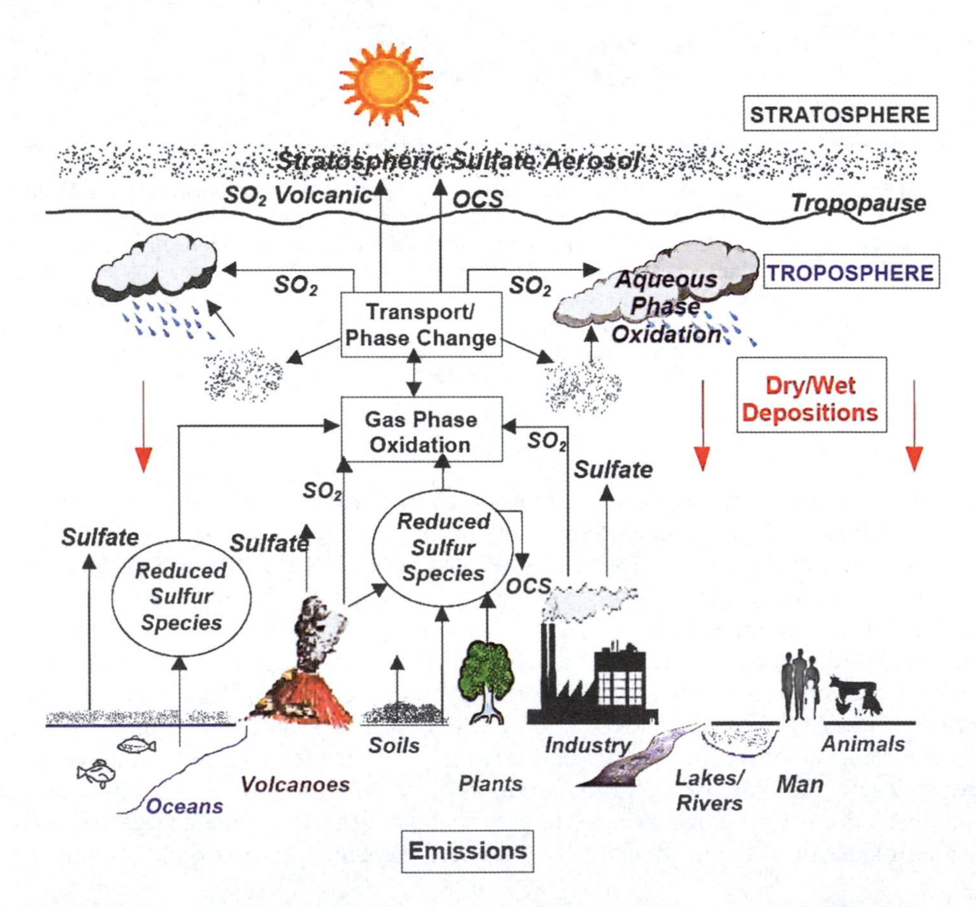

Figure 6.6 The sulfur cycle. Sulfur exists in nature in oxidation states from −2 to +6. Natural emissions are both geologic (*e.g.* from volcanoes) and biologic. These emissions are mainly, but not entirely, of the more reduced compounds (*e.g.* H_2S, $(CH_3)_2S$), but SO_2, CS_2 and COS are also emitted. Man adds very markedly to the releases, especially of SO_2. In the atmosphere, the reduced compounds are oxidized to SO_2, and ultimately to SO_3, in both homogeneous and heterogeneous processes. The SO_3 can become hydrolysed and incorporated in cloud or rain droplets as sulfuric acid. Return of the dissolved acid to the surface completes the cycle. Source: Figure 1.1 in thesis submitted by M. Albu, Bergische Universität Wuppertal, 2008.

compounds. Atmospheric mixing ratios for the sulfur-containing compounds are rather small. Sulfur dioxide is present at a globally averaged mixing ratio of 167 ppt (parts per trillion or parts per million million) corresponding to a total atmospheric load of 0.9×10^9 kg S. Hydrogen sulfide is very variable, typical global mixing ratios lying between zero and 100 ppt ($\equiv$ 0 to 0.6×10^9 kg S). Lifetimes for H_2S must be only a few days even for the highest concentrations observed, and are probably more usually a few hours. Similar remarks apply to SO_2. The mechanisms for the rapid destruction of these species are uncertain, but may involve reaction with OH, or heterogeneous oxidation steps.

Two more sulfur-containing gases, carbonyl sulfide (COS) and carbon disulfide (CS_2), are present in the troposphere, and it has been suggested that these are the main precursors of SO_2 and sulfate aerosol. Carbonyl sulfide is thought to be the most abundant (500 ppt) sulfur gas in the troposphere. Its distribution with latitude is uniform, consistent with a rather long lifetime of more than one year. While carbon disulfide is much less abundant, and highly variable (2–120 ppt) because its atmospheric lifetime is only a few weeks, tropospheric oxidation appears to produce COS, so that CS_2 contributes to the total COS available. Most sulfur-containing gases emitted into the troposphere from natural or artificial sources are too reactive or too soluble to reach the stratosphere, but COS is an important exception. Apart from volcanic injection, tropospheric COS is the main source of stratospheric sulfur, and the stratospheric sulfate layer thus depends on it. The layer influences temperature structure in the lower stratosphere and may have some effect on ozone concentrations (see Section 9.5.5). Oceans are a global source of COS, contributing about 0.15×10^9 kg yr^{-1}. Ash collected from the eruption of Mount St Helens (18 May 1980) gave off large amounts of COS and CS_2 at room temperature, although these gases were much less concentrated than H_2S in the gaseous part of the plume. Interestingly, SO_2 became the dominant gas only after the eruption of 15 June. The eruption of Mount Pinatubo (1991) injected enormous quantities of SO_2 into the stratosphere that within months was converted into H_2SO_4 aerosol and multiplied (temporarily) by a factor of 100 the aerosol surface area (see Section 9.5.5).

Dimethyl sulfide (DMS) is an important sulfur-carrying gas that has natural sources. Large quantities are produced by algae in the oceans. For example, in spring and summer, algae along the coasts of the North Sea produce enough DMS during April and May to make a contribution after oxidation of up to 25 per cent of the H_2SO_4 burden carried in the troposphere over some parts of Europe. Further discussion of DMS and its influences on the environment is provided in Section 7.2.4, and the mechanism of the oxidation of DMS is discussed in Section 8.4.

CHAPTER 7

Life and the Atmosphere

Recall how different the atmosphere of Earth is from that of our two neighbours, Venus and Mars. Figure 1.2 provided a dramatic demonstration of the effect that life has on the atmosphere. The CO_2 that is the dominant constituent (95 per cent or more) of the atmospheres of the other two planets has become a minor constituent of Earth's atmosphere. On the other hand, N_2 and O_2 make up 99 per cent of the 'dry' components of our atmosphere, yet N_2 is present in our neighbours' atmospheres at levels of just a few per cent, and there is much less again of O_2.

As explained in Section 3.2, the prebiogenic atmospheres of the three planets probably consisted of CO_2, N_2 and H_2O, together with a little of the reduced gases CO and H_2, with similar initial compositions on all three planets. The presence of life on Earth seems an obvious candidate for association with our peculiar present-day atmosphere. But is it that living organisms have created the unusual atmosphere, or is it that the atmosphere has allowed the existence and evolution of life? As it turns out, the influences work in both directions, and the interactions and feedbacks are amongst the most fascinating aspects of our extraordinary planet. It is these matters with which the present chapter is concerned. We start with an examination of how our atmosphere might have looked in the *absence* of life, using the properties of our neighbours' atmospheres as a guide.

7.1 ATMOSPHERES ON PLANETS WITHOUT LIFE

Life may be responsible for the unexpected composition of Earth's atmosphere, but we are likely to wonder about the fate of the CO_2 found in the atmospheres of Venus and Mars but missing from ours, and about the origins of the N_2 and O_2 in our atmosphere. A simplified answer can be given very quickly. Most of the CO_2 on Venus is in the atmosphere, most of the CO_2 on Earth is chemically combined in carbonate rocks (with some dissolved in the oceans), and most of the CO_2 on Mars is present as the low-temperature solid. To a reasonable approximation, the total CO_2 (gas, solid, in combination, and dissolved) is the same fraction of the planetary mass for all three planets under discussion. How about the N_2? Nitrogen makes up only 3.5 per cent of the Venusian atmosphere, but, on the other hand, the surface pressure is 93 Earth atmospheres. Thus, the *mass* of N_2 in that atmosphere is certainly not smaller than that in ours (78.1 per cent of 1 atm). Mars, on the other hand, has only 2.7 per cent of nitrogen in its rather thin atmosphere, whose pressure is about 0.006 of an Earth atmosphere. It appears, then, that Mars shows the unexpected behaviour in

Atmospheric Chemistry
By Ann M. Holloway and Richard P. Wayne

Published by the Royal Society of Chemistry, www.rsc.org

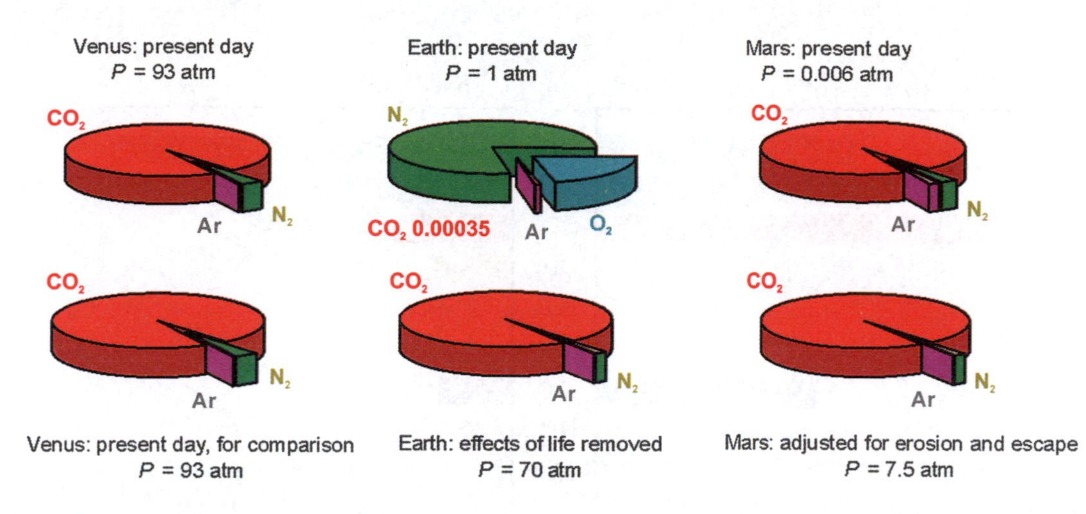

Figure 7.1 Abundances of gases in the atmospheres of Venus, Earth and Mars. The pie charts show the fractional abundances of the dominant gases, while those of other key components are shown in parentheses. The lower set of charts show what would happen on Earth if the effects of life were removed, and on Mars if the atmosphere were adjusted for loss by erosion and escape. For comparison, the top set reproduces what has already been presented in Figure 1.2 for the present-day compositions. From R.P. Wayne, *The Evolution of the Earth's Atmosphere* in *Atmospheric Science for Environmental Scientists*, ed. C.N. Hewitt and A.V. Jackson, Wiley-Blackwell, Oxford, 2009.

terms of N_2, and not Earth, with the Martian atmosphere being strongly depleted compared with either Earth or Venus. The explanation is that almost all the N_2 that would have made up the early atmosphere of Mars has escaped to space. We shall return to that escape shortly.

It is instructive to work out what the graphs of Figure 1.2 would look like if there had been no life on Earth, and if there had been no escape to space in the case of Mars. Figure 7.1 presents the results of one calculation, and shows the charts of Figure 1.2 again for comparison. Carbon dioxide would be the dominant (96–98 per cent) atmospheric constituent for all three planets. Nitrogen would be a minor gas, making up very similar proportions of about 2–4 per cent of the three atmospheres. Compare the pie charts in the two rows, and the dramatic differences brought about by the individual planetary histories is clearly evident. Figure 7.2 looks in more detail at the composition of Earth's atmosphere as it is with life present, and as it would be if life were now absent. The differences in N_2, O_2 and CO_2 already noted are, or course, reflected in the new figure. Bear in mind that the scale is logarithmic, so that the bars for CO_2, for example, represent nearly 3000 times more on the left-hand side (no life) compared with the right (with life). But now information is displayed about the gases present in our atmosphere at smaller concentrations. CH_4, H_2, NH_3 and HCl would almost all be virtually absent without life, while N_2O is more than 1000 times more abundant with life present than without it. Only CO would be several thousand times more abundant without life. With life present, the chemical composition of the atmosphere is far from equilibrium. The atmosphere resembles a low-temperature combustion system, with both fuels and oxidant present simultaneously. The entropy reduction implied by this statement is achieved by the biota using solar energy to drive the fuels and oxidants apart.

A more complete picture of how the compositions of the three atmospheres differ draws together information about some physical properties of the planets (distance from the Sun, surface temperature and pressure, and the velocity that atoms or molecules need to escape the gravitational attraction) and about the chemical composition and behaviour of the solids, gases, and possibly liquids, that are found. Because the principles involved illustrate so vividly important aspects of atmospheric behaviour, it is worth looking at them very briefly.

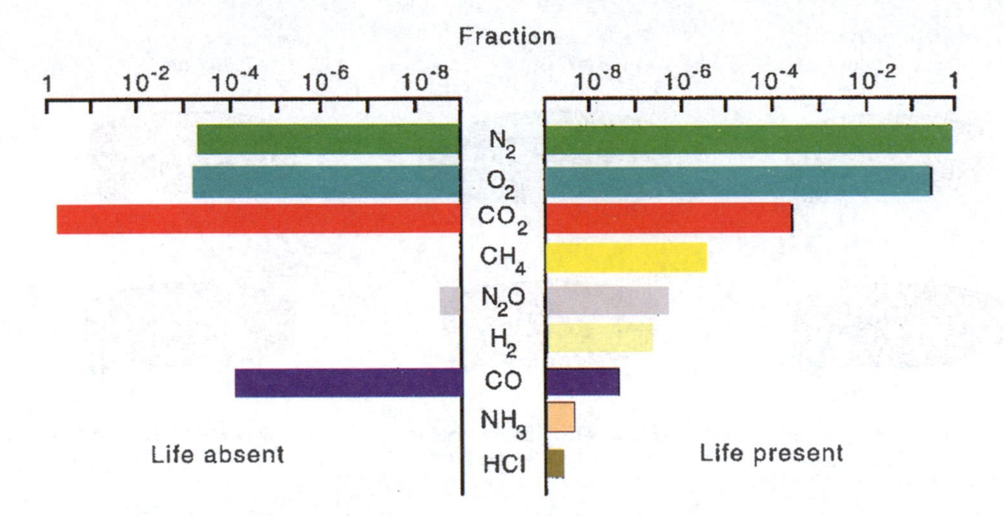

Figure 7.2 Life's influence on Earth's atmosphere. The diagram shows, on a logarithmic scale, the mixing ratios for the major gases and some significant trace species found in our atmosphere in the presence of life and those expected in its absence. Diagram devised by Professor Peter Liss.

Distances from the Sun have a primary influence on planetary temperatures, the *inverse-square law* predicting that the flux of solar energy reaching a planet will drop off as the square of the distance; reflection and scattering by the atmosphere and surface, and trapping of radiation ('greenhouse warming') by atmospheric constituents modify the temperatures expected on the basis of solar flux alone. The mass of a planet affects the gravitational attraction it exerts, and that, together with the temperature and composition, determines the atmospheric pressure.

7.1.1 Water and Venus

Venus is closer to the Sun than us; other things being equal, it should be warmed more by the Sun's energy than we are. In reality, the heating is much greater than expected because of the greenhouse warming exerted by the 93 atm of CO_2. Water on the planet would have evaporated into the atmosphere, and water is itself a powerful greenhouse gas. As the temperature rose, yet more water entered the atmosphere, adding more again to the greenhouse warming. This supposed behaviour in the Venusian past is referred to as a *runaway greenhouse effect*, and is an example of a phenomenon showing *positive feedback*. Almost all the Venusian water would have been in the atmosphere, and the temperatures were too high for it to condense. The rain of Earth and snow of Mars are absent from Venus. Chemistry now intervenes. Water is photodissociated by short-wavelength ultraviolet radiation; the sequence of steps can be represented by the equation

$$2H_2O + 3h\nu \rightarrow 2H_2 + O_2. \tag{7.1}$$

Outgassing might be expected to release materials to the Venusian atmosphere with oxidation states similar to those for terrestrial volcanic gases ($[CO]/[CO_2] \sim 10^{-2}$; $[H_2]/[H_2O] \sim 10^{-2}$). However, at high Venusian temperatures, the gas-phase reaction between CO and H_2O and the loss of H_2O in heterogeneous reactions at the planetary surface would have increased the H_2 content relative to H_2O. Molecular hydrogen would thus have been the dominant gas in the early upper atmosphere of Venus. H and H_2 are light enough to escape to space at a significant rate (on a geological timescale) from Venus, and over the 4.5 Gyr since the planet was formed, most of the H_2 has been lost. The Venusian atmosphere is now very dry, and there is certainly no liquid water on the surface at a temperature of about 740 K! The escape rate calculated for loss of H_2 would have exhausted the equivalent of Earth's oceans in about 280 million years.

7.1.2 Nitrogen and Mars

If gases can escape from Venus, it is clear that they can escape much more readily from the much smaller Mars. The escape velocity is less than half that for Venus, so that the *energy* needed for escape from Mars is under one-quarter needed to escape from Venus. Added to that, Martian atmospheric chemistry favours several processes that can yield fragments that carry much kinetic energy. For example, the neutralization of N_2^+ ions by combination with electrons

$$N_2^+ + e \rightarrow N + N \quad (7.2)$$

is typical of reactions occurring in the Martian ionosphere. It is rapid because the ionization energy of N_2 can be carried off as translational energy by the two fragment (neutral) N atoms. The energy corresponds to 279 kJ mol^{-1} in each of the N atoms, while the escape velocity of 5 km s^{-1} is equivalent to 168 kJ mol^{-1}. There is thus more than enough energy released in reaction (7.2) for the N atoms to escape completely from Martian gravitation. An interesting observation from the Viking landers that arrived on Mars in 1976 corroborates the notion that nitrogen has been escaping from the atmosphere since the planet was formed. The ratio of abundances of the isotopes ^{15}N and ^{14}N was determined by mass spectrometry. The ratio $^{15}N/^{14}N$ showed an enrichment by a factor of about 1.6 compared with the terrestrial abundance (the ratio is 3.66×10^{-3} on Earth, and 5.94×10^{-3} on Mars). Assuming that the Martian N_2 initially possessed the terrestrial composition, the enrichment could be brought about by preferential escape from Mars of the lighter ^{14}N isotope over geological time. The preference for loss of ^{14}N over ^{15}N is only small, so that large total amounts of nitrogen must have been lost during the planet's history. Calculations of that amount, together with allowances for sinks of nitrogen such as fixation by lightning, are the source of the 'adjusted' abundances of N_2 given along with Figure 7.1.

7.1.3 Carbonates and Nitrates

We have established that liquid water cannot have been present on the surface of Venus from early in the history of the solar system. For Mars, the situation is much less clear, because imaging and other science conducted by satellites and landers is very suggestive of liquid water at some time in the past. However, such water as exists now on the surface and in the atmosphere is in the form of ice and snow and the vapour in equilibrium with them. The absence of liquid H_2O precludes the weathering of silicates discussed in Sections 6.1 and 6.2. Thus, on Venus all CO_2 released from the interior of the planet remains in the atmosphere. CO_2 on Mars can condense out as solid at the winter poles, where temperatures can drop to as low as 130 K, and it may be that the pressure at the surface of Mars is determined by the equilibrium vapour pressure of CO_2 at the mean temperature of the polar CO_2-snow cover. Only on Earth has most of the CO_2 been incorporated into carbonate sedimentary deposits by the conversion of silicates or the deposition of the shells of living organisms. The presence of liquid water is the key, a point that we shall focus on soon.

One might wonder why similar considerations do not apply to nitrogen in Earth's environment. Why has the bulk of the atmospheric nitrogen not been turned into solid nitrates in rock formations below the surface? The answer lies in a fact of nature, which is that the nitrates of many elements (and specifically of calcium and magnesium) are highly soluble in water, while the carbonates are virtually insoluble. Nitrate ions in solution keep the combined nitrogen available for those processes that can convert it back into atmospheric N_2.

7.2 AN ATMOSPHERE ON A PLANET WITH LIFE: EARTH

7.2.1 Water and Carbon Dioxide

The principles set out in the preceding section do not explicitly invoke living organisms to explain the composition of Earth's atmosphere. But we have emphasized one notable feature of our environment, and that is the presence of large quantities of liquid water. Without that water, life could not exist and CO_2 would not have been removed from the atmosphere by the formation of carbonate rocks. Of course, without the appropriate living organisms, the calcite from their shells would not been deposited, and that contribution to the total solid carbonate is quite significant. More subtly, although the inorganic weathering of silicates by CO_2 (Section 6.1, reaction (6.7)) appears to be abiological, the rates can be modulated by biological influences. Partial pressures of CO_2 in the soil where weathering occurs are 10–40 times higher than the atmospheric pressure, and these high partial pressures are maintained by soil bacteria. Thus, when the relevant organisms had developed, conversion rates could be greatly enhanced. Most of the weathering that occurred after life appeared would have been of this kind. Water can be seen to be needed twice in the efficient conversion of silicates to carbonates: first, it is the medium without which the process would not occur, and second it makes possible the life that accelerates the chemistry.

One curious and important fact is that our water seems never to have *completely* frozen over or *completely* evaporated since the geological and biological records begin, say 3.5 Gyr (3.5×10^9 yr) ago. Both the geochemistry of the rocks and the persistence of life itself point to this conclusion. Mean surface temperatures have probably never departed from the range of 5 to 50 °C, and may have been highest at very early periods. Why we call this fact 'curious' is that there have been several influences that ought to have pushed temperatures far beyond these limits. For example, the luminosity of the Sun has changed over these billions of years, and so has Earth's orbit around it. The standard model of evolution of solar intensity has created the *faint young Sun paradox*, because it translates into a decrease of Earth's effective temperature by 8 per cent, low enough to keep sea-water frozen for ~2 Gyr, if the atmosphere possessed its present-day composition and structure. The paradox may be only illusory if atmospheric behaviour was different 3–4 Gyr ago from what it is today. Explanations proposed include changes in albedo or increases in the greenhouse efficiency. Alterations in clouds could exert a *negative-feedback* effect. Negative feedback in this context means that some process or processes act in a way that works against a stimulus given to a system, diminishing its effect and stabilizing the system. In the case of clouds and temperature, negative feedback is likely to occur because lower temperatures would mean decreased cloud cover and reduced reflection away of solar radiation. The temperature drop expected in the absence of feedback is thus opposed to some extent. Water vapour makes the largest contribution to the greenhouse effect in the contemporary atmosphere, but it is unlikely to be the agent of long-term temperature control. Its relatively high freezing and boiling points render its blanketing effect prone to destabilizing positive feedbacks by increasing ice and snow albedo at low temperatures (further reducing temperature), but by increasing water-vapour content at high temperatures (and yet further increasing the greenhouse effect).

Whatever greenhouse gas or other mechanism kept the Earth warm, it must have been smoothly reduced to avoid exceeding the high-temperature limit for life. Carbon dioxide seems the most likely greenhouse gas to have exerted thermostatic control of our climate, although in the earliest stages after the appearance of life, methane might also have been a critical contributor. Negative-feedback mechanisms can be identified for both CO_2 and CH_4. In the case of CO_2, non-biological control might include acceleration of the weathering of silicate minerals to carbonate deposits in response to increased temperatures. However, as we just discussed, present-day weathering is biologically determined, and the biota both sense and amplify temperature changes. Rates of weathering can double for every 10 °C increase in temperature. Any attempt by a stimulus such as increased luminosity to increase surface temperatures will be met by greater losses of atmospheric CO_2 and a consequent decrease in greenhouse warming; this negative feedback thus acts as a thermostat to stabilize atmospheric temperatures, with the living organisms themselves functioning as thermometers!

7.2.2 Oxygen and Ozone

Of all the gases in Earth's atmosphere, oxygen is the one most obviously associated with life on the planet. Observations of the oxidation states of mineral deposits in rocks of age greater than 2.2 Gyr, coupled with models of the rates of photochemical formation of O_2 from CO_2 and H_2O, indicate that the prebiological atmospheric concentration near the planetary surface was unlikely to be more than 5×10^{-9} of its present value. Two factors, in particular, limit the production rates. First, the short-wavelength UV required to split the only two abundant precursors that contain oxygen is itself strongly absorbed by O_2. As O_2 levels began to build up, they would have greatly reduced the residual intensities available for photolysis: this is the process of *shadowing*. Second, the coproduct of photolysis is mainly H (the CO from CO_2 reacts with OH from H_2O to regenerate CO_2 and liberate H), which may go on to combine to H_2. But any of the H or H_2 that does not escape to space will rapidly react with the O_2 formed, thus undoing the entire operation. All that has happened is that solar ultraviolet radiation has been converted to heat through the intermediacy of a chemical cycle, and not much O_2 has appeared!

Photosynthesis (see Section 6.2) is now the dominant source of O_2 in the atmosphere, and the oxygen cycles through the biosphere in roughly 3000 years (Section 6.3), a period negligible on geological timescales.

It is important to realize that substantial concentrations of oxygen can build up in the atmosphere only if the carbohydrate formed in the photosynthetic process is removed from contact with the atmospheric oxygen by some form of burial. Without such burial, spontaneous oxidation would rapidly reverse the changes brought about by photosynthesis. This requirement is analogous to the need for H and H_2 escape in the inorganic mechanism. In the contemporary atmosphere, marine organic sediment deposition buries about $0.12 \times 10^{12}\,\text{kg yr}^{-1}$ of carbon, and so releases about $0.32 \times 10^{12}\,\text{kg yr}^{-1}$ of O_2. At that rate, atmospheric oxygen could therefore double in concentration in about 4×10^6 yr. There are balancing processes, including geological weathering (*e.g.* of elemental carbon to CO_2, sulfide rocks to sulfate, and iron(II) rocks to iron(III)) and oxidation of reduced volcanic gases (*e.g.* H_2 and CO). Nevertheless, over geological time marked variations in atmospheric O_2 are likely to have occurred.

Prebiological oxygen concentrations in the palaeoatmosphere are of importance in two ways connected with the emergence of life. Organic molecules are susceptible to thermal oxidation and photo-oxidation, and are unlikely to have accumulated in large quantities in an oxidizing atmosphere. Living organisms are photochemically sensitive to radiation at $\lambda \leq 290$ nm. Surface life-forms known to us depend on an ultraviolet screen provided by atmospheric oxygen and its photochemical derivative, ozone, because DNA and nucleic acids are readily destroyed. Biological evolution therefore seems to have proceeded in parallel with the changes in our atmosphere from an oxygen-deficient to an oxygen-rich one.

Ozone has already been introduced in this book several times as one of the most important atmospheric trace gases, and the molecule will link several of the subsequent chapters as we discuss its chemical effects. In the context of life on Earth, the shielding of organisms living on the surface from solar UV by O_3 is one of the most critical features of our atmosphere, as explained in Section 5.5. It would be reasonable to say that atmospheric O_2 is most essential to life because it is the precursor of O_3; many organisms are *anoxic* and live more happily without oxygen than with it in their immediate surroundings.

It is worth noting that, although the tiny prebiological concentrations of oxygen preclude the existence of a useful oxygen and ozone shield against ultraviolet radiation, the low concentrations were probably essential in the early stages of the synthesis of complex organic molecules that became the basis of life. Living organisms are known to develop mechanisms and structures that protect against oxidative degradation, and so are able to survive in atmospheres containing large amounts of oxygen. At about 1 per cent of the present-day concentration, organisms can derive energy from glucose by respiration rather than by anaerobic fermentation, and they gain an energy advantage of a factor of 16. However, the fact remains that oxygen is toxic, and organisms have to trade off the energy advantage against the need to protect themselves from the oxidant.

One other feature of atmospheric ozone chemistry is striking in relation to the biota on Earth. Stratospheric ozone concentrations are determined by the rate of loss competing with the rate of production, and we have already explained (Section 6.1) that the losses of odd oxygen (ozone and atomic oxygen) can be represented by the equation

$$O + O_3 \rightarrow O_2 + O_2 \tag{7.3}$$

However, the direct reaction is dominated by catalytic cycles involving species written as X and XO. This catalysis was exemplified in Section 6.1 by a cycle in which the identity of X was NO, and XO thus NO_2. As will become apparent in Section 9.3, where catalytic ozone loss is discussed in greater depth, several other cycles and catalysts are known to operate in the stratosphere. The three most important catalysts X are shown in Table 7.1, along with the major precursors or sources of the catalyst (see also Section 9.3.1). Entries in bold italic are primarily biogenic: methane, CH_4, for example, is produced by the decay of vegetation, by termites, by fermentation in the stomachs of animals, and so on. Figure 7.3 shows the relative contributions of different natural sources to the biogenic CH_4, and demonstrates how important the wetlands are. N_2O, nitrous oxide, is liberated from soils as a result of incomplete biological nitrification or denitrification. Most of the methyl chloride (CH_3Cl) comes from the oceans, where it is formed microbiologically from the chloride ions of dissolved seasalt. Some is liberated as a result of the burning of vegetation, and some is emitted in volcanic eruptions. Man has added immensely to the atmospheric burden of chlorine (see Section 11.2), but at this point in our exposition we just note that Man is one form of living organism.

What is particularly striking is that life is directly involved not just in the processes that lead to the formation but also in those that control losses of stratospheric O_3. The O_2 precursor of O_3 is generated by photosynthesis, and the catalysts that remove O_3 are also largely biogenic in origin. Several mechanisms are thus in place by which living organisms could alter the amount of ozone in the stratospheric layer that protects them from damage by solar UV. We leave this topic for the moment at a stage where, as for the control of temperature by CO_2, we are *hinting* at the possibility that the biota could modulate the amount of UV reaching the surface *for their own benefit*.

Table 7.1 Catalyst species involved in cyclic destruction of stratospheric ozone.

X	*XO*	*Precursors or* ***sources of X***	
OH	HO_2	***CH_4***	H_2O
NO	NO_2	***N_2O***	Lightning
Cl	ClO	***CH_3Cl***	*Man!*

Precursor species in bold italic have their origins mainly or entirely in biological processes. Man is a major contributor to atmospheric chlorine as a consequence of the manufacture and subsequent release of chlorinated hydrocarbons, and especially of CFCs (chlorofluorocarbons).

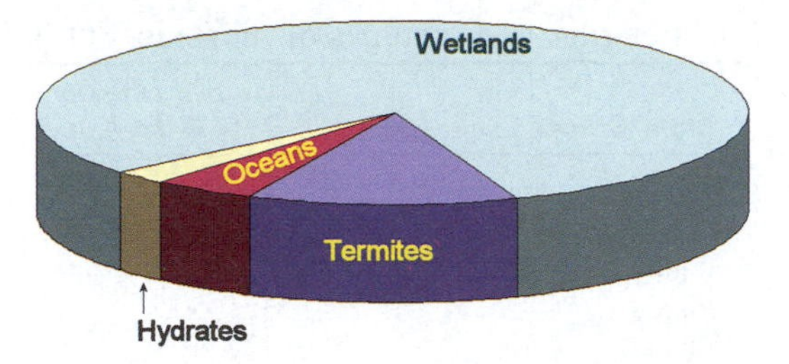

Figure 7.3 The relative contribution of the most important natural sources to global atmospheric methane emissions. Total natural emissions of CH_4 from these sources are roughly $203 \times 10^9\,kg\,yr^{-1}$. Data are mid-points of ranges given in Table 7.6 of *Climate Change 2007 – The Physical Science Basis*, Fourth Assessment Report of the Intergovernmental Panel on Climate Change, Cambridge University Press, 2007.

7.2.3 Hydrocarbons and Other Organic Compounds

The methane mentioned in connection with stratospheric ozone is just one of 1000 or more *volatile organic compounds* (VOCs) released to the atmosphere biogenically. Of the biogenic hydrocarbons, methane is by far the most abundant in terms of mixing ratio. But in terms of emission rates, even more *non-methane hydrocarbons* (NMHCs) and partially oxidized VOCs are released to the atmosphere. These compounds are much more reactive in the atmosphere than is methane, so that the concentrations of individual species are much smaller. However, the high reactivity also means that the biogenic VOCs make a significant, as well as fascinating, contribution to atmospheric chemistry.

Vegetation is known to release a wide range of hydrocarbons to the atmosphere, some of them of a fairly exotic nature. The list of natural VOCs known to enter the atmosphere now extends to well over 1000 compounds. Most of these compounds are emitted primarily from terrestrial plants. Except for methane (and dimethyl sulfide: see next section), animals, soils, sediments, and the oceans are much weaker sources. Soils and sediments possess a high capacity for microbial oxidation of organic species (it is estimated that only 10 per cent of the methane formed escapes such oxidation), which accounts in part for the weakness of their emissions of VOCs. Forests cover more than 40 per cent of all land surfaces, and the release of VOCs, and especially NMHCs, from them plays an important role in shaping tropospheric chemistry. Table 7.2 gives an indication of the sources and emission strengths of important biogenic VOCs. The table shows that very substantial quantities of isoprene and the terpenes are produced by plants. The terpenes are isoprenoid structures. Isoprene (C_5H_8) itself, and the related compound 2-methylbut-2-en-3-ol (MBO: C_5H_9OH) are classed as hemiterpenes; monoterpenes ($C_{10}H_{16}$) and sesquiterpenes ($C_{15}H_{24}$) are the volatile or semivolatile terpenes emitted by plants. Figure 7.4 shows the structures of some typical examples of these NMHCs. The monoterpene family contains about 1000 members, and the sesquiterpenes more than 6500, so the potential number of minor VOCs emitted to the atmosphere is very great indeed.

Oxygenated compounds include the C_1 family (CH_3OH, HCHO, HCOOH) and the corresponding C_2 species (C_2H_5OH, CH_3CHO, CH_3COOH). The only C_3 compound known to be released is acetone (CH_3COCH_3), but biogenic formation of this ketone is of particular interest, because surprisingly high concentrations (0.5 ppb) have been found in the free troposphere at northern mid-latitudes, with rather less (0.2 ppb) at southern latitudes.

Volatile C_6 compounds produced include several isomers of hexenol and hexenal; (3*Z*)-hexenyl acetate is also liberated from some plants. A mixture of these compounds is said to give rise to the characteristic odour of green tea, and the 'green' smell of cut grass or damaged leaves is also attributed to combinations of some of the aldehydes and alcohols. Indeed, (2*E*)-hexenal is

Table 7.2 Sources, emission strengths, and lifetimes of biogenic VOCs.[a]

Compound	*Main natural sources*	*Annual emission (10^9 kg carbon yr^{-1})*	*Scale of lifetime*
Methane[b]	See text	152	Years
Dimethyl sulfide	Oceans	15–30	Hours to days
Ethylene (ethene)	Plants, soils, oceans	8–25	Days
Isoprene	Plants	175–503	Hours
Terpenes	Plants	127–480	Hours
Other reactive VOCs[c]	Plants	~260	Hours
Less reactive VOCs[d]	Plants, soils	~260	Days to months

[a]Based on a table presented by R. Fall, in C. N. Hewitt, (ed.) *Reactive Hydrocarbons in the Atmosphere*, Academic Press, San Diego, 1999.
[b]The source strength for methane is the one reported in Table 3.2 multiplied by 12/16 (RMMs of C and CH_4)
[c]For example, acetaldehyde, 2-methylbut-3-en-2-ol (MBO), hexenal family.
[d]For example, methanol, ethanol, formic acid, acetic acid, acetone.

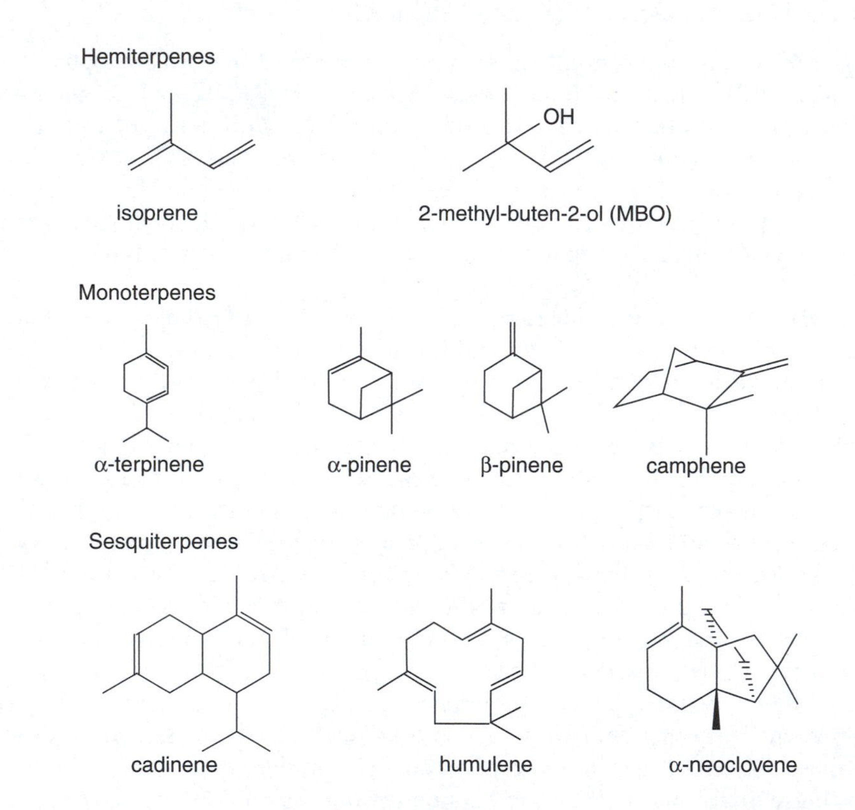

Figure 7.4 Structures of isoprene and typical terpenes.

sometimes called 'leaf aldehyde' and (3*Z*)-hexenol referred to as 'leaf alcohol'. Hexenyl acetate is one of the few volatile esters shown to be liberated by vegetation; another is methyl salicylate, which is generated by diseased tobacco plants.

Plants thus have the metabolic potential to generate and emit a large variety of VOCs. Figure 7.5 illustrates the concept in the form of a hypothetical 'VOC tree' that emits all the most important VOCs that we have discussed. The probable tissues and compartments for the generation of different compounds are indicated in the boxes to each side of the tree, together with a suggestion of the associated production mechanism.

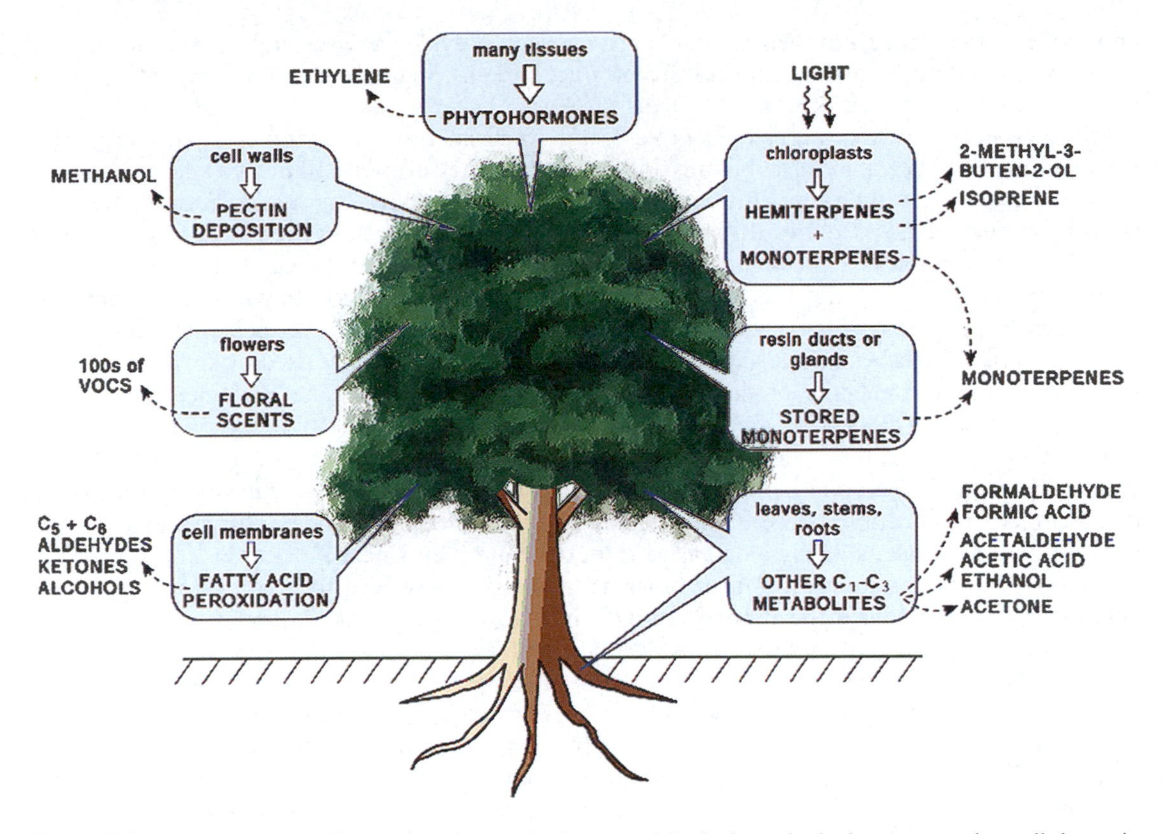

Figure 7.5 The VOC Tree, illustrating the metabolic potential of a hypothetical tree to produce all the main known VOCs emitted by plants. Based on a diagram of R. Fall presented in *Reactive Hydrocarbons in the Atmosphere*, ed. C.N. Hewitt, Academic Press, San Diego, 1999.

Why do plants produce VOCs on a large scale and in wide variety? Strongly emitting plants typically convert one or two per cent of the CO_2 that they have fixed by photosynthesis into the VOCs, and the biochemical energy required to convert the CO_2 into VOC is considerable. It would seem likely that there would be some advantage to the plant in expending this energy. Floral scents are, perhaps, fairly obviously designed to attract pollinating insects, and other VOCs play the same role. Attracting predators of herbivores would be another similar use, and some VOCs are repellants of the herbivores themselves. Many of the VOCs that have been identified also possess antibacterial, antiviral, or antifungal activity. The C_6 alcohols and aldehydes, for example, possess antibiotic properties, and their rapid formation—within less than a minute—after leaf injury probably reflects the plant's attempt to prevent infection by bacteria that reside on the leaf surfaces.

Ethylene (ethene) has been shown to play a most important role in controlling plant growth and development, and it behaves as a volatile plant hormone. Seed germination, flowering, fruit ripening, senescence of flowers and leaves, and sex determination are all affected by exposure to ethylene. Furthermore, ethylene production is greatly enhanced in response to stress: infection, physical wounding, and exposure to some chemicals can all increase the production rate by up to 400-fold. In general, it seems that stress-ethylene serves to induce protective responses in affected plants.

The function or functions of isoprene, which is probably the VOC emitted by plants in greatest quantity, is rather uncertain. One suggestion is that isoprene interferes with interactions between the host plant and competing plant species, or between the plant and insects. Another idea is that isoprene serves as a volatile mediator of membrane fluidity, in order to protect the photosynthetic apparatus of the plant from thermal damage. One piece of evidence that supports this view is that leaves of plants grown at low temperature produce little isoprene, but rapidly generate isoprene when exposed to high temperatures.

Monoterpenes have several ecological roles that have been established with some certainty. They provide direct defence against herbivores and pathogens; they attract pollinators and predators of herbivores; and they interfere with competing plants. One interesting role of monoterpenes in conifers is evident in its outward manifestations to anyone who has stood in a pine forest. A wounded conifer releases *oleoresin*, a pleasant smelling sticky substance that is made up of monoterpenes and diterpene acids, both of which are toxic to insects and to pathogenic fungi. The monoterpene mixture (collectively called *turpentine*) acts as a solvent for the acids. Evaporation of the turpentine then leaves behind a hardened barrier of *rosin* (a substance used by string-instrument musicians to treat the hairs of their bows) that seals the wound. The biological role of the sesquiterpenes is less clear. They are emitted from plants in response to leaf injury, and they may act as *signalling molecules*: their release may guide a predator to the herbivore, say a caterpillar that is causing the injury. The concept of volatile molecules being able to convey information may even extend to 'communication' between one plant and another. Methyl salicylate, for example, may be such a signal molecule, eliciting a defensive response from plants near a diseased, emitting, plant.

Having examined the production of biogenic VOCs, we now turn to a consideration of their behaviour once they reach the atmosphere. One noteworthy feature is the involvement of several compounds in the generation of aerosols. Oxidation of biogenic compounds initiated by reaction with OH, NO_3, or O_3 often leads to the formation of oxygenated or nitrated products with vapour pressures lower than those of the starting reactants. The product compounds may then partition to the aerosol phase. For example, a major product of the oxidation of α-pinene is pinonic acid. Pinonic acid has only a small vapour pressure, and therefore undergoes gas-to-particle formation to yield an organic aerosol. The volatility of some of the sesquiterpenes may be so low that they could condense as the emitted gases rise and encounter lower temperatures, even without any chemical conversion. Organic aerosols can influence climate by scattering solar radiation back to space (net cooling) and by absorption of thermal radiation from the surface of the Earth (net warming).

7.2.4 Dimethyl Sulfide

In Sections 1.5 and 6.5, we mentioned the compound dimethyl sulfide (DMS, $(CH_3)_2S$) as an example of a biogenically produced sulfur-bearing trace gas. Marine DMS constitutes by far the largest *natural* source of sulfur compounds in the atmosphere; the compound is produced by metabolic processes in certain algae. Oceanic biological activity is thought to contribute $12–58 \times 10^9$ kg sulfur yr^{-1}, and the land biosphere only $0.1–7 \times 10^9$ kg sulfur yr^{-1}. Volcanoes provide a flux of $3–9 \times 10^9$ kg sulfur yr^{-1}. The species that result from the atmospheric oxidation of DMS are of considerable significance. If oxidation of DMS leads to SO_2, then after further oxidation SO_4^{2-} can be formed in the aqueous phase (see Section 8.4.1). One of the most important properties of sulfate aerosol particles is that they can act as *cloud-condensation nuclei* (CCNs) (Sections 1.3 and 3.3.1), so that DMS has the potential for influencing large-scale climate change. In addition, its oxidation to SO_2 and then to SO_3 and H_2SO_4 could be a substantial (up to 25 per cent) natural contributor to total acid deposition (see also Section 11.1.5).

Field measurements, laboratory studies, and models all currently indicate that SO_2 is the major oxidation product from DMS under most atmospheric conditions, and that DMS is the main source of non-seasalt sulfate. This conclusion suggests a hypothesis that DMS is a significant source of CCN, and thus of cloudiness. The DMS can then potentially influence the climate by determining the amount of solar radiation reflected to space. For DMS to act in this way, it must increase the *number* of CCN particles, and not just the total mass of sulfate aerosol. Such an increase could arise either by the creation of new particles or by causing the growth of pre-existing smaller particles to a size at which they can act as CCN. Direct evidence for the contribution to particle and CCN formation is provided, for example, by measurements made over the South Atlantic. Strong correlations exist between the marine DMS emission flux and the concentrations both of total aerosol particles and of CCN. Backscattering by clouds influences both illumination and temperature at the surface. Both these factors may affect the rate at which DMS is generated. Once again, the elements of a 'planetary thermostat' are in place. (There is some speculation about the possible role of biogenic VOC emissions as natural precursors of cloud-condensation nuclei, thus providing an analogue for continental areas to the cloud-formation mechanism postulated for dimethyl sulfide in marine zones.)

7.3 BIOGEOCHEMISTRY, FEEDBACKS AND GAIA

Section 7.2 should have made it abundantly clear that the biota on Earth have modified its atmosphere in several ways. Let us summarize the influences that we have noted: the living organisms affect

- chemical composition of the atmosphere;
- surface pressure;
- surface temperature;
- cloudiness;
- intensity of surface illumination;
- wavelength distribution of solar radiation at the surface.

These effects do not stand in isolation from one another, but are quite obviously linked. The biological removal of CO_2 determines the pressure and the temperature. Both the temperature and some aspects of the composition (*e.g.* DMS) influence the cloudiness, which in turn bears on the surface solar intensity and the temperature. Meanwhile, the O_2 from photosynthesis is a precursor of O_3. Both the O_2 and the O_3 absorb solar ultraviolet radiation, greatly modifying the spectral distribution of the radiation that does reach the surface. Chemical reactions in the troposphere are often driven by ozone as well, and the reaction paths and rates of the chemistry that occurs in the atmosphere are dependent on the temperature.

Such interactions thus mean that the biota can exert an impact on their physical environment by bringing about atmospheric changes directly or indirectly. The study of the ways in which living organisms alter the chemistry of the Earth, along with the study of the interactions and feedbacks, is often referred to as *biogeochemistry*. But the effects that living organisms have on their surroundings are only part of the story, because the surroundings themselves may be more or less favourable for the existence of the organisms; this side of the science has been termed *geophysiology* by some. Of paramount importance is the continued presence of liquid water on Earth since life first appeared. It is hard to imagine how living organisms could survive, let alone reproduce and evolve, without that liquid. And, as we have emphasized earlier, water is also essential for the processes of photosynthesis and of weathering that remove CO_2 from the atmosphere.

At this stage, we should observe that an increase in temperature accelerates the rates of both photosynthesis and biological weathering, so that CO_2 will tend to be removed from the atmosphere more rapidly. In consequence, radiation trapping (greenhouse warming) by CO_2 will be diminished, so that the increase in temperature is opposed. This is another example of negative feedback (see Section 7.2.1). The outcome is that any factor that would be expected to lead to a rise in temperature—the Earth's orbit getting closer to the Sun, for example—produces a smaller rise in temperature than would be expected if there were no feedback. Something like this stabilizing effect is likely to have kept the Earth's water within the relatively narrow range of 100 °C between freezing and boiling. Over nearly 4 billion years, somewhere on the planet there has been liquid water despite changes in solar intensity and orbit that might have been expected to produce much wider temperature excursions. Of course, we are all familiar with this type of feedback in *control systems* that are designed to keep a potentially variable quantity constant at some chosen value. In the domestic context, the quantity might be the temperature of a house. Some device or *sensor* measures the quantity: in our example, the measuring device could be a thermometer. The actual temperature in the house is compared with the desired temperature, and the difference produces an *error signal*. The error signal is now *amplified* enough that it can be used to drive a heater if the temperature is lower than desired. With any luck, the temperature of the house will be maintained close to the desired value, and the control is *locked*. Returning to the natural environment again, we see that the temperature sensors are the living organisms themselves, with the rates of photosynthesis and biological weathering being the 'signal'. Effective amplification exists in the system because of the high sensitivity of these rates to temperature changes. Carbon dioxide concentrations respond to this (amplified) signal to trap less or more solar infrared radiation, and thus heat or cool the atmosphere to maintain temperatures within the limited range. Some researchers even suggest that insufficient damping in the climate system is responsible for the swings into and out of ice ages that appear to have occurred with a period of roughly 10^5 years.

Many other processes in nature exhibit negative-feedback behaviour. We can see that the increased cloudiness associated with biological DMS release is another potential negative-feedback mechanism that can stabilize temperatures. However, nature also affords examples of *positive feedback*. One example often cited is that of snow lying on the branches of the trees in a pine forest. The white snow reflects much of the light and the infrared energy from the Sun back into space. If the temperature increases enough to melt a small amount of the snow, great quantities may fall off the branches, exposing the dark-green and sunlight-absorbing needles. As a consequence, the temperature increases further, melting yet more snow, and so on. The effect of melting a small amount of snow has been to expose great areas of low-reflectivity material; the behaviour of snow on downward-pointing branches is acting as an amplifier. The comparison is between the area exposed as a result of tree-snow melting for a given solar-energy input with the area of, say, grass exposed on a flat surface. In this case, unlike the one of biological control of CO_2, the outcome is that a small temperature increase is converted into a large one, and the system is inherently *unstable*. This example is introduced here because there are cases of potential positive feedback that might be significant in relation to climate change. A particular case concerns the hydrates of methane mentioned earlier in this chapter. The increases of temperature expected from Man's production of CO_2 (Section 11.4) could shift the equilibrium between the hydrates and the gaseous CH_4 to cause the release of the gas to the atmosphere; since CH_4 is a very powerful greenhouse gas in its own right, there is a positive feedback for the heating, which could increase until all the stored CH_4 had been used up. To say the least, the effect would be a disaster.

More than a quarter of a century ago, James Lovelock had suggested that one way of determining if another planet harboured life would be to examine the major and minor gases in its atmosphere. If the gases were in chemical equilibrium with each other and their surroundings, then the planet was lifeless. But if there were a composition far from equilibrium, as in Earth's atmosphere, that would be an indication that life in some form had brought about the disequilibrium. These ideas led Lovelock and his collaborator, biologist Lynn Margulis, to promote their *Gaia hypothesis*. The name *Gaia* refers to the ancient Greek goddess of the Earth, and calls on the postulate that the climate and the chemical composition of the Earth's surface and atmosphere are kept at an optimum by and for the biosphere. Gaia is most emphatically not some New Age religion, as some may have thought, but rather a highly imaginative concept that links what we have called biogeochemistry and geophysiology. From receiving a somewhat lukewarm reception from scientists whose approach was perhaps too specialized, the hypothesis has gained widespread recognition as one that illuminates a profitable path to thinking about the interactions, and that may well reflect the way that nature really operates. Biogeochemistry shows us that the composition of Earth's atmosphere is displaced from equilibrium because of interactions with the biosphere. Lovelock and Margulis go further, and see the interaction between life and the atmosphere as so intense that the atmosphere can be regarded as an extension of the biosphere. The atmosphere is not living, but is a construction maintained by the biota. This view has Gaia as a tightly coupled system, whose constituents are the biota, the atmosphere, the oceans, and the surface rock, with the system providing self-regulation of important properties like climate and chemical composition.

The relationship between composition and the biosphere is seen as analogous to that between the circulatory blood system and the animal to which it belongs. In the case of the atmosphere, if highly improbable arrangements (equivalent to low entropy) extend beyond the boundaries of living entities so as to include also their planetary environment, then the environment and life taken together can be considered to constitute a single larger entity. It is the 'operating system' of life and its environment that is called Gaia. Chemistry, pressures, and temperatures are all regulated. Indeed, following the line of thought further, the expected efficiency of evolution would mean that every trace gas in the atmosphere had a purpose as a chemical information carrier. As we have seen, each of the many exotic trace gases emitted by plants appears to have a 'purpose'. Control systems in general require feedback mechanisms and amplifiers. In nature, several biogeochemical amplifiers can be identified: some involving CO_2, N_2O and DMS have been described earlier in this chapter. Those who mistrust Gaia, or at least her hypothesis, take the more cautious view that the close links between atmosphere, oceans, and biosphere do not necessarily imply the existence of an adaptive control system. Rather, they believe that the ocean–atmosphere system has adjusted to biological activity such as photosynthesis, and that the biosphere has responded by optimizing the use of available free energy. The Gaian counter-argument asks why living organisms, which generally exhibit an economy of function, should produce, for example, trace gases such as N_2O or CH_4, unless there is some evolutionary advantage to the organism. The potential control of atmospheric composition and climate by these trace gases is then seen, in the Gaian context, as their regulatory role. In the engineering control-system analogy, the difference between Gaian and non-Gaian views is concerned with whether or not there are closed feedback loops.

The concept of closed-loop control is of more than passing interest in predicting the atmosphere's response to natural or anthropogenic disturbance, a matter likely to be of increasing, and perhaps overwhelming, concern as the 21st century unfolds. But it may be that the greatest value of the Gaia hypothesis to us is that it has taught us to take a wider view of the relevant sciences than was previously the case, and to look closely for interactions between the coupled systems in which chemicals in the atmosphere provide the communicating link.

7.4 ATMOSPHERIC OXYGEN AND THE EVOLUTION OF LIFE

We have seen in this chapter how the biota have modified our atmosphere from its probable prebiological composition, pressure and temperature. Several other aspects of atmospheric behaviour are attributed to chemistry influenced by living organisms. We have also seen how atmospheric properties can make the environment less (or more) hostile for life. Of particular note in this respect is the filtering of ultraviolet radiation from the Sun by photosynthetically-generated O_2 and its product O_3, which together allow organisms to survive when exposed to the Sun on the solid surface of the planet and near the surfaces of its oceans.

What we have *not* done so far is to provide any timescale for the rise in atmospheric oxygen concentrations since life first emerged. There is obviously great interest in seeing if the changes in atmospheric $[O_2]$ over billions of years can be linked to the evolution of life over the same period.

Perhaps even more fascinating is the question of finding a possible link between the development of atmospheric protection from short-wavelength ultraviolet radiation and the evolution of life. It is to a brief consideration of these questions that we now turn.

The rise in [O_2] from its prebiotic levels has generally been linked to the geological timescale either on the basis of the stratigraphic record of oxidized and reduced mineral deposits (including information about isotope ratios) or from fossil evidence combined with estimates of the oxygen requirements of ancient organisms. Evolution of the climate system is likely to have influenced the composition and mineralogy of sedimentary rocks. Attempts have been made to use palaeoclimatological evidence to suggest the history of [CO_2] in the Earth's atmosphere. It has become increasingly apparent that changes in atmospheric [O_2] and [CO_2] must be studied alongside each other, and that it is necessary to have a reliable estimate of past [CO_2] in order to correctly infer past values of [O_2].

Living organisms appear to have been present on our planet from a very early stage, and perhaps from as long ago as 3.85 Gyr BP (the abbreviation 'BP' stands for 'before present'). From that time on, there has thus been a potential photosynthetic source of O_2 that could lead to an increase of atmospheric concentrations over their very low prebiotic levels. The information available from geology includes the finding that '*red beds*', which contain some iron in the higher oxidation state, III, are absent before roughly 2 Gyr BP, and that reduced minerals, such as uraninite, were generally formed before this date. *Banded-iron formations* (BIFs), which contain iron (II) rather than iron (III), were also formed only up until about 1.85 Gyr BP. Much evidence has now accumulated to confirm that there was a large change in atmospheric O_2 level just before 2 Gyr BP. The indications are that the atmospheric pressure of O_2 was at $\sim 5 \times 10^{-4}$ atm, or 2.5×10^{-3} of the present atmospheric level (PAL) before that date. Nevertheless, the presence of red beds, containing some Fe(III), suggests that by about 2.2 Gyr BP there was a substantial (but unquantified) amount of O_2 present in the atmosphere.

The reason for the apparently sudden change in atmospheric O_2 concentrations at about 2 Gyr BP remains a matter for speculation. Isotope studies[i] suggest an abnormally high rate of organic carbon deposition between 2.22 and 2.06 Gyr BP, consistent with a rapidly increasing rate of O_2 production by photosynthesis. The transition in O_2 concentrations may thus reflect a time when the net rate of photosynthetic production of O_2 became greater than the rate of oxygen consumption in reaction with volcanic H_2 and other reduced gases. Perhaps, also, the transition was accentuated when the reduced and lower-oxidation state mineral sinks on the surface had finally been almost all consumed.

One line of evidence about the presence of life involves geologically stable *biomarker molecules* that can act as 'molecular fossils'. Biomarker lipids discovered in northwestern Australia have been shown to be almost certainly contemporaneous with the 2.7 Gyr-old shales in which they are found, and thus provide fairly secure evidence for the presence of life. Some of the compounds are formed only by cyanobacteria. Photosynthetic bacterial cells of this kind initially changed the atmosphere from its oxygen content of $< 10^{-8}$ PAL towards much higher concentrations. The earliest cells lacked a nucleus, and are classified as *prokaryotic*: bacteria fall into this category, and can operate

[i] Isotope ratios in mineral deposits are a source of information about the rise in oxygen levels from their prebiotic levels. Simple physical chemistry predicts that both the positions of chemical equilibria and the rates of chemical change may depend on the isotopic masses of the participating species. There is thus the possibility that chemical reactions could enrich the products of the process in the heavier isotope (and consequently deplete the reactants) or *vice versa*. Of course, the global isotopic ratios cannot change, since elements are neither created nor destroyed (except in the case of radioactive decay or radiogenic formation). However, if an ancient deposit is laid down as a result of chemistry (including biological chemistry) that is isotope dependent, and that deposit is resistant to isotope-exchange processes, then it may retain a signature of the way it was formed in a difference in its isotopic composition compared with the global mean. Photosynthesis preferentially utilizes $^{12}CO_2$ rather than $^{13}CO_2$, so that plant remains are slightly depleted in ^{13}C compared with the global average. There is a large excursion in the amount of the C isotope between 2.22 and 2.06 Gyr BP, which was probably related to an abnormally high rate of organic-carbon deposition, and thus of O_2 production.

anaerobically. Larger cells, which are almost certainly eukaryotic, or possessing a nucleus, are found in the fossil record. It was formerly thought that eukaryotic organisms first appeared about 1.4 Gyr BP, although it has been proposed that the discovery of the corkscrew-shaped organism *Grypania* puts the fossil record of eukaryotic cells back to 2.1 Gyr BP. The importance of this finding for the interpretation of atmospheric evolution is that almost all eukaryotic cells require large quantities of oxygen ($\sim$0.01 PAL) to function. Cell division is preceded by a clustering and splitting of the chromosomes within the nucleus (mitosis), a process dependent on the protein actomyosin that cannot form in the absence of oxygen.

One important biological indicator at the 'transition stage' for O_2 at 2 Gyr BP is the appearance of enlarged thick-walled cells on filaments of cyanobacteria found as fossils. The significance of this observation is that cell structures were developing in response to the need to protect the organisms against the quantities of O_2 by then building up in the atmosphere. Nitrogenase, the enzyme that reduces N_2 to NH_3 or amines, is poisoned by O_2, and cyanobacteria have evolved complex mechanisms to protect their nitrogenase. Some contemporary cyanobacteria have developed specialized cells in which the N_2 fixation occurs, very similar to those found in the ancient fossils. Other cyanobacteria photosynthesize O_2 by day, and reduce N_2 by night, or remove N_2 during the morning and release O_2 during the afternoon. The extraordinary conclusion is that these most ancient of identified fossils are of organisms that were already surprisingly advanced: life must certainly have been established well before these fossils were formed.

As the atmospheric oxygen reached about 10^{-2} PAL, a biological revolution occurred, because eukaryotic cells, and then animal and plant life, appeared. Respiration and large-scale photosynthesis became of importance, enough free oxygen became available for the fibrous protein collagen to be formed, and the scene was set for the appearance of *metazoans*, or multicelled species. About 550 million years ago, the Cambrian period opened. According to earlier ideas, this period heralded an 'evolutionary explosion'. In many ways, the real significance of the Cambrian period is that the first animals with clear external skeletons are preserved as fossils whose identity has been recognized for centuries, while remains of earlier life-forms had not yet been discovered or understood. Metazoan fossils from the preceding 120 million years, the 'Ediacarian' period, are now known. Many of these are from species resembling jellyfish. Such organisms can absorb their oxygen through the external surfaces at concentrations of about 7 per cent of PAL. A reasonable estimate for when this level of oxygen was reached in the atmosphere can thus be set at about 670 Myr BP. The relatively impervious surface coverings of the Cambrian metazoans suggest that 120 million years later the oxygen concentration was approaching 10^{-1} PAL. Following the opening of the Cambrian, the complexity of life is known to have multiplied rapidly and the foundations for all modern phyla were laid. 'Advanced' life forms (*i.e.* non-microscopic) were found ashore by the Silurian age (420 Myr BP), and by the Early Devonian, only 30 Myr later, great forests had appeared. Soon afterwards, amphibian vertebrates ventured onto dry land.

Measurements of the carbon isotope ^{13}C even suggest an overshoot of atmospheric O_2 levels beyond 1 PAL peaking at roughly 300 Myr BP, and the presence of giant insects (especially dragonflies) only around that time has been explained in terms of the elevated oxygen. Several arguments link the rise of O_2 to the rise of trees and other vascular plants from about 390 Myr BP, as just mentioned. For example, these plants synthesize lignin as their '$(CH_2O)_n$' (see equation (6.16)), and this material is relatively resistant to biodegradation. As a result, more of the organic carbon becomes buried and separated from the liberated O_2, and concentrations of the O_2 can rise. Large plants can exert another influence on atmospheric composition. They possess deep and extensive root systems that greatly enhance the rate of chemical weathering. The increased rates of weathering and of photosynthesis will both lead to a decrease in atmospheric $[CO_2]$, and thus to reduced greenhouse warming. The lowered $[CO_2]$ is believed in this way to have helped initiate the glaciation at about 290 Myr BP, which was the longest and most extensive of the past 550 Myr. It may therefore be that the rise of large land plants during the Devonian was nearly as important to the evolution of the atmosphere (and life) as was the development of microbial photosynthesis some 2 Gyr earlier.

Another large rise in atmospheric [O_2] appears to have occurred since about 205 Myr BP at the boundary between the Triassic and Jurassic Periods. Extensive margins that developed along the Atlantic Ocean during this period may have provided a long-term storehouse for organic matter, once again separating the fuel ($(CH_2O)_n$) from the oxidant (O_2). The rise of O_2 over 150 Myr since the beginning of the Triassic almost certainly contributed to the evolution of large animals.

We have emphasized already (Section 5.5 and 7.2.2) that ozone formed from oxygen in the atmosphere acts, together with the O_2 itself, as a screen for biologically-damaging ultraviolet radiation. Our next task will be to see how and when the rise in atmospheric [O_2], and thus [O_3], enabled new developments in living organisms. There is one preliminary point to make. Until the atmospheric attenuation of ultraviolet intensity became sufficient, photosynthetic organisms such as the eukaryotic phytoplankton would not have been present in the oceans, and the rates both of carbon burial and of oxygen generation would have been less than they are at present. The atmospheric shield would have become effective for oceanic organisms with [O_2] in the range 0.01–0.1 PAL.

As explained in Section 7.2, the only species capable of attenuating the Sun's radiation in the critical wavelength region of 230–290 nm is ozone, at least in the present-day atmosphere. The starting point for many discussions of the relationship between biological evolution and the composition of the palaeoatmosphere is the suggestion that life was able to evolve in response to increasing protection from solar ultraviolet radiation as the atmospheric ozone shield developed. One idea is that life would initially develop in stagnant pools where liquid water of 10 m depth or more would be able to filter out the damaging radiation. At this stage, life in the oceans would be unlikely since organisms would be brought too close to the surface to survive. As the atmospheric content of oxygen, and thus of protective ozone, increased, life could migrate from the safety of the pools and lakes to the oceans and, finally, to dry land. Accompanying these changes would be a greatly enhanced photosynthetic and, indeed, evolutionary activity. Marine biota certainly seem to have paved the way, in terms of modification of the atmosphere, for the evolution of the terrestrial biota. Life was abundant long before the Cambrian, and the apparent 'evolutionary explosion' at the dawn of the Cambrian was illusory, as the discussion earlier in the present section will have made clear. All the same, complexity and diversity did increase rapidly after the opening of the Cambrian, and the possibility cannot be discounted that increased mobility was linked to atmospheric ozone reaching levels that afforded sufficient protection.

A major question attached to the interpretation just presented is whether the biological evolutionary events were linked causally to the atmospheric changes that undoubtedly occurred. If they were, then some kind of feedback mechanism may have been in operation, since the atmospheric evolution was certainly mediated by the biota. Resolution of this question will require further information: in the first place, it is necessary to put the time history of growth of O_2 and O_3 in the atmosphere on a firmer footing than it seems to be at present.

Whatever shield is needed by the developing organisms, it is still difficult to specify the flux of radiation that might be dangerous to micro-organisms. Damage would occur primarily at the few hours near midday, when the Sun is most nearly overhead. For organisms whose generation span is one day or less, the maximum exposure is thus about 4 h of high-intensity light. Experiments on genetic damage to corn pollen suggest that the maximum acceptable dose over the period of 4 h would be $0.1\,J\,m^{-2}\,s^{-1}$ for $\lambda \leq 302$ nm. An adequate ultraviolet screen would be provided by an ozone column density of 7×10^{18} molecule cm^{-2}. Other calculations indicate that, at these densities, roughly 10 per cent of a colony of cyanobacteria would survive. If a survival fraction of 10^{-3} is acceptable, then the critical column density of ozone is 4.5×10^{18} molecule cm^{-2} for the most sensitive organisms. Atmospheric photochemical models can be used to calculate the ozone concentrations that accompanied smaller O_2 levels in the early atmosphere. According to one modelling exercise, the critical column densities of O_3 of 7×10^{18} molecule cm^{-2} would require [O_2] near 0.1 PAL. Although other models give rather lower values for the critical [O_2], it is abundantly clear that there would be insufficient ozone in the atmosphere to afford adequate protection for land-based life during the initial stages of oxygen growth. Enough ozone would not be present until the opening of the Cambrian (say 550 Myr BP) if the [O_2] required is 0.1 PAL.

Figure 7.6 summarizes the material presented so far in this section. The growth in surface [O_2] is identified on this plot by markers denoting the geochemical and fossil evidence. Column ozone abundances calculated using the model just described are also displayed in the figure.

It is also interesting to examine the emergence of life onto dry land in relation to the growth in protection by atmospheric ozone. For [O_2] $\leq$ 0.01 PAL, a layer of water of thickness > 4–5 m is needed for additional protection; by [O_2] ~0.1 PAL, water is probably no longer required. Does the transition to [O_2] $\geq$ 0.1 PAL then really explain the appearance of life on dry land? A link of this kind would certainly fit in with the concepts of the Gaia hypothesis (see Section 7.3). Shelled organisms require dissolved oxygen that would be in equilibrium with > 0.1 PAL in the atmosphere, so that the critical level of O_3 for biological protection would have already been passed

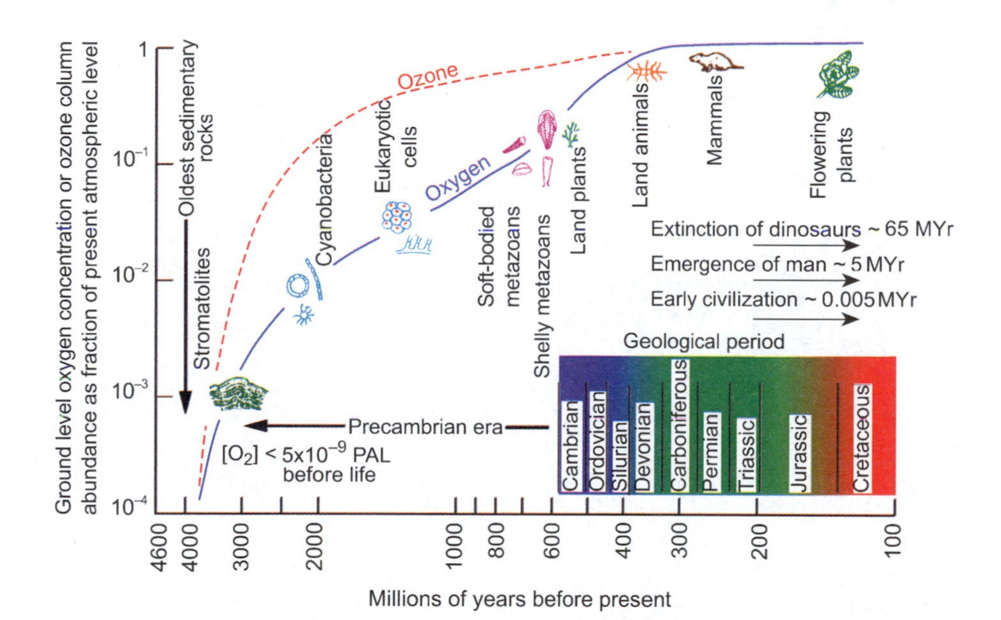

Figure 7.6 Evolution of oxygen, ozone, and life on Earth. In the absence of life, surface oxygen concentrations are unlikely to have exceeded ~5×10^{-9} of the present value. The build-up of oxygen to its present level is largely a result of photosynthesis. Early organisms would have found high oxygen concentrations toxic, but eukaryotic (nucleated) cells require at least several per cent of the present level for their respiration. Soft-bodied metazoans could have survived at similar oxygen levels, but the reduced surface oxygen uptake area available once the species had developed shells must mean that the concentration was approaching one-tenth of its current value about 570 Myr ago. Considerations such as these are used in drawing up the oxygen growth curve. Ozone concentrations can be derived from a photochemical model. Life could not have become established on land until there was enough ozone to afford protection from solar ultraviolet radiation. From R.P. Wayne, *Chemistry of Atmospheres*, 3rd edn, Oxford University Press, 2000 with permission from Oxford University Press.

when the organisms appeared abundantly in the Cambrian period (550 Myr BP). According to the arguments presented in this chapter, even in the worst case there must have been an adequate atmospheric filter by 550 Myr BP at the latest to protect life on dry land from solar ultraviolet radiation. Yet life was apparently not firmly established on land until some 170 Myr later, towards the end of the Silurian. Thus the possibility exists that the development of the ozone shield was not directly linked to the spread of life onto land. The connection between the emergence of life out of water and the filtering of ultraviolet radiation by the atmosphere certainly remains tantalizing.

CHAPTER 8

Chemistry in the Troposphere

8.1 INTRODUCTION

In earlier chapters, we have described the wide variety of chemical species that are found in the Earth's troposphere. About 90 per cent of the total atmospheric mass resides in the troposphere, and the bulk of the minor trace-gas burden is also found there. Some of the species are liberated from the surface or interior by purely geological sources, gently over long periods or violently in volcanic eruptions (*e.g.* H_2S, SO_2), and some are generated in thunderstorms (*e.g.* CO, NO). But many of the substances are present because they or their precursors are generated by living organisms (*e.g.* CH_4, N_2O). Man is a special case, because human activities can lead to intentional or accidental releases of large quantities of gaseous or aerosol materials, some of which have no 'natural' source (*e.g.* $CFCl_3$): these matters are discussed separately in Chapter 11. We have pointed out, also, the propensity for the compounds emitted from all the sources to contain elements in their lower oxidation states; the compounds are often relatively insoluble in water. Tropospheric chemical change thus often involves oxidation steps, and the partially and fully oxidized compounds are much more likely to be highly soluble (*e.g.* CO_2 from CH_4 and other hydrocarbons, N_2O_5 from nitrogen compounds, SO_3 from sulfur species) and to yield acids in solution (H_2CO_3, HNO_3 and H_2SO_4 in the three examples just given).

It is the purpose of the present chapter to examine the chemical steps that are believed to account for the transformations observed. Identification of the reaction pathways followed and their relative importance represents the kernel of an understanding of atmospheric chemistry. Determination of the rates at which processes occur is then essential in the construction of models of atmospheric chemistry (see Chapter 4).

Hydroxyl radicals dominate the daytime chemistry of the troposphere in the same way that oxygen atoms and ozone dominate the chemistry of the stratosphere. The high reactivity of the OH radical with respect to a wide range of species leads to oxidation and chemical conversion of most trace constituents that have an appreciable physical lifetime in the troposphere. Free-radical chain reactions oxidize hydrogen, methane and other hydrocarbons, and carbon monoxide to CO_2 and H_2O. The reactions thus constitute a low-temperature combustion system. Other species, notably NO_x (see Section 4.5) and sulfur compounds, participate in the reactions to modify the course of the combustion processes. Species that survive both physical loss and chemical conversion in the troposphere (*e.g.* N_2O, some CH_4, and CH_3Cl) are transported to the stratosphere where they yield

Atmospheric Chemistry
By Ann M. Holloway and Richard P. Wayne

Published by the Royal Society of Chemistry, www.rsc.org

the NO_x, HO_x, and ClO_x radicals that destroy ozone in the catalytic cycles to be discussed in Chapter 9. Hydroxyl-radical chemistry in the troposphere provides an efficient chemical scavenging mechanism for both natural and Man-made trace constituents, and has a major influence not only on tropospheric composition, but also on stratospheric behaviour.

At night, the nitrate radical, NO_3, takes over from OH as the dominant oxidant in the troposphere. Although NO_3 is generally much less reactive than OH, its peak tropospheric concentration is higher, so that it plays an important role in atmospheric chemical transformations. The diurnal impact of OH and NO_3 is complementary, because OH is generated photochemically only during the day, while NO_3 is readily photolysed, and so can survive only at night.

Radical-chain processes in the troposphere are photochemically driven, although stratospheric ozone limits the solar radiation at the Earth's surface to wavelengths longer than 280 nm (Section 5.5). It is thus necessary to characterize carefully the flux of solar photons as a function of wavelength as these photons pass through the atmosphere and are absorbed and scattered. The most important species that are photochemically labile at $\lambda > 280$ nm are O_3, NO_2, and HCHO (formaldehyde). As we shall see in Section 8.3, all three of these species can indirectly lead to OH (or HO_2) formation, and thus initiate the oxidation chains. Ozone photolysis is a critical step, since the other photolytic processes owe either their origin or their importance to it. Although only 10 per cent of the total atmospheric ozone is found in the troposphere (see Section 5.4), all *primary* initiation of oxidation chains in the natural atmosphere depends on that ozone.

8.1.1 Boundary Layers

We noted in Section 1.2 that the lowest part of the atmosphere, that in contact with the surface, is a *boundary layer* (a *marine boundary layer*, MBL, over the oceans). So far, we have had no strong reason in terms of chemistry to make a distinction between the boundary layer and the *free troposphere* that lies above it. Any chemical differences between the boundary layer and the free troposphere are associated with the higher concentrations of a possibly extended range of species, and with much greater turbulent mixing, in the boundary layer. In this context, the boundary layer is that lowest part of the atmosphere that communicates with the sources and sinks of chemical species and is influenced by those species on a timescale of tens of minutes.

The boundary layer arises as a consequence of turbulence near the surface being confined by a layer of high stability (a *capping inversion*). This is the meteorology explained in Section 2.3.3 and illustrated in Figure 2.5. At night, the surface of the solid Earth is typically at a lower temperature than the air lying overhead, so that in fine weather there is nothing to drive convection. Conversely, during the day, when the ground is heated, vigorous convection currents are set up, so that the cool and calm night is replaced by a warm and gusty day. Because of these diurnal changes, and because the induced turbulence will also be influenced by the physical features of the underlying land, the thickness of the boundary layer is highly variable with place and time. Usually, it is roughly 1–2 km thick, but can be as little as a few tens of metres. The MBL is generally thinner, except near to shore, since the ocean temperature does not rise significantly during the day. However, if cold air is advected over a warmer water surface an unstable, turbulent MBL can develop.

Inside the boundary layer, mixing is efficient, of course, but the capping inversion can trap materials released from the surface within the layer. These effects may be particularly important where there are local releases of relatively short-lived species. This is the situation that often arises with urban air pollution (see Section 11.1), and the result is that instead of the pollutants being diluted in the entire troposphere, they are confined at higher concentration near the ground. It is exactly the combination of physical geography (the Pacific to the west, high mountains everywhere else) and meteorology (strong inversions) that make the Los Angeles basin an ideal reactor in which vehicular emissions can be turned into photochemical smog (Section 11.1.6).

8.2 OXIDATION AND TRANSFORMATION OF VOLATILE ORGANIC COMPOUNDS (VOCS)

Our more intimate consideration of the details of tropospheric chemistry starts by looking at how hydrocarbons are oxidized. Methane is both the simplest hydrocarbon and the most abundant in the troposphere, so it is the obvious first example to illustrate the nature of the processes. Later, more complex volatile organic compounds (VOCs) are introduced into the discussion.

At atmospheric temperatures, a saturated hydrocarbon such as CH_4 does not react directly with O_2. Rather, oxidation is a *radical-chain* process possessing the conventional main steps of *initiation*, *propagation* and *termination*. Our first topic is therefore the initiation step(s).

8.2.1 Photochemical Chain Initiation by Hydroxyl Radicals

In the real troposphere, several species are present that are capable of absorbing solar radiation and hence initiating radical-chain oxidation. More insight may be gained if we take as our starting point the artificial situation where no CH_4 has yet been oxidized. Ozone is then the photochemical precursor of hydroxyl radicals. As we discussed at some length in Chapter 5, ozone is photolysed at wavelengths less than ˜ 310 nm to yield an excited, ^{1}D, oxygen atom that is energetically capable of reacting with water vapour to yield OH

$$O_3 + h\nu \rightarrow O^*(^1D) + O_2^*(^1\Delta_g) \tag{8.1}$$

$$O^*(^1D) + H_2O \rightarrow OH + OH. \tag{8.2}$$

Since H_2O is itself a minor component of the atmosphere, reaction in process (8.2) is a minor fate of $O^*(^1D)$ atoms compared with quenching

$$O^*(^1D) + M \rightarrow O + M \qquad M = N_2,\ O_2. \tag{8.3}$$

However, virtually all ground-state O atoms will regenerate ozone

$$O + O_2 + M \rightarrow O_3 + M, \tag{8.4}$$

so that quenching does not constitute a loss of odd oxygen. Although the recycling of atomic oxygen maintains the mass balance of odd oxygen, the rate of OH production and of chain initiation does depend on the relative rates of reactions (8.2) and (8.3), and, in a similar way, on the quantum yield for $O(^1D)$ production in reaction (8.1) as a function of wavelength.

Ozone of stratospheric origin (see Section 5.3) can well be that needed for $O(^1D)$ production, but if NO_2 is also present in the troposphere, then NO_2 photolysis (at $\lambda < 400$ nm)

$$NO_2 + h\nu \rightarrow O + NO, \tag{8.5}$$

followed by the combination reaction (8.4), can provide a tropospheric source. As we shall see shortly, NO itself can be oxidized back to NO_2, so that the formation of O_3 is not stoicheiometrically limited by the number of NO_2 molecules initially present.

8.2.2 Oxidation Steps for CO and CH_4

Hydroxyl radicals formed in reaction (8.2) react mainly with CO and with CH_4

$$OH + CO \rightarrow H + CO_2, \tag{8.6}$$

$$OH + CH_4 \rightarrow CH_3 + H_2O. \tag{8.7}$$

Roughly 70 per cent of the OH reacts with CO, and 30 per cent with CH_4, in the unpolluted atmosphere. We note that in both processes an active species is formed that is capable of adding molecular oxygen to produce a different radical

$$H + O_2 + M \rightarrow HO_2 + M \tag{8.8}$$

$$CH_3 + O_2 + M \rightarrow CH_3O_2 + M. \tag{8.9}$$

These radicals are *peroxy* radicals, generically written as RO_2 (HO_2 is *hydroperoxy* and CH_3O_2 is *methylperoxy*, the first in the series of *alkylperoxy* radicals). The atmospheric fates of the peroxy radicals depend on the local concentration of the oxides of nitrogen (NO_x). In tropospheric regions where NO concentrations are very low, the peroxy radicals HO_2 and CH_3O_2 are consumed mainly in the reactions

$$HO_2 + HO_2 \rightarrow H_2O_2 + O_2 \tag{8.10}$$

$$CH_3O_2 + HO_2 \rightarrow CH_3OOH + O_2, \tag{8.11}$$

(Self-reaction of two CH_3O_2 radicals is rather slow at ambient temperatures.) One fate of the hydrogen peroxide (H_2O_2) and the methyl hydroperoxide (CH_3OOH) is that they can dissolve in cloud droplets and be removed from the troposphere in the form of rain. In this respect, therefore, reactions (8.10) and (8.11) are loss or terminating steps, although alternative fates of the peroxides, such as photolysis or reaction with OH, may regenerate radicals.

If oxides of nitrogen are present, then a quite different course of events can follow the formation of the peroxy radicals in reactions (8.8) and (8.9). Peroxy radicals react rapidly with NO

$$HO_2 + NO \rightarrow OH + NO_2 \tag{8.12}$$

$$CH_3O_2 + NO \rightarrow CH_3O + NO_2. \tag{8.13}$$

Reaction (8.12) regenerates OH, while (8.13) produces an *alkoxy* radical, RO, in this case *methoxy*, that can in turn react with O_2,

$$CH_3O + O_2 \rightarrow HCHO + HO_2, \tag{8.14}$$

to yield formaldehyde. Formaldehyde itself is photochemically labile; the major photolytic pathway at $\lambda < 338$ nm produces two radical fragments

$$HCHO + h\nu \rightarrow H + HCO, \tag{8.15}$$

and both radicals re-enter the HO_x chain, *via* reaction (8.8) for H, and for HCO, *via* the reaction

$$HCO + O_2 \rightarrow CO + HO_2. \tag{8.16}$$

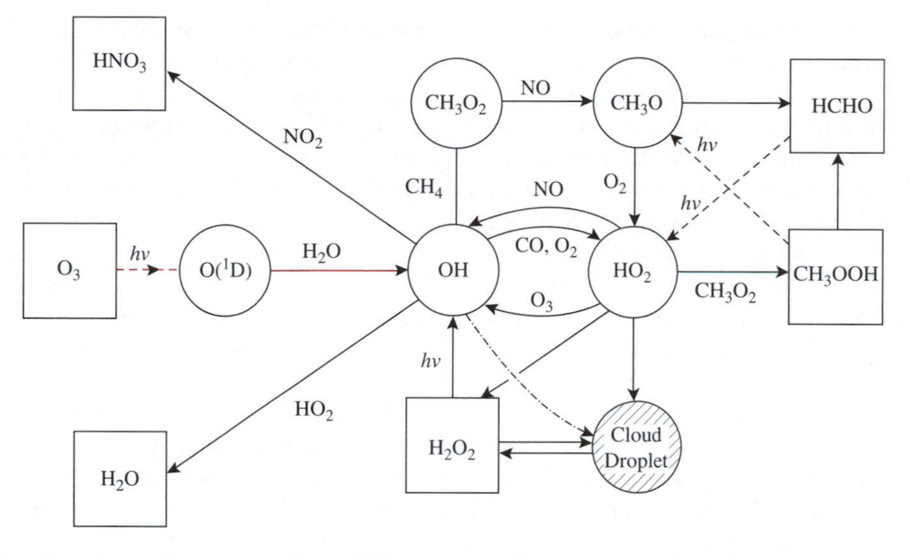

Figure 8.1 Chemistry of the troposphere. This schematic diagram emphasizes the processes that create and destroy the OH and HO_2 radicals that are key intermediates in the oxidation steps.

The essential feature is the conversion of NO to NO_2 while preserving active radicals capable of oxidizing further molecules of CO and CH_4 or converting more NO to NO_2. This formation of NO_2 is essential to the formation of ozone in the troposphere, as we shall discuss in Section 8.2.3.

Figure 8.1 presents in the form of a 'flow diagram' the chemistry discussed so far, with the emphasis laid on processes creating and destroying the radical intermediates OH and HO_2.

8.2.3 Tropospheric Ozone Production

Photolysis of NO_2 is the only known way of producing ozone in the troposphere. Reactions (8.12) or (8.13) achieve not only the conversion of peroxy radicals (HO_2, CH_3O_2) to oxy radicals (OH, CH_3O), but also the oxidation of NO to NO_2. They thus provide the link required for catalytic generation of O_3 *via* reactions (8.5) followed by (8.4). One cyclic process for tropospheric ozone production can now be written

$$NO_2 + h\nu \rightarrow O + NO \tag{8.5}$$

$$O + O_2 + M \rightarrow O_3 + M \tag{8.4}$$

$$OH + CO \rightarrow H + CO_2 \tag{8.6}$$

$$H + O_2 + M \rightarrow HO_2 + M \tag{8.8}$$

$$HO_2 + NO \rightarrow OH + NO_2 \tag{8.12}$$

Net $$CO + 2O_2 + h\nu \rightarrow CO_2 + O_3$$

Very similar chain reactions can be written that involve the CH_3O_2 species. In all cases, NO_x must be present. Below a certain critical value of the ratio [NO]/[O_3], HO_2 reacts primarily with O_3

$$HO_2 + O_3 \rightarrow OH + 2O_2 \tag{8.17}$$

rather than with NO in reaction (8.12), and ozone loss can dominate over generation. Regions of the Earth characterized by extremely low concentrations of NO, such as the remote Pacific, are thus likely to provide a net photochemical sink for odd oxygen, while the continental boundary layer at mid-latitudes, where concentrations of NO are relatively high, is likely to provide a net source. So long as NO_x is available, production of ozone in the troposphere is ultimately limited by the supply of CO, CH_4, and other hydrocarbons. Each molecule of CO can generate one molecule of O_3, while it is estimated that as many as 3.5 O_3 molecules could be formed from the oxidation of each CH_4.

Tropospheric loss processes for O_3 include photolysis in the visible and ultraviolet regions, and reaction with species such as NO_2, HO_2, and unsaturated hydrocarbons. Model calculations suggest that the globally averaged tropospheric sources and sinks for ozone are roughly in balance. This view of the *in-situ* balance is consistent with estimates that the surface sink is equal to the stratospheric injection rate.

The tropospheric abundance of ozone is obviously critically important in determining the oxidizing capacity of the troposphere, because O_3 is the primary source of OH radicals (as well as being an oxidizing species in its own right). Background concentrations of NO_x and of the peroxy radicals derived from CO, CH_4, and other hydrocarbons determine the rate of formation of ozone. Future trends of tropospheric oxidizing capacity will thus be tied closely to future atmospheric burdens of all these species.

8.2.4 The Importance of NO_x

The discussion of the previous two sections has shown how important NO_x, and especially NO, is to the overall oxidation rates and to the ozone distribution in the troposphere. The influence of NO_x is emphasized in Figure 8.2, which is a restatement of the essential steps of Figure 8.1 in another form. The left-hand semicircle shows the oxidation steps that do not require the presence of NO. In this case, the sequence of transformations following the heavy arrows leads from O_3 to CH_3OOH. If, however, NO is present, then the transformations of the right-hand semicircle close the loop, with regeneration of OH and the concomitant production of two molecules of NO_2 (and thus, potentially, of O_3).

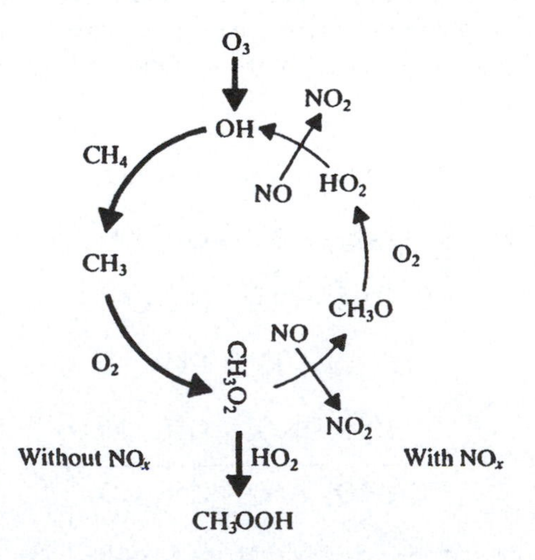

Figure 8.2 Essential steps in tropospheric methane oxidation. The heavier arrows on the left-hand side of the diagram indicate the steps that can occur in the absence of NO_x. With NO_x present, the processes on the right-hand side can close a loop, with regeneration of OH and oxidation of NO to NO_2. From a diagram devised by M.E. Jenkin.

Model estimates of tropospheric ozone production obviously depend critically on the altitude–concentration profiles adopted for NO_x, so that we must consider the atmospheric budgets for these species. Natural sources of NO_x include microbial actions in the soil, which produce NO as well as N_2O (see Section 6.4). Controlled laboratory experiments suggest that NO might even be a slightly more abundant initial product than N_2O, but NO has a far shorter chemical lifetime. Oxidation of biogenic NH_3, initiated by OH radicals, would be another significant natural source of NO_x. Lightning appears to be responsible for less than 10 per cent of the total NO_x budget, while biomass burning and high-temperature combustion (*i.e.* 'anthropogenic' sources) are perhaps responsible for 50 per cent globally. The ratio of $[NO]/[NO_2]$ depends on the rate of photolysis (reaction (8.5)) and the rates of NO oxidation in processes (8.12), and (8.13) and its analogues for other VOCs, together with the reaction

$$NO + O_3 \rightarrow NO_2 + O_2. \tag{8.18}$$

This latter reaction has a significant activation energy ($\sim 13\,kJ\,mol^{-1}$) so that $[NO]/[NO_2]$ is larger at higher tropospheric altitudes where the temperature is lower.

It is important to note that all the sources mentioned, except lightning, release NO_x to the boundary layer. Since the major loss for NO_x involves nitric acid formation,

$$OH + NO_2 + M \rightarrow HNO_3 + M, \tag{8.19}$$

followed by wet deposition, the NO_x release to the free troposphere may be appreciably smaller than the surface source strengths suggest. Some workers have argued that injection of NO_x from the stratosphere could provide a more important source for the upper troposphere than the surface emissions.

8.2.5 Alkanes and Alkenes

The oxidation steps that we have written for methane obviously have their analogues for higher hydrocarbons, but in all cases depend on the switch between RO_2 (peroxy-) radicals and RO (oxy-) radicals in the interaction with NO, represented for the general case as

$$RO_2 + NO \rightarrow RO + NO_2. \tag{8.20}$$

We turn now to a rather more detailed consideration of the oxidation of simple alkanes and alkenes. To keep the discussion reasonably compact, we shall assume that the initial attack is always by hydroxyl radicals, although it must be understood that attack by NO_3 is important at night (Section 8.3.1), and that O_3 (Section 8.3.2), HO_2, and O can play a minor role.

Alkanes are lost in the troposphere almost exclusively by reaction with OH, with H-atom abstraction from the C–H bond yielding an alkyl, R, radical. The radicals react initially with O_2

$$R + O_2 \rightarrow RO_2, \tag{8.21}$$

a third body probably being needed in the case of $R = CH_3$ (*cf.* reaction (8.9)). For ethane and propane, the subsequent steps are probably

$$RO_2 + NO \rightarrow RO + NO_2 \qquad (8.20)$$

$$RO \rightarrow R' + R''CHO \tag{8.22}$$

$$RO + O_2 \rightarrow R'R''CO + HO_2 \tag{8.23}$$

where R′ and R″ are daughter alkyl radicals or groups. The aldehydes and ketones (R″CHO, R′R″CO) can be photolysed, as we have seen in Section 8.2.2 for formaldehyde in reaction (8.15), or become further oxidized in thermal reactions. For example, OH abstracts H from acetaldehyde to form CH_3CO, which itself adds oxygen to yield the acetylperoxy radical, $CH_3CO.O_2$

$$CH_3CHO + OH \rightarrow CH_3CO + H_2O, \tag{8.24}$$

$$CH_3CO + O_2 \rightarrow CH_3CO.O_2. \tag{8.25}$$

As we shall see in Section 8.2.6, one important reaction of $CH_3CO.O_2$ is addition of NO_2, but the usual (*cf.* reactions (8.12), (8.13) and (8.20)) oxidation of NO is possible, and in that case the resulting $CH_3CO.O$ radical can fragment

$$CH_3CO.O_2 + NO \rightarrow CH_3CO.O + NO_2 \tag{8.26}$$

$$CH_3CO.O \rightarrow CH_3 + CO_2. \tag{8.27}$$

In the reactions described, the higher alkyl radicals are degraded to lower ones, and ultimately, *via* the reactions of CH_3, to CO and CO_2.

Ethene and propene, C_2H_4 and C_3H_6, appear to react with OH predominantly by an addition mechanism

$$OH + C_2H_4 \rightarrow C_2H_4OH \tag{8.28}$$

$$OH + CH_3CHCH_2 \rightarrow CH_3CHCH_2OH. \tag{8.29}$$

Both reactions show a negative temperature coefficient of rate, characteristic of an addition process, and a collisionally stabilized adduct has been observed explicitly in the case of reaction (8.29). These alkene–OH adducts appear to undergo reactions analogous to those of alkyl radicals. Thus, for terminal addition of OH to propene, the major atmospheric oxidation steps are

$$CH_3CHCH_2OH + O_2 \rightarrow CH_3CH(O_2)CH_2OH, \tag{8.30}$$

$$CH_3CH(O_2)CH_2OH + NO \rightarrow CH_3CH(O)CH_2OH + NO_2, \tag{8.31}$$

$$CH_3CH(O)CH_2OH \rightarrow CH_3CHO + CH_2OH, \tag{8.32}$$

which should be compared with reactions (8.21), (8.20) and (8.22) for the R, RO_2, and RO radicals. In the atmosphere, the CH_2OH reacts exclusively with O_2 to yield $HCHO + HO_2$

$$CH_2OH + O_2 \rightarrow HCHO + HO_2, \tag{8.33}$$

so that the products from attack of OH on propene are the aldehydes HCHO and CH_3CHO, which are ultimately oxidized or photolysed as described earlier. The HO_2 radical product of the process reacts further with NO in reaction (8.12) to regenerate OH, emphasizing that the oxidation scheme is a chain process. Chain carriers are not consumed overall, even though C_3H_6 has been oxidized to CH_3CHO and HCHO. Of course, two molecules of NO are converted to NO_2 for each OH radical cycle, and that is a matter of importance for ozone production.

8.2.6 PAN

Measurements in regions remote from anthropogenic sources have generally been thought to be representative of NO_x in the 'natural' atmosphere, because the chemical lifetime of NO_x is small compared with the times taken for geographical redistribution. The situation is rather more involved than appears at first sight, because a reservoir species for NO_x is now known that can extend the lifetime of NO_x and provide a source when others are absent. This reservoir species is an adduct of the peroxyacetyl radical, $CH_3CO.O_2$, with nitrogen dioxide, of formula $CH_3CO.O_2NO_2$. It is known universally as peroxyacetyl nitrate, or PAN, although the compound is not an ester of HNO_3 so the 'nitrate' in the name is not really appropriate. A proper systematic name for PAN might be ethane peroxoic nitric anhydride as the compound is a mixed anhydride of two acids.

PAN has been recognized for many years as a component of photochemical air pollution (Section 11.1.6), but is now known to be present also in the clean air of remote oceanic regions. Concentrations of 10 to 400 parts in 10^{12} of PAN have been observed over the Pacific Ocean, where NO_2 itself constitutes less than 30 parts in 10^{12} of the air. The importance of PAN is that it is in thermal equilibrium with its precursors,

$$CH_3CO.O_2 + NO_2 \rightleftharpoons CH_3CO.O_2NO_2, \qquad (8.34)$$

and that the equilibrium is shifted to the right-hand side at lower temperatures. Above the boundary layer, temperatures are sufficiently low for PAN to be relatively stable, but the molecule is unstable close to the surface. PAN will release NO_2 as it is transferred from cooler to warmer regions.

Formation of PAN, and of other peroxyacyl nitrates, requires the initial generation of the peroxyacyl radicals. As we saw in Section 8.2.5, many hydrocarbons, both saturated and unsaturated, produce aldehydes as oxidation intermediates, and it is attack of OH on these carbonyl compounds that yields the acyl, and ultimately the peroxyacyl, radicals. Several other oxidation mechanisms can lead to alkoxy radicals, RO. So long as R contains more than two carbon atoms, oxidation can yield a carbonyl fragment R′CO and thence, *via* reactions analogous to (8.24) and (8.25), a peroxyacyl species. PAN is therefore a compound expected when hydrocarbons are oxidized in the presence of NO_2. It is now apparent that the species must be included with NO, NO_2, and HNO_3 when considering NO_x or NO_y budgets and transport in the troposphere.

8.2.7 Oxidation of More Complex VOCs

The discussion of Section 7.2 will have shown that the VOCs released by nature to the atmosphere are not just confined to the simple alkanes and alkenes considered in Section 8.2.5. Rather, these VOCs include much more complex hydrocarbons, alcohols, aldehydes, and so on. To this list, Man's activities add yet more organic compounds such as aromatic hydrocarbons (especially benzene, toluene and the isomers of xylene), as will become apparent in Chapter 11. Nevertheless, the general principles explained earlier in the present chapter remain the basis for the interpretation of the atmospheric oxidation of all these VOCs. Figure 8.3 brings together in pictorial form the steps for OH-initiated oxidation of a generalized VOC. Steps now included are the formation of aldehydes and ketones in the degradation of the alkoxy radicals; the production of peroxides, acids and alcohols from peroxy radicals; and the diversion of peroxy radicals into the RO_2NO_2 temporary reservoirs (see previous section). The focus is now on the oxidation of the VOC and on the organic species rather than on the cycling of the inorganic radicals emphasized in Figures 8.1 and 8.2.

As pointed out in Section 7.2, oxidation of biogenic VOCs often leads to the formation of oxygenated or nitrated products with vapour pressures lower than those of the starting reactants, and thus to the formation of particles (aerosols). The 'blue hazes' that are formed in summer over certain forested areas, such as the Smoky Mountains (USA) or the Blue Mountains (Australia), probably owe their origin to VOCs emitted by the vegetation. Reaction with ozone generated in the

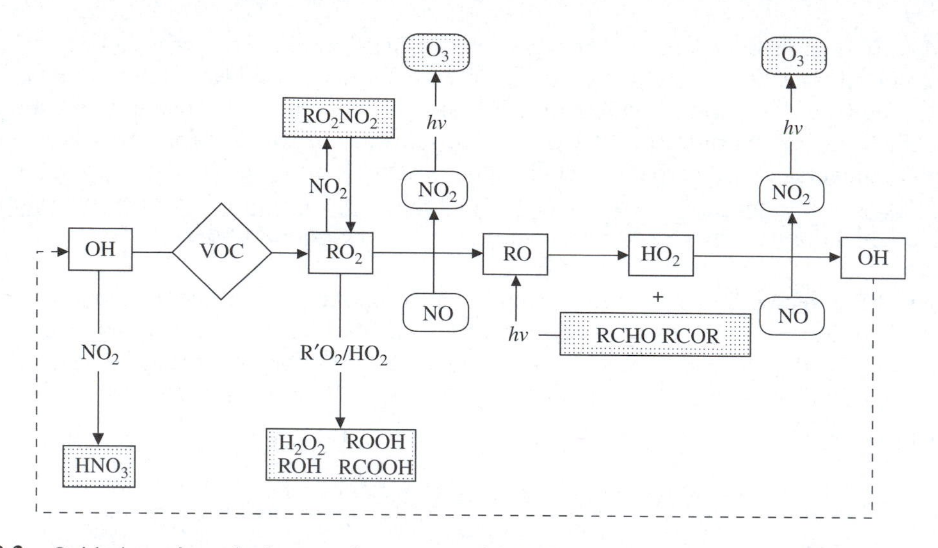

Figure 8.3 Oxidation of a volatile organic compound (VOC) during the daytime. Oxidation is initiated by attack of OH radicals on the VOC, followed by the addition of O_2 to form a peroxy radical, RO_2. From *Chemical Processes in Atmospheric Oxidation*, ed. G. Le Bras, Springer-Verlag, Berlin, 1997.

troposphere (Section 8.2.3) or originating in the stratosphere (Section 5.3) leads to the formation of aerosol in a kind of natural photochemical smog (Section 11.1.6). Indeed, parts of California experienced something like photochemical smog long before the invention of the motor car, and it may well be that terpenes from the citrus groves interacted with ozone to produce the hazes. A rather fine laboratory demonstration mimics this behaviour. A piece of orange peel is squeezed near a glass jar containing ozonized oxygen: as fine jets of the orange oils encounter the ozone, a multitude of white streamers develops in the jar.

The gas-phase atmospheric chemistry of the biogenic VOCs proceeds along the lines already explained in Sections 8.2.2 and 8.2.5. Initiation can involve attack by OH, NO_3 or O_3. As before, we focus on OH-initiation here, but will return in Section 8.3 to NO_3 and O_3. The more complex VOCs undergo essentially similar reactions, but the presence of different functional groups attached to each end of the unsaturated bond, and possibly the existence of more than one such double bond, increase the number of distinguishable reaction sites and thus reaction pathways. For example, isoprene, whose formula we shall represent as $CH_2{=}C(CH_3)CH{=}CH_2$ for the present purposes, has four chemically different carbon atoms attached to the double bonds. Thus attack by, say, OH can yield four different radical adducts. It might be supposed that, if each of these radicals were represented as R, the next stages could be represented by four sequences of the type

$$R \xrightarrow{+O_2} RO_2 \xrightarrow{+NO} RO \xrightarrow{+O_2} HO_2 + HCHO + \text{carbonyl}. \tag{8.35}$$

However, two of the initial adducts are allylic, and can undergo isomerization. Each of these radicals thus produces *two* distinguishable peroxy radicals, RO_2, on addition of oxygen, making a total of six subsequent pathways. Remember, also, that RO_2 can react in the atmosphere not only with NO, as suggested by equation (8.35), but also with HO_2 and all other RO_2 species present, and the complexities of the chemistry begin to emerge. Figure 8.4 is a simplified diagram of the steps that can occur in OH-initiated oxidation of isoprene, and it will give some idea of the steps that can follow the initial attack. The products observed in the laboratory include, in addition to

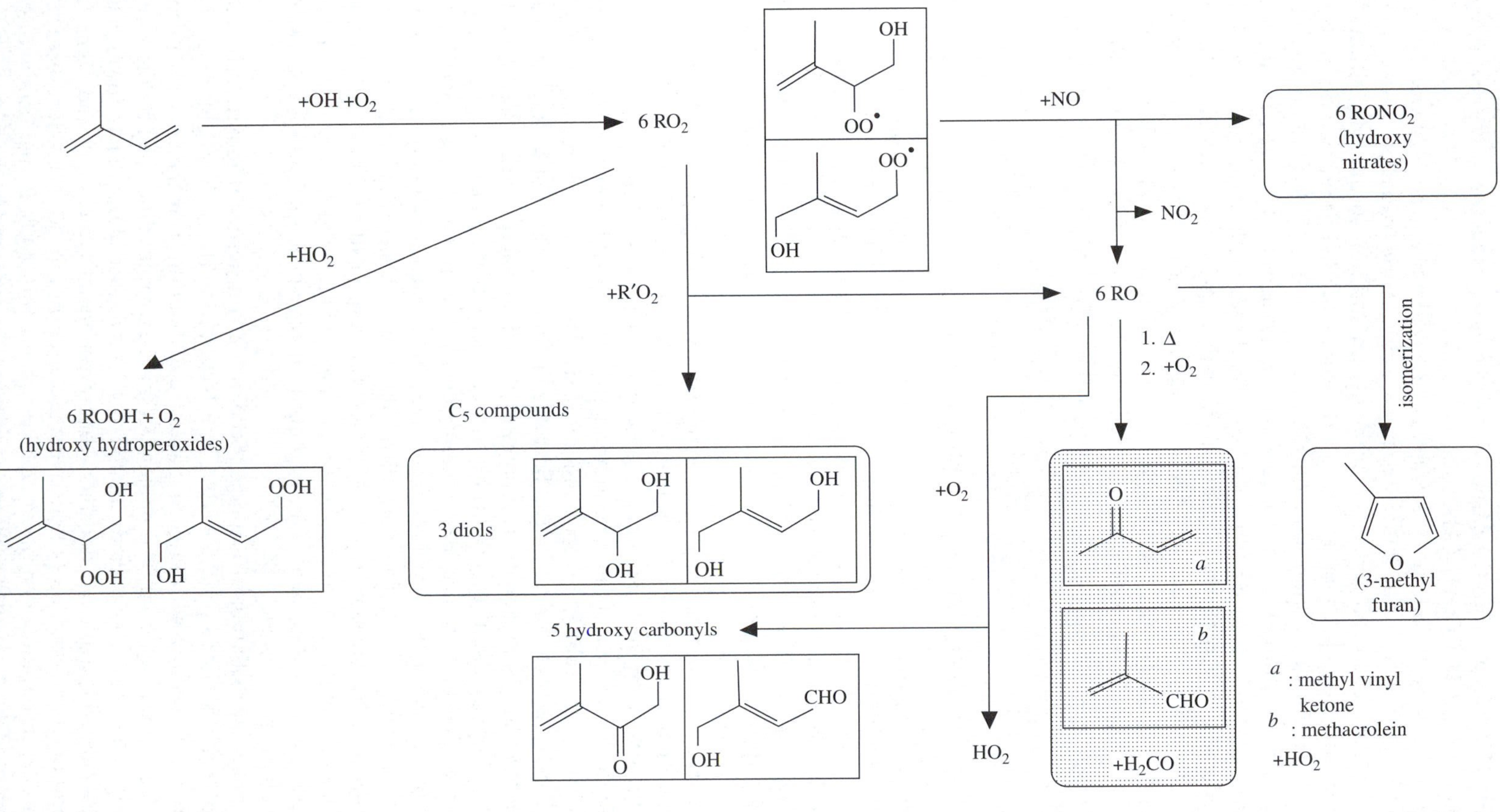

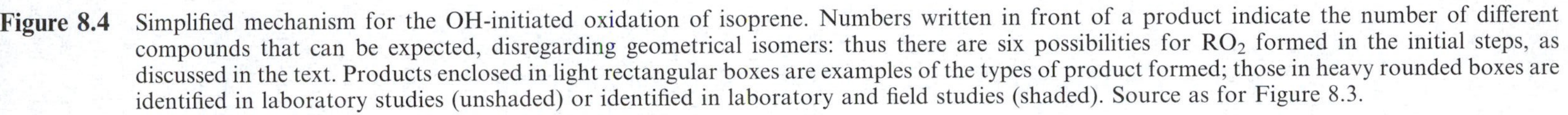
Figure 8.4 Simplified mechanism for the OH-initiated oxidation of isoprene. Numbers written in front of a product indicate the number of different compounds that can be expected, disregarding geometrical isomers: thus there are six possibilities for RO_2 formed in the initial steps, as discussed in the text. Products enclosed in light rectangular boxes are examples of the types of product formed; those in heavy rounded boxes are identified in laboratory studies (unshaded) or identified in laboratory and field studies (shaded). Source as for Figure 8.3.

formaldehyde, HCHO, the unsaturated carbonyl compounds methacrolein, $CH_2{=}C(CH_3)CHO$, and methyl vinyl ketone, $CH_3COCH{=}CH_2$.

Let us conclude the present section by examining briefly the chemical behaviour of aromatic hydrocarbons in the atmosphere. The initial step in the atmospheric oxidation is mainly attack by OH during the daytime. The process involves primarily an addition to the ring, although, if the molecule possesses an alkyl side-chain, H-abstraction from the alkyl group can occur as a minor process. Figure 8.5 attempts to show the complex chemistry that follows. Pathway (1) is the addition process, while pathway (2) is a minor abstraction reaction. A series of reactions occur, many without the intervention of NO_x, that lead to several saturated and unsaturated dicarbonyl compounds. Product **A** is a muconaldehyde (substituted hexa-2,4-diene-1,6-dial). Several such muconaldehydes have been synthesized, and on reaction with OH they yield unsaturated 1,4-dicarbonyl compounds, glyoxal, methyl glyoxal, and maleic anhydride, which are those products also seen in the reaction of OH with toluene in the presence of O_2. Many of the aromatic oxidation products, and especially the unsaturated 1,6-dicarbonyl species, are known to be toxic, with both carcinogenic and mutagenic properties.

8.3 INITIATION OF OXIDATION BY NO_3 AND BY O_3

Initiation of radical-chain steps by the hydroxyl radical, OH, has been the starting point for the oxidation processes discussed so far in this chapter. However, other initiators can contribute to the chemistry, and for some VOCs may even dominate as the agents of attack. In this section, we discuss initiation by the nitrate radical, NO_3 (Section 8.3.1), and by ozone, O_3 (Section 8.3.2). Halogen atoms and related radicals are also known to act as chain initiators in certain special cases and environments; these processes will be considered along with other aspects of tropospheric halogen chemistry in Section 8.5.

It should be remembered that the major source of OH involves photolysis of ozone by sunlight, so that most initiation of oxidation by OH is limited to the hours of daylight. Conversely, the NO_3 radical is photodissociated by sunlight, so its role as an initiator is confined to the night. O_3 itself may be present in the atmosphere at all times (although its concentration may change between day and night), and it can act as an initiator on a 24-hour basis. The relative importance of initiation of oxidation by OH, NO_3, and O_3 can be judged from the lifetimes of different VOCs with respect to reaction with the three attacking agents. Estimates of such lifetimes for rural sites in the UK are shown in Table 8.1 Absence of an entry indicates that attack by that species is negligible, so that it immediately emerges that only unsaturated compounds are susceptible to atmospheric attack by O_3. For many of those compounds, however, it is evident that reaction with O_3 makes a significant contribution to the overall reaction. The more alkyl-substituted alkenes react most rapidly with O_3; 2-butene and 2-methyl-2-butene have *shorter* lifetimes with respect to reaction with O_3 than with OH. It is worth noting, however, that for these two compounds, and a few others, reaction with NO_3 at night is the most important of all. One example is dimethyl sulfide, DMS, to be discussed in Section 8.4, and another is isoprene.

Although we looked at oxidation of isoprene (Section 8.2.7) in terms of initiation by attack of OH, inspection of Table 8.1 shows that reaction with NO_3 during the night might be rather more important, and that the reaction with O_3 also makes a contribution. For the terpenes themselves, reactions with NO_3 and O_3 may even be dominant. A good example is afforded by the compound terpinoline, for which lifetimes have been calculated of 49 min towards OH, 17 min towards O_3, and just 7 minutes towards NO_3, with OH and NO_3 concentrations at typical day and night 12-hour average values, and O_3 at a reasonable 24-hour average concentration. At face value, these numbers mean that 64 per cent of terpinoline in the atmosphere reacts with NO_3, 27 per cent with O_3, and only nine per cent with OH. It is worth pausing at this point to recollect that the lifetime of methane

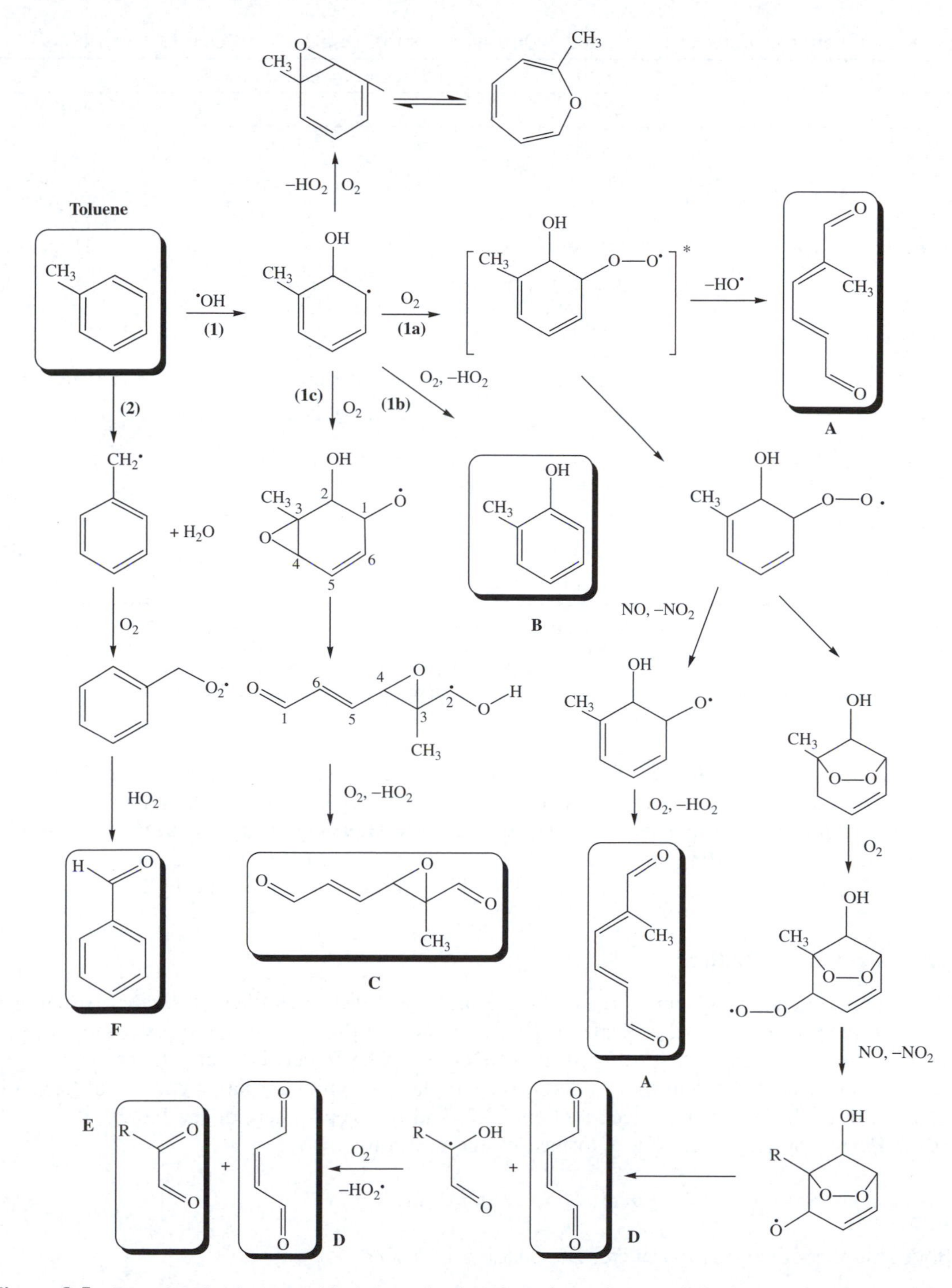

Figure 8.5 Postulated primary oxidation mechanism for toluene for attack by OH on the ortho- position. Products are shown in boxes, and are: A, 2-methyl-hexa-2,4-diene-1,6-dial; B, *o*-cresol; C, epoxide; D, butenedial; methylgyoxal; F, benzaldehyde. Based on a figure presented by C.O. Balteretu, E.I. Lichtman, A.B. Hadler and M.J. Elrod, *J. Phys. Chem. A*, 2009, **113**, 221.

Table 8.1 Chemical lifetimes of selected VOCs with respect to attack by OH, O_3, and NO_3.[a]

VOC	*Lifetime against attack by*		
	OH	*O_3*	*NO_3*
Methane	3.0 years		
Ethane	29 days		65 years
Propane	6.3 days		5.6 years
Butane	2.9 days		1.9 years
2-Methyl butane	1.9 days		11 months
Ethene	20 hours	9.7 days	5.2 months
Propene	6.6 hours	1.5 days	3.5 days
1-Butene	5.5 hours	1.6 days	2.5 days
2-Butene	2.9 hours	2.4 hours	2.1 hours
2-Methyl-2-butene	2.0 hours	0.9 hours	0.09 hours
Isoprene	1.7 hours	1.2 days	1.2 hours
Benzene	5.7 days		
Toluene	1.2 days		1.3 years
o-Xylene	12 hours		2.9 months
m-Xylene	7.1 hours		4.7 months
p-Xylene	12 hours		2.4 months
Dimethyl sulfide	1.5 days		0.7 hours[b]

[a]Data from *Ozone in the United Kingdom*, Department of the Environment, London, 1997. Lifetimes are for UK rural sites at assumed ambient levels of oxidant concentration ($OH = 1.6 \times 10^6$, $O_3 = 7.5 \times 10^{11}$, and $NO_3 = 3.5 \times 10^8$ molecule cm^{-3}).
[b]In the marine troposphere, where DMS oxidation is most important, concentrations of NO_3 might be three times less than those used for the calculations; the lifetime of DMS would then be roughly two hours. See Section 8.4.

with respect to attack by OH is of the order of *years*, and to reflect how the presence of the double bonds in the terpene have enhanced its reactivity. Other terpenes show different absolute and relative reactivities towards the oxidants OH, NO_3, and O_3. However, whatever the detail, it is quite evident that an understanding of the atmospheric chemistry of the biogenic VOCs must encompass an unravelling of the reaction paths for attack by NO_3 and O_3 as well as that by OH.

8.3.1 The Nitrate Radical

Radical intermediates in chemical reactions are most positively identified through their optical absorption spectra. The nitrate radical, NO_3, was observed in this way over 100 years ago, in 1881, and may thus have been the first transient radical species to be detected directly. Over the past few decades, it has become evident that the NO_3 radical plays a significant part in chemical transformations in both the stratosphere (see Section 9.3.2) and the troposphere of the Earth.

In the Earth's atmosphere, NO_3 is formed by the reaction

$$NO_2 + O_3 \rightarrow NO_3 + O_2 \tag{8.36}$$

in both stratosphere and troposphere. Dissociation of N_2O_5

$$N_2O_5 + M \rightarrow NO_3 + NO_2 + M \tag{8.37}$$

is apparently an additional source, but since N_2O_5 is formed by the reaction

$$NO_3 + NO_2 + M \rightarrow N_2O_5 + M \tag{8.38}$$

it is ultimately dependent on the occurrence of reaction (8.36). Dinitrogen pentoxide, N_2O_5, is an important product in its own right, since it can react heterogeneously with H_2O (Section 8.6) to yield HNO_3, and thus contribute to atmospheric acidification (see Section 11.1.5). Incidentally, NO_3 does not itself appear to react with H_2O, although it does react with aqueous negative ions to yield nitrate ions and other products.

During the day, the NO_3 radical is rapidly photolysed, but at night NO_3 has available a multitude of organic compounds with which it could react. Two main kinds of initial step can be envisaged: hydrogen abstraction, and addition to unsaturated bonds, as typified by the reactions

$$NO_3 + RH \rightarrow HNO_3 + R \tag{8.39}$$

$$NO_3 + {>}C{=}C{<} \rightarrow {>}C(ONO_2)\dot{C}{<} \tag{8.40}$$

Nitric acid is thus a direct product of hydrogen abstraction by the radical, and reaction (8.39) may make some contribution to its generation in the troposphere. Furthermore, the radical produced in reaction (8.39) is likely to add O_2 in air to form a peroxy radical, RO_2. In the special case when RH is formaldehyde, HCHO, the HCO radical will ultimately yield the HO_2 radical; for other aldehyde precursors, the acyl products of reaction (8.39) yield acylperoxy radicals, $R.CO.O_2$, and are thus potential sources of peroxyacylnitrates.

The initial adduct formed in reaction (8.40) can eliminate NO_2 to yield an epoxide. However, the species is a radical, and in the presence of air it is therefore expected to add oxygen, and the peroxy radical formed then can participate in reactions that lead to the end products. For example, for the radical derived from propene, the reaction is

$$CH_3\dot{C}HCH_2(ONO_2) + O_2 \rightarrow CH_3CH(OO\bullet)CH_2(ONO_2) \tag{8.41}$$

Several products are observed in the system when O_2 and NO_x are present, including CH_3CHO, HCHO, 1,2-propanediol dinitrate (PDDN), nitroxyperoxypropyl nitrate (NPPN), and α-(nitrooxy)-acetone

CH_3—CH(O_2NO)—CH_2(ONO_2)	CH_3—CH(O_2NO)—CH_2($OONO_2$)	$CH_3C(=O)CH_2ONO_2$
PDDN	**NPPN**	**α-(nitrooxy)acetone**

The possible formation of α-(nitrooxy)acetone in ambient air is of particular interest, since the compound has been reported to be a mutagen. The other products are also of concern because dinitrates have been shown to produce several adverse health reactions.

In the general case, the secondary reactions have the form

$$RO_2 + NO_3 \rightarrow RO + NO_2 + O_2 \tag{8.42}$$

$$RO + O_2 \rightarrow R'R''CO + HO_2 \tag{8.23}$$

Figure 8.6 provides a pictorial summary of night-time oxidation of a VOC in a manner identical to that given for the day in Figure 8.3. An important additional point is now made clear: an HO_2 radical is formed in the reaction (8.23), so that NO_3-initiated night-time chemistry is a source of

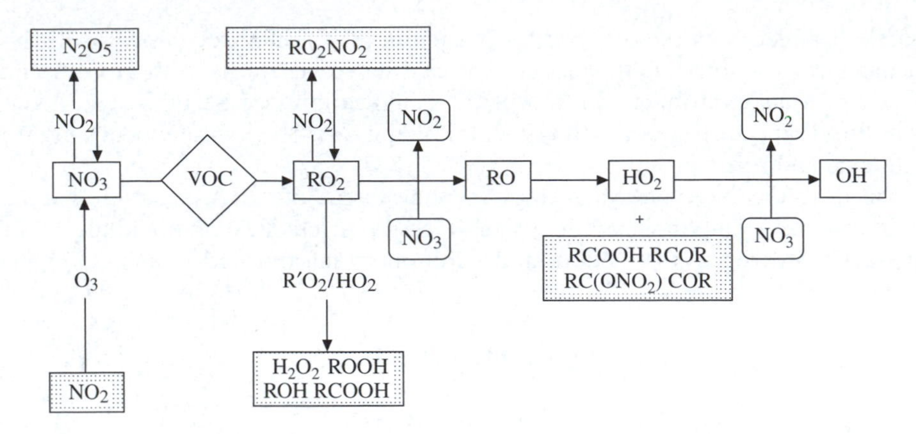

Figure 8.6 Oxidation of a volatile organic compound at night. Oxidation is initiated by attack of NO_3 radicals on the VOC, followed by the addition of O_2 to form a peroxy radical, RO_2 (*cf.* Figure 8.3). Source as for Figure 8.3.

both RO_2 and HO_2 radicals. Interactions of NO_3 with such radicals may thus be of atmospheric significance, particularly since NO, so important in the reactions of the peroxy radicals in reactions (8.12) and (8.20), usually falls to low concentrations during the night. Indeed, it is now apparent that the reactions might even constitute a night-time source of OH radicals, species that were hitherto regarded as solely of daytime photochemical origin. Two reactions can convert the HO_2 product of reaction (8.23) to OH

$$HO_2 + O_3 \rightarrow OH + 2O_2 \tag{8.17}$$

$$HO_2 + NO_3 \rightarrow OH + NO + O_2, \tag{8.43}$$

the relative importance of the two steps depending on the amount of NO_x present, which determines the ratio of $[NO_3]$ to $[O_3]$. OH radicals are generated at the expense of O_3 and NO_3. Evidence is accumulating to confirm the validity of these concepts. Significant concentrations of peroxy radicals have been detected at night, and, depending on the exact nature of the observations, there is support both for the formation of RO_2 initiated by the reactions of NO_3, and for the interaction between the RO_2 and NO_3.

8.3.2 Ozone

As indicated in Table 8.1, unsaturated VOCs such as alkenes and dienes may be attacked in the atmosphere not only by OH and NO_3, but by O_3 itself. Organic radicals, and even OH, may be formed as products, and a chain oxidation can follow the initial step. In circumstances when solar intensity is low (resulting in low [OH]) or NO_x levels are small (resulting in low $[NO_3]$), the direct reactions with O_3 can make a significant contribution to the loss of VOC.

Figure 8.7 is the analogue for attack by O_3 of those for initiation by OH (Figure 8.3) and NO_3 (Figure 8.6). It shows in diagrammatic form the essential steps in the 'ozonolysis' of a simple alkene. Reaction starts by the addition of O_3 to the double bond to form an ozonide. This ozonide

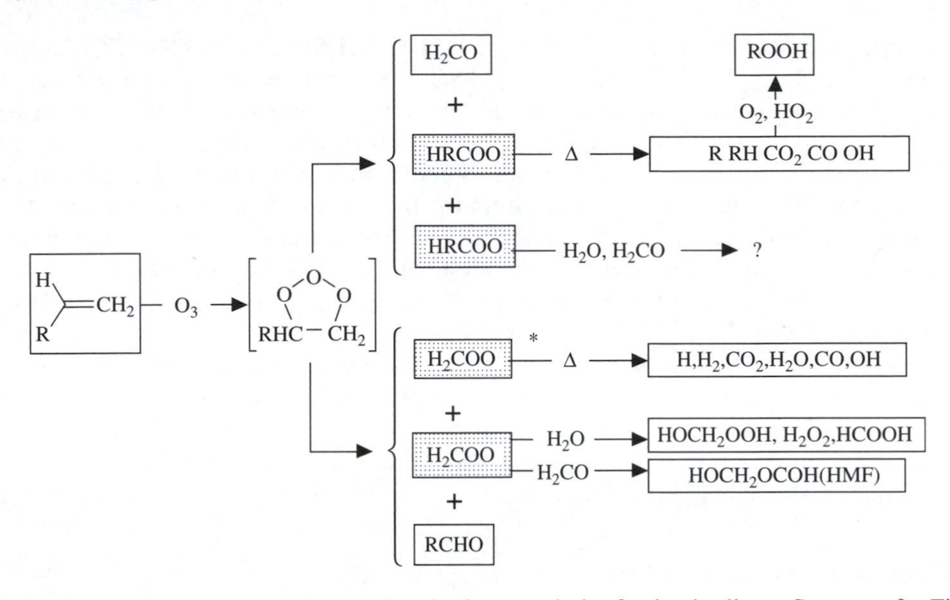

Figure 8.7 Flow diagram showing the reactions in the ozonolysis of a simple alkene. Source as for Figure 8.3.

is energy rich, and rapidly decomposes. One suggestion is that a biradical (*Criegee radical*) is formed, together with a carbonyl compound, as illustrated by the reaction of O_3 with propene

$$O_3 + CH_2{=}CHCH_3 \rightarrow CH_3\dot{C}HO\dot{O} + HCHO \tag{8.44a}$$

$$\rightarrow \dot{C}H_2O\dot{O} + CH_3CHO \tag{8.44b}$$

The Criegee biradicals are formed *excited* (energy-rich) in these reactions. A substantial fraction of the excited radicals are stabilized in the atmosphere as a result of collisional removal of their energy, and can isomerize and react with water to form carboxylic acids, aldehydes, and hydroperoxides. But the internal energy of the remaining radicals makes possible several decomposition pathways, some of which lead to new radicals (CH_3, OH and H in the case of the Criegee product of reaction (8.44a)). Thus, the ozonolysis reactions may produce organic free radicals that can form RO_2 by addition of O_2, as well as HO_x (either OH, or HO_2 after addition of O_2 to H). Laboratory experiments demonstrate the formation of OH in the interaction of ozone with alkenes, and the yield of OH increases with increasing alkyl substitution of the double bond. Thus, propene, 2-butene, and 2-methyl-2-butene yield 30, 60, and 90 per cent of OH on reaction. Since these are just the alkenes that also react most rapidly with O_3 (see Table 8.1), there is evidently the potential for significant HO_x production in the troposphere when such alkenes are present. The formation of organic radicals, ultimately in the form of RO_2, also has interesting consequences. We know already that such radicals are agents in tropospheric O_3 production (Section 8.2.3), so that, depending on the ambient conditions, initiation of oxidation by the ozone interaction can lead to net ozone *production*.

8.4 SULFUR CHEMISTRY

The sources and source strengths of the compounds of sulfur have been discussed several times in this book (Sections 1.5, 6.5 and 7.2.4). The general picture is that biogenic processes tend to emit reduced compounds such as H_2S, CS_2, COS, CH_3SH, CH_3SCH_3 and CH_3SSCH_3. Volcanoes release both reduced (H_2S, COS) and oxidized (SO_2) compounds to the atmosphere; different volcanoes (and sometimes different eruptions of the same volcanoes) emit different mixes of the gases. Man adds huge quantities of sulfur to the natural sources, mostly in the form of sulfur dioxide (SO_2) from the burning of fossil fuel (and some from biomass burning). The natural sulfur compounds are mostly oxidized first to SO_2 within the troposphere. The exception is the long-lived COS, which may reach the stratosphere and be the dominant source of stratospheric sulfate aerosol during periods of volcanic quiescence. Dimethyl sulfide (CH_3SCH_3), DMS, is the largest contributor to the biogenic emissions, and its oxidation may have significant environmental consequences by altering cloudiness, as discussed in Section 7.2.4.

In the present section, we address specifically the chemical processes that give rise to the observed oxidation products. Since oxidation of the reduced species leads first to SO_2, which is also one of the species emitted, it makes sense to divide the discussion into two parts. First, there are the steps that convert sulfur in oxidation state –2 (*e.g.* CH_3SCH_3) through six oxidation states to +4 (SO_2). Then, there follows the oxidation of SO_2 to state +6 in SO_3 (or its hydrolysis product H_2SO_4).

8.4.1 Oxidation from State –2 to +4: the Steps to SO_2

For the inorganic reduced-sulfur species, oxidation is driven by the hydroxyl radical, as with so many tropospheric processes. In each case, SH radicals are formed

$$OH + CS_2 \rightarrow COS + SH \tag{8.45}$$

$$OH + COS \rightarrow CO_2 + SH \tag{8.46}$$

$$OH + H_2S \rightarrow H_2O + SH. \tag{8.47}$$

Reaction (8.45) with CS_2 probably proceeds *via* an intermediate complex $HOCS_2$ that can react with O_2 to yield COS + SH. In the presence of O_2, O_3, or NO_2, the SH radicals are then further oxidized to SO_2, *via* the intermediate radicals SO and possibly HSO and S atoms. The oxidation steps probably include the pair of reactions

$$SH + O_2 \rightarrow OH + SO \tag{8.48}$$

$$SO + O_2 \rightarrow SO_2 + O, \tag{8.49}$$

which interestingly regenerate the OH used in the primary step as well as bringing about the oxidation to SO_2. Another oxidation sequence that can follow the formation of SH proceeds with atomic S as an intermediate

$$SH + SH \rightarrow H_2S + S \tag{8.50}$$

$$S + O_2 \rightarrow SO + O. \tag{8.51}$$

The disproportionation reaction (8.50) creates one S atom that can be oxidized in reaction (8.51) followed by (8.49); it also provides a pathway for the formation of H_2S from the more oxidized SH radicals. Reaction of SH with OH, HO_2, HCHO, H_2O_2, and CH_3OOH likewise generates H_2S, so there exist photochemical pathways that link CS_2 and COS to H_2S. Figure 8.8 illustrates the main transformations that lead to SO_2 from reduced inorganic sulfur compounds.

Oxidation of the organic sulfur compounds can be initiated by attack either of OH or of NO_3. For CH_3SCH_3 (DMS), a typical atmospheric lifetime against attack by NO_3 at night is less than

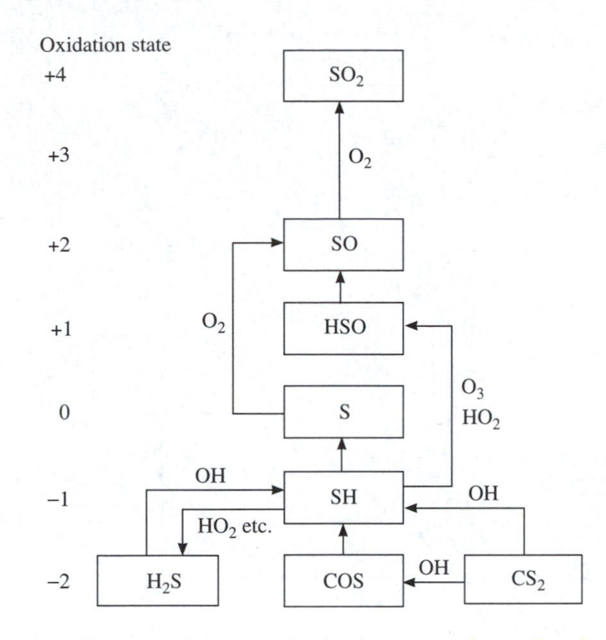

Figure 8.8 Conversions of sulfur-containing species in the troposphere, showing the progression towards more oxidized compounds.

0.7 h, as against 1.5 days for daytime attack by OH radicals (Table 8.1). CH_3SSCH_3 seems to react predominantly with OH, although NO_3 still contributes about 10 per cent to the loss of the compound.

The types of oxidation process that the organic sulfides undergo can be illustrated with reference to oxidation of DMS initiated by reaction with hydroxyl radicals. Figure 8.9 is intended to provide a simplified guide to the description of the reactions that follows. Products observed are dimethyl sulfoxide (DMSO), COS, SO_2, H_2SO_4, methane sulfonic acid (MSA), and dimethyl sulfone ($DMSO_2$). Initial attack of OH on DMS proceeds *via* two parallel steps

$$OH + CH_3SCH_3 \rightarrow H_2O + CH_3SCH_2 \qquad (8.52a)$$

$$OH + CH_3SCH_3 \rightarrow CH_3S(OH)CH_3. \qquad (8.52b)$$

Channel (8.52a) involves hydrogen abstraction to yield a radical that behaves much as an alkyl radical, and channel (8.52b) is addition of OH to form an adduct. The figure shows that these two pathways lead to different final products, and it is noteworthy that the branching ratio is highly temperature dependent: abstraction is favoured at higher temperatures and addition at lower temperatures. Thus, at $T = 285$ K the two channels are of equal importance, but by $T = 298$K channel (8.52a) makes up about 80 per cent of the reaction. Relative product yields will therefore be dependent on atmospheric temperatures.

In the presence of air, initial products from both channels react further with O_2. The peroxy radical derived from channel (8.52a), $CH_3SCH_2O_2$, reacts with NO as though it were a simple alkylperoxy radical, the oxy radical fragmenting to a methyl thiyl radical (CH_3S) and formaldehyde. As with other RO_2 radicals, if NO_x is low (as it may be in the marine boundary layer) reactions with HO_2 occur (*cf.* reaction (8.11)). A series of competing reactions now follows that lead to the products COS, MSA and SO_2. The addition channel of the primary interaction, reaction (8.52b), is responsible for the formation of yet further sulfur-containing species found in

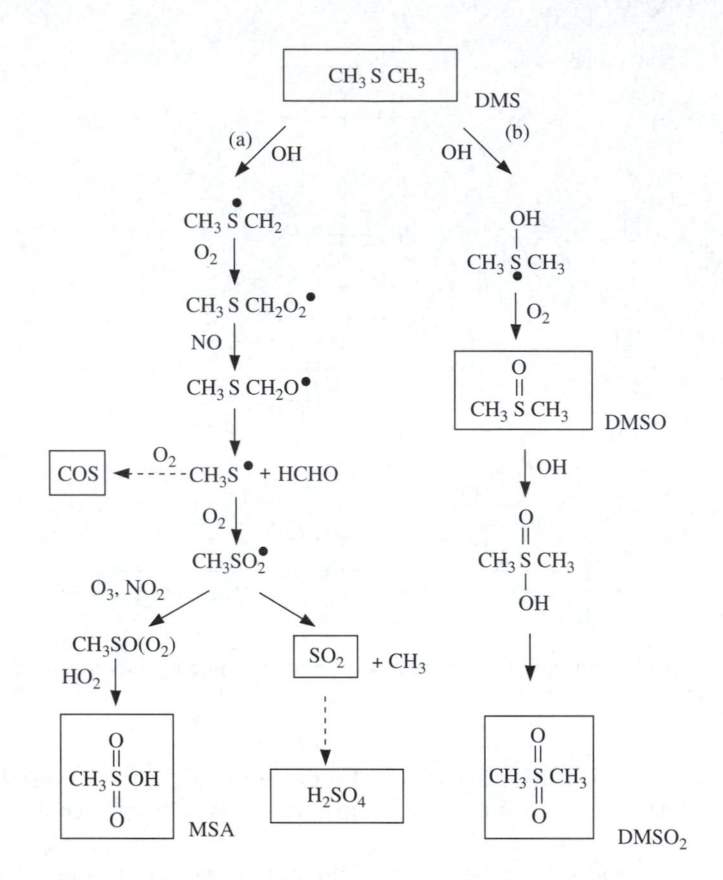

Figure 8.9 Simplified pathways for the atmospheric oxidation of dimethyl sulfide (DMS) initiated by reaction with OH. Species shown in boxes are observed atmospheric products: DMSO = dimethyl sulfoxide; $DMSO_2$ = dimethyl sulfone; MSA = methyl sulfonic acid.

the atmosphere, such as dimethyl sulfoxide (DMSO) and dimethyl sulfone ($DMSO_2$). Plausible routes are shown in the diagram.

The reaction of NO_3 with DMS is more efficient than many interactions of this radical with organic compounds, and it possesses a small negative temperature coefficient, which is indicative of an addition mechanism. However, HNO_3 is rapidly eliminated from the adduct first formed, so that the products are HNO_3 and the radical CH_3SCH_2, which participates in the reactions described in the preceding paragraph. The subsequent oxidation steps are thus identical to those described for the abstraction channel of OH-initiated oxidation of DMS. The rapid reaction of NO_3 may be especially important during high-latitude winter, even in remote regions where NO_x is relatively low. It is not at present clear if reactions of NO_3 are responsible for observed cyclic seasonal variations in the ratio of non-seasalt sulfate to MSA, which show a pronounced maximum in winter.

8.4.2 Oxidation from State +4 to +6: the Steps from SO_2 to H_2SO_4

In the troposphere, SO_2 is almost all oxidized to H_2SO_4 in the form of aerosol, and the atmospheric sulfur cycle is closed by wet precipitation of the sulfuric acid. The aerosol particles, once formed, are rapidly incorporated into water droplets, and the particles may, indeed, act as condensation nuclei. Reactions on and in droplets are thus evidently of central importance in the atmospheric chemistry of SO_2. Gas-phase, aqueous-phase, and surface oxidation steps all seem to contribute to the overall conversion of SO_2 to H_2SO_4. Reactions involving more than one phase are referred to as *heterogeneous* as opposed to *homogeneous* processes occurring solely within the gas phase. In view of the widespread nature of heterogeneous chemistry in the troposphere, it is appropriate to devote a separate section (Section 8.6) to a consideration of the general principles that apply. For the present purposes, we just accept that heterogeneous reactions take place, and that often the chemical changes that take place are different from the homogeneous ones, or at least may occur more rapidly.

Meteorological conditions such as the relative humidity and the presence or absence of clouds or fogs are likely to control the relative importance of the homogeneous and heterogeneous processes. Enhanced oxidation rates at high relative humidities or when condensed water is present indicate the participation of liquid-phase reactions. Rates of wet and dry deposition of SO_2 itself are five to ten times faster than the rate of homogeneous oxidation in the boundary layer, thus suggesting that heterogeneous conversion of SO_2 to H_2SO_4 must normally constitute the major oxidation path. Some reaction may occur on solid aerosol surfaces, but within a cloud most of the oxidation takes place in the liquid phase. However, there are uncertainties about the rates for many of the reactions possible in the gas and liquid phases. In highly polluted environments, species present other than sulfur compounds may act as oxidants, and various metals can act as catalysts. Quantitative estimates of the contributions of the different routes to SO_2 oxidation are not, therefore, available. Instead, we must examine the processes most likely to bring about homogeneous and heterogeneous oxidation.

In the gas phase, the ubiquitous hydroxyl radical seems once again to initiate the dominant oxidation route by addition to SO_2

$$OH + SO_2 + M \rightarrow HOSO_2 + M. \tag{8.53}$$

The $HOSO_2$ radical is further oxidized to H_2SO_4, although the mechanism is probably more complex than just the addition of a further OH radical. Several schemes have been proposed, of which one of the most interesting starts with the addition of O_2 to form HSO_5 ($HOSO_2O_2$). HSO_5 has some of the properties of other peroxy radicals, but it may also become hydrated with n water molecules

$$HSO_5 + nH_2O \rightarrow HSO_5(H_2O)_n \tag{8.54}$$

Hydrated HSO_5 is seen as a strong oxidizing agent that can convert SO_2 to SO_3 in a reaction that is described as 'quasiheterogeneous' because, with large n, the hydrated radicals are virtually aerosol particles. Sulfuric acid formation is possible, as indicated by the equation

$$HSO_5(H_2O)_n + SO_2 \rightarrow HSO_4(H_2O)_nSO_3 \rightarrow HSO_4(H_2O)_{n-m} + H_2SO_4(H_2O)_{m-1}, \tag{8.55}$$

m of the original n molecules of water being used to hydrolyse and hydrate the SO_3. Reactions analogous to equation (8.55) can also oxidize NO and HO_2. Nitric acid is a potential product of the reaction with NO, a matter of considerable interest since nitric and sulfuric acids are frequently found together in polluted environments (see Section 11.1). The hydrated HSO_4 radical that appears as a product in reaction (8.55) is likely to be a highly reactive species that could react with itself, or with HO_2 or NO_2.

We turn now to a consideration of the *heterogeneous* oxidation from the S(IV) oxidation state of SO_2 to S(VI) of SO_4^{2-} in the aqueous droplets that constitute aerosols, clouds, fogs, and rain. Solution of SO_2 in water proceeds first through hydration of the neutral SO_2 and then by ionization to yield HSO_3^- and SO_3^{2-} ions

$$SO_2 + H_2O \rightleftharpoons H^+ + HSO_3^- \rightleftharpoons H^+ + SO_3^{2-} \tag{8.56}$$

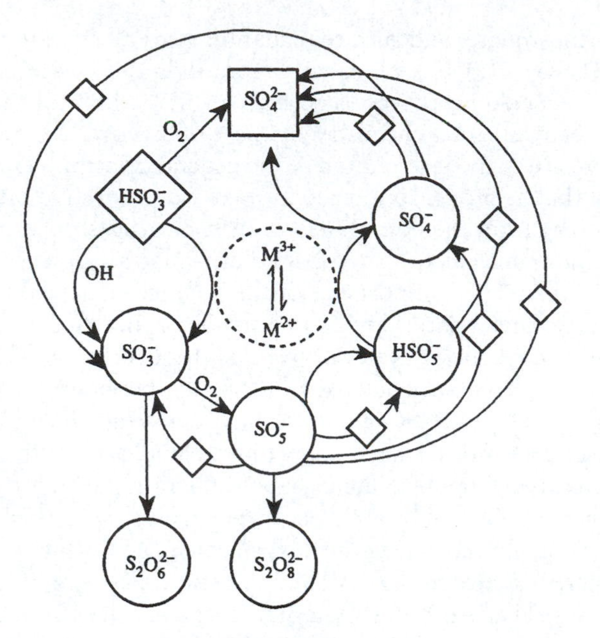

Figure 8.10 Generalized scheme for the oxidation of sulfur (IV) in aqueous solution. Reaction pathways marked with a diamond shape consume HSO_3^- as one of the reactants. The dotted circle and arrow links in the centre represent the catalytic pathways opened by the presence of transition metal ions. Based on a diagram presented by P. Warneck in *Heterogeneous and Liquid Phase Processes*, ed. P. Warneck, Springer-Verlag, Berlin, 1996.

The HSO_3^- and SO_3^{2-} ions are finally oxidized to SO_4^{2-} by hydrogen peroxide and ozone dissolved from the gas phase. Molecular oxygen plays a more equivocal role, since it is not clear if any oxidation occurs at all in the absence of catalysts; several metals have, however, been shown to promote the reaction.

In evaluating different routes for the oxidation of S(IV) to S(VI), consideration must always be given to the availability of sufficient oxidant to effect the conversion. For example, in urban environments, where SO_2 concentrations can be around 100 ppb, it seems that, however *efficient* an oxidant H_2O_2 might be, it is almost certainly present in quantities only enough to oxidize a small fraction of the SO_2. Even O_3 concentrations might be inadequate for the purpose. An obvious step is to look for cyclic processes in the aqueous phase capable of bringing about the oxidation. Free radicals in the liquid phase, such as OH and HO_2, may be important in promoting the reactions. The radicals can be scavenged from the gas phase, or formed by reactions in the droplets, as will be described in Section 8.6. Aqueous OH reacts with HSO_3^- and SO_3^{2-} to yield SO_4^{2-}; radical-ion intermediates such as SO_3^-, SO_4^-, and SO_5^- are believed to be involved. Figure 8.10 is an attempt to present diagrammatically this rather complex chemistry. Ignore initially the large dashed circle in the middle of the diagram. The diamond shape contains the starting S(IV) compound, HSO_3^-, while the square contains the oxidized S(VI) product, SO_4^{2-}; the circles contain the various intermediates, as well as two other terminal ions, $S_2O_6^{2-}$ and $S_2O_8^{2-}$. The lines are the available chemical pathways. A diamond shape on a line means that HSO_3^- is the coreactant in the process, so that each of the seven reactions so marked allows yet another S(IV) compound to enter the oxidation chain. Several cyclic paths can be constructed, although a critical link seems to be between SO_4^- and SO_3^- ions, represented in the diagram by the outermost left-hand link.

Much evidence has accumulated to show that transition metals catalyse the oxidation of S(IV) compounds, and such metals are often abundant in atmospheric particles present in just those polluted environments where SO_2 concentrations are also high. Catalytic pathways in the radical-ion oxidation mechanism have now been discovered, and they are indicated schematically as the inner part of Figure 8.10, where the variable-valence metal ions are shown within the dotted circle, and the arrows show the species linked through the catalyst.

In the real atmosphere, a combination of the different proposed mechanisms may bring about the oxidation of SO_2, and each may be of dominant importance under different conditions. Whatever the oxidation route, and be it gas phase or aqueous phase, H^+ ions become dissolved in cloud water and lower its pH. Precipitation thus brings with it acidity as well as dissolved sulfate.

8.5 'REACTIVE' INORGANIC HALOGENS AND THEIR OXIDES

Organic halogenated compounds are found in wide variety in the atmosphere. Many of them are of anthropogenic origin, but a few are natural. For example, it has long been recognized that the methyl halides, CH_3Cl, CH_3Br, and CH_3I, are present: they are produced predominantly in the oceans. However, until the 1980s it was believed that inorganic halogen atoms and radicals were significant only in the stratosphere, where they are of the greatest importance in catalytic cycles that destroy ozone (see Section 7.2.2 and Chapter 9). Tropospheric processes energetic enough to liberate reactive species to the gas phase were unknown at that time. Since then, however, the increasing understanding of heterogeneous chemistry gained from studies of the stratosphere has led speculation that *inorganic* halogens in the form of the atomic and molecular elements (*e.g.* Br and Br_2) and their radical oxides (*e.g.* BrO) could also participate in tropospheric chemistry.

Vast quantities of inorganic halides are present as particles in the atmosphere, and we shall see shortly that chemical mechanisms exist that can effect the transformation of these particles into 'active' reactants. Wave action in marine areas generates small airborne droplets of sea-water that either remain as concentrated aqueous solutions, or evaporate to leave suspended particles of the solids. Seasalt itself contains 55.7 per cent of Cl, 0.19 per cent of Br, and 2×10^{-5} per cent of I by weight. Polar regions seem to be particularly effective in providing sources of inorganic halogens, a topic to which we shall return. Salt flats, and inland bodies of salt water, such as the Dead Sea, provide additional sources of salt-containing particles, which are generated by wind action and erosion of crusts. Some volcanic eruptions, such as that of El Chichón, generate a brief pulse of salt emissions. Figure 8.11 shows schematically the ways in which the precursors of reactive inorganic halogens can enter the lower atmosphere. Note that the sources are mostly natural, and largely associated with halogens in ocean water.

Figure 8.11 goes on to indicate the chemistry that leads from the oceanic precursors to the reactive atoms and radicals. The molecular halogens (Cl_2, Br_2, I_2), interhalogens (*e.g.* BrCl) and some other halogen compounds (*e.g.* $ClNO_2$) can be readily photolysed by the wavelengths of solar radiation reaching the troposphere

$$Br_2 + h\nu \rightarrow Br + Br \tag{8.57}$$

$$BrCl + h\nu \rightarrow Br + Cl \tag{8.58}$$

$$ClNO_2 + h\nu \rightarrow Cl + NO_2, \tag{8.59}$$

so that several of the processes include steps such as these. Working from the right-hand side of the diagram, the mechanisms identified are

(i) photolysis of gas-phase precursors;
(ii) acid displacement;
(iii) acidic ion chemistry and photolysis.

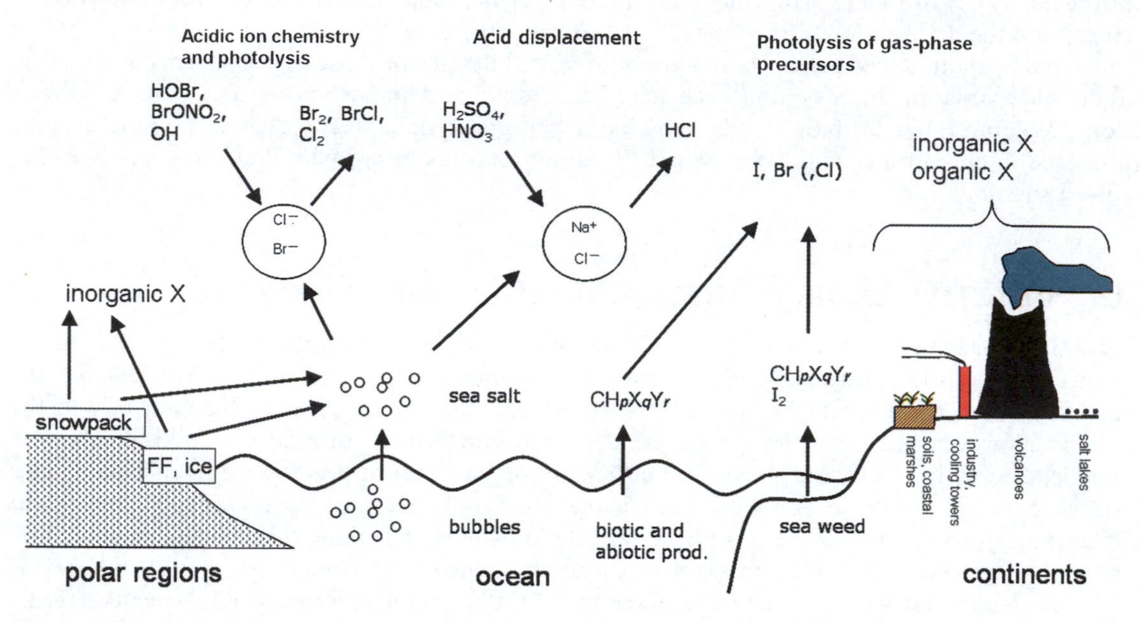

Figure 8.11 Release mechanisms for reactive halogens in the troposphere. Adapted from R. von Glasow, *IGAC Newsletter* 2008, **39**, 2.

As an example of process (i), CH_3I can be photolysed by the radiation that penetrates to the troposphere to generate an iodine atom

$$CH_3I + h\nu \rightarrow CH_3 + I. \tag{8.60}$$

Other biogenic iodides and mixed halides such as CH_2I_2 and CH_2BrI behave in the same way. The reaction

$$I + O_3 \rightarrow IO + O_2 \tag{8.61}$$

is then a source of the IO radical, which is quite widespread in occurrence: it has been detected from space, and found from the Antarctic (at very high concentrations), to the Cape Verde islands and Mace Head (Ireland). At Mace Head, IO has been measured at concentrations of up to 6 ppt (6×10^{-12}). The IO follows solar radiation intensity very closely, with concentrations below the level of detectability during the night. This correlation suggests a photochemical origin of the radicals, most probably from the precursors CH_2I_2 and CH_2BrI, which were observed at elevated concentrations.

Acid displacement, mechanism (ii), may be a major source of active halogens. It has been known for at least five decades that seasalt particles are deficient in chloride and bromide ions relative to sodium, suggesting that reactions with atmospheric gases (acids, oxides of nitrogen, or other oxidants) may have converted condensed-phase halides into gas-phase halogen species. Heterogeneous reactions such as

$$NaCl_{(s)} + N_2O_{5\,(g)} \rightarrow ClNO_{2\,(g)} + NaNO_{3\,(s)} \tag{8.62}$$

occurring on the particles generate nitryl chloride ($ClNO_2$), which is readily photolysed to liberate Cl atoms. Figure 8.12 shows in more detail one reaction scheme proposed to explain the formation

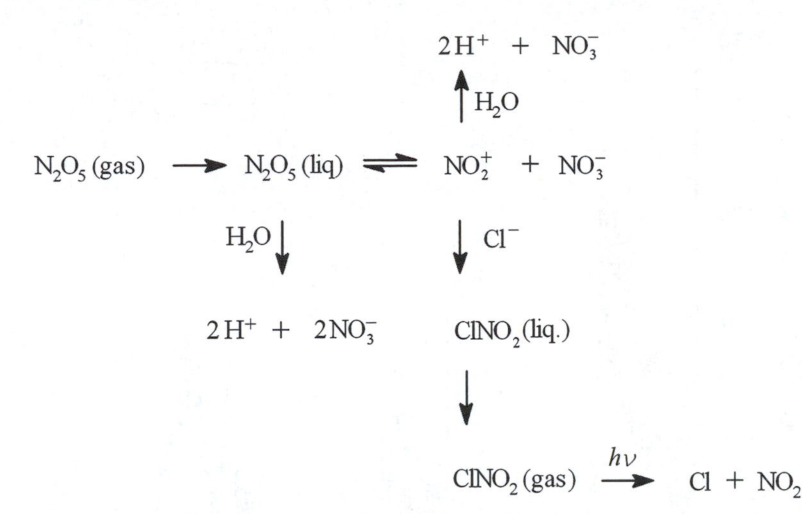

Figure 8.12 Reaction scheme for the formation of $ClNO_2$, and then Cl, from N_2O_5 and droplets of NaCl solution.

of $ClNO_2$ and, *via* reaction (8.59), free Cl atoms: there are several steps involving ionic reactions within the droplet.

Reaction (8.62) has been shown to occur in the laboratory, and its occurrence in the atmosphere as well appears to be nicely confirmed by simultaneous *in-situ* detection of $ClNO_2$ and N_2O_5 made from a research vessel travelling off the northeastern coast of the USA. Figure 8.13 shows results obtained during a period in March 2008 in the International Chemistry in the Arctic Lower Troposphere (ICEALOT) campaign (although the conditions were by no means 'Arctic' here, but at about 42.3°N, with air arriving from the direction of New York!) Not only is the compound $ClNO_2$ found to be present in the atmosphere, but the close correlation between its concentrations and those of N_2O_5 are highly suggestive of reaction (8.62) being the source of the $ClNO_2$.

As we shall see later in this section, there is evidence of bursts of high levels of halogen activation in polar regions. Variants of mechanism (iii) have been called on to explain this behaviour. The special feature of the suggested schemes is the occurrence of heterogeneous ionic reactions such as

$$HOBr + Br^- + H^+ \rightarrow Br_2 + H_2O \tag{8.63a}$$

$$HOBr + Cl^- + H^+ \rightarrow BrCl + H_2O. \tag{8.63b}$$

The Br_2 or BrCl are formed initially in the aqueous phase, but enter the gas phase, where they undergo photolysis in reactions (8.57) or (8.58) to liberate free halogen atoms. Well-established gas-phase reactions involving O_3 and HO_2 can be used to write a coherent cyclic scheme. Here is one that employs formation of Br_2 in reaction (8.63a)

$$Br + O_3 \rightarrow BrO + O_2 \tag{8.64}$$

$$BrO + HO_2 \rightarrow HOBr + O_2 \tag{8.65}$$

$$HOBr + Br^- + H^+ \rightarrow Br_2 + H_2O \tag{8.63a}$$

$$Br_2 + h\nu \rightarrow Br + Br \tag{8.57}$$

$$\text{Net} \qquad HO_2 + H^+ + O_3 + Br^- + h\nu \rightarrow Br + 2O_2 + H_2O.$$

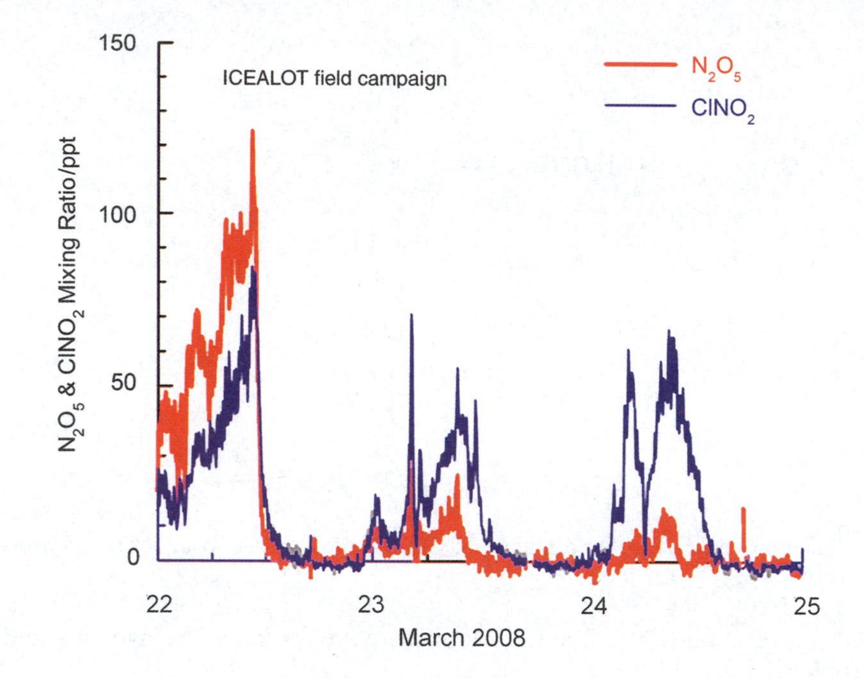

Figure 8.13 The $ClNO_2$ (blue) and N_2O_5 (red) mixing ratios measured from 22 March to 24 March 2008 during the ICEALOT field campaign.

Several interesting features are apparent on examining this scheme. It is a cyclic process because the Br atoms that initiate the sequence are regenerated at the end, and acidic conditions are required. More remarkably, it is not just one Br atom that is formed, but *two*, in reactions (8.63a) + (8.57), for every *one* that went in to reaction (8.64). That is, there are more Br atoms at the end than at the beginning; the net effect, as shown, is the generation of Br atoms. Since each 'new' Br can initiate its own cycle, there is the potential for exponential growth both of rate of reaction and of concentration of Br in the gas phase. This situation is analogous to that in the explosive H_2–O_2 reaction, and the bromine cycle just described has been termed the *bromine explosion*.

This is the appropriate point at which to return to the tropospheric consequences of reactive halogen chemistry. One of them, the *ozone-depletion event* (ODE), is probably related to the 'bromine explosion' just described. Sudden and large ozone depletions can occur in geographically rather localized areas of polar regions. Figure 8.14 shows one such event in which ozone concentrations dropped to near zero over a few hours, and remained low over several days. Several spectroscopic studies, and especially those using satellite imaging, link ODEs to locally high concentrations of BrO, so that it is not unreasonable to suppose that it is bromine explosions that produce the ODEs, with ozone being consumed in reaction (8.64). As to why these events occur specifically in polar regions, it seems that the freezing of sea water makes more halide ions enter the atmosphere than are provided by seasalt spray over open oceans. The measurements of BrO point to first-year sea-ice as an important region for bromine activation. Two hypotheses have been presented for the mechanism of bromide salt transport from the ocean to reactive bromine in the atmosphere, both of which are associated with young ice. One possibility is that snow contaminated with salts, which are prevalent on first-year sea-ice, is the surface from which bromine is released. Direct mass spectroscopic evidence of release of Br_2 and BrCl from snow, and depletions of Br^- in coastal snow, support the salty-snow hypothesis. Freezing sea-water separates ice from brine, which is a concentrated salt solution. Some of this brine is forced to the new ice surface, causing high initial surface salinities. As the ice ages, the surface salinity decreases. Snow on the sea-ice is

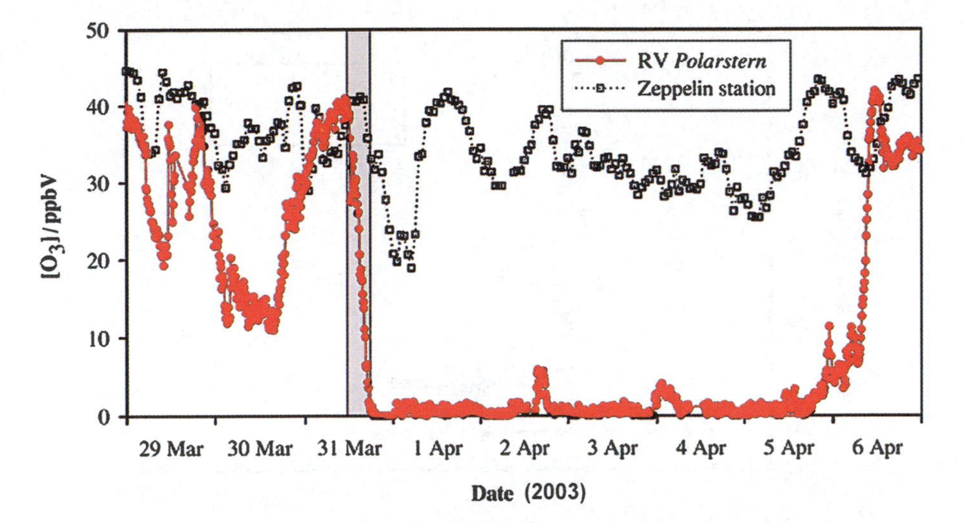

Figure 8.14 An ozone-depletion event measured (red points and line) on board the Polarstern in the Arctic during spring 2003 in the vicinity of new ice fields. Data from a land station at Ny-Ålesund, Spitsbergen, are shown in grey for comparison, illustrating that the observed ozone depletion is occurring locally rather than being due to long-range transport of ozone-free air. The grey bar indicates a drop in ozone from 40 ppb to close to zero in approximately 7 h. From H.-W. Jacobi *et al.*, *J. Geophys. Res.* 2006, **111**, D15309.

contaminated by salts when brine wicks up the snow or the wind scours the snow to re-expose saline surfaces or salt aerosols deposited on the snow. Sea-ice that survives the summer has a much lower surface salinity because the brine drains away during summer melting. Multiyear sea-ice and land surfaces therefore contain less atmospherically accessible salt and are less likely to be reactive halogen sources than first-year sea-ice. An alternative is that *frost flowers*, which are themselves associated with young ice, may provide the surface from which bromine is released. Frost flowers form when small nodules in the forming ice act as condensation nuclei for vapour deposition of angular ice crystals that look like delicate flowers. The flowers wick brine from the sea-ice surface through the action of surface tension, producing highly saline crystals with large specific surface areas.

Ozone depletion and halogen chemistry have a significant impact on VOC and OVOC photochemistry. Light hydrocarbons are rapidly consumed (mostly by Cl atoms) during ODEs. The lifetime of propane, for example, is about 14 days under normal conditions when oxidation is initiated by OH. However, during ozone-depletion events, the high [Cl] coupled with the high reactivity towards propane (8800 times greater for Cl compared with OH) results in a propane lifetime of just 8 h. Reactive alkanes, such as butane and *n*- and *i*-pentane, are removed nearly completely during ODEs. A singularly telling piece of evidence that links elevated halogen atom levels with the Arctic tropospheric ozone depletions is reproduced in Figure 8.15. Concentrations of *i*-butane, *n*-butane, and propane were measured as they decayed as a result of chemical reaction, and in the absence of new emissions or physical dilution. Referring to the three hydrocarbons as *i*, *n*, and *p*, the rate constants for reaction with OH are roughly in the ratios $i{:}n{:}p = 2{:}2{:}1$, while for reaction with Cl the rate-constant ratios are approximately $i{:}n{:}p = 1{:}2{:}1$. If, then, loss of the hydrocarbons is primarily in reaction with OH, it might be expected that $[i]/[n]$ would remain constant, while $[i]/[p]$ decreased. On the other hand, if reaction with Cl were largely responsible for chemical loss of the hydrocarbons, $[i]/[p]$ would remain constant, while $[n]/[i]$ would decrease. The figure shows that the experimental data divide themselves into two distinct groups. When there is a normal ozone concentration at Alert, the behaviour is that of OH-controlled

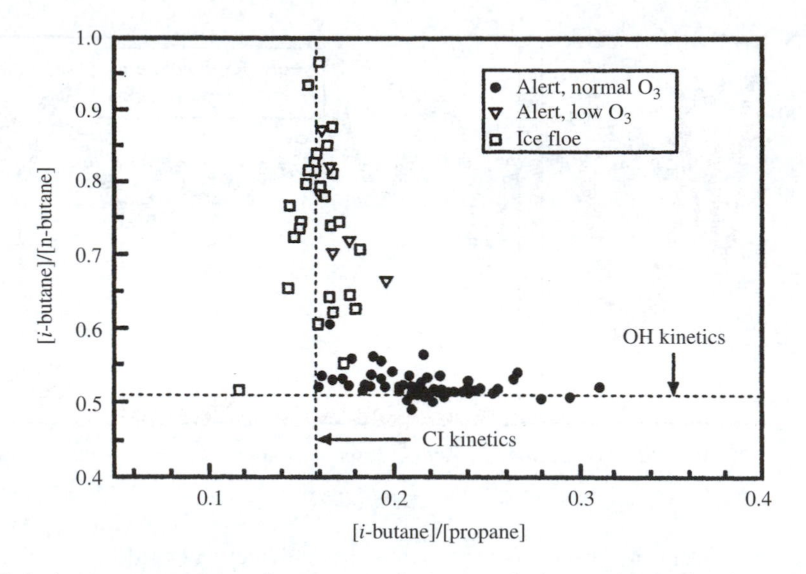

Figure 8.15 Relative concentrations of propane, *n*-butane, and *i*-butane in the Arctic troposphere at Alert (Canada) and on an ice floe 150 km north of Alert. As presented by B.J. Finlayson-Pitts and J.N. Pitts, Jr., *Science* 1997, **276**, 1045.

concentrations, while when O_3 is low at Alert or on the ice floe, the loss of hydrocarbon is essentially that expected if losses were dominated by reaction with Cl. There is thus good evidence that Cl concentrations are elevated at the times of O_3 loss, and it does not require too great an act of faith to believe that the elevated Cl, and species associated with it, might itself be responsible for the O_3 destruction.

Possibly more important than the rapid destruction of alkanes, alkenes and most aromatic compounds is the production of OVOC oxidation products in the stable polar marine boundary layer. For example, during ozone depletion events, there is a substantial consumption of propane, and a concomitant increase in acetone. Alkane oxidation alone cannot account for the increase in gas phase OVOCs during ODEs, and it may be that some of the species, such as HCHO, are additionally produced in and then emitted from, the snow pack.

Before leaving the subject of polar bromine activation, there are two further effects of enhanced BrO concentrations that should be noted. *Mercury depletion events* in the Arctic have been observed, and they seem to be related to bromine chemistry in a similar way to the ODEs. The relevance of the mercury events is to Arctic life, because largely-inert gaseous elemental mercury is converted into bioavailable mercury. Sulfur chemistry may also be modified by the presence of the activated halogens. Some measurements suggest that the oxidation of dimethyl sulfide (DMS: see Section 8.4.1) under coastal Antarctic conditions is dominated by the reaction of BrO with DMS. It is expected that BrO reactions would follow pathways analogous to the addition route (b) on the right-hand side of Figure 8.9, and that DMSO would be the almost exclusive initial product. But substantial amounts of methane sulfonic acid (MSA) are observed in the polar regions associated with elevated [BrO]. That finding implies that there is a route, not shown on Figure 8.9, that leads from DMSO to MSA, because the abstraction pathway to MSA (far left of the diagram) depends on reactions with NO_x, which is not present in the remote polar troposphere. Gas-phase conversion seems to be too slow, so the conversion may be another one that occurs in or on aqueous droplets. The processes are a matter of particular interest in relation to the formation of cloud-condensation nuclei, and their potential rôle in climate control (Section 7.2.4). As pointed out in Section 7.2.4, one requirement for DMS to act in promoting cloudiness is that it must produce *new* CCNs to increase their number, and not just add to pre-existing nuclei. If the dominant reaction with BrO through the addition channel leads to products that do not nucleate, it may be that high polar [BrO] could produce cloud feedbacks that oppose the expected increase of cloud albedo caused by DMS emission.

Inorganic halogens participate in tropospheric chemistry in non-polar oceanic regions, and, in particular, in the *marine boundary layer* (MBL). The strongest source of bromine and chlorine in the MBL is seasalt aerosol resulting from bubble bursting, with a smaller fraction released from biogenic organic halogens. Iodine compounds, on the other hand, are largely released as organic compounds and molecular iodine from micro- and macroalgae that accumulate iodine from the ocean. Globally, seasalt aerosol is, along with dust, the strongest particle source in terms of mass. Models show that BrO at mixing ratios of 0.5–4 ppt can have a significant impact on sulfur budgets in the MBL. In aerosol particles and droplets, HOBr and HOCl oxidize S(IV) to S(VI), and in the gas phase BrO adds to DMS, potentially leading to different oxidation products and the predominant growth of *existing* particles rather than the production of H_2SO_4 as a precursor of *new* particle formation. Direct attack of Cl atoms on hydrocarbons offers another potential influence of the halogens on MBL chemistry: it has been estimated that the Cl reactions in the Southern Hemisphere could account for several per cent of CH_4 loss.

Ozone photochemistry is also strongly influenced by the presence of the halogens. Measurements made on the Cape Verde islands showed very high levels of both BrO (2.5 ppt) and IO (1.4 ppt). Local sources should be minimal in this location, so that the concentrations may be representative of the MBL. At such concentrations, ozone photochemistry is completely dominated by the halogen-mediated reactions. Iodine chemistry shows some particularly interesting features, since the formation of IO in reaction (8.61) can be followed by the combination reaction

$$IO + IO \rightarrow 2I + O_2 \qquad (8.66)$$

to provide a catalytic route to ozone destruction that might play a minor part in ozone loss in the troposphere. However, the dependence of the rate of the reaction on the *square* of the relatively small [IO], and the existence of alternative channels for the interaction of two IO radicals (to yield $I_2 + O_2$ or an IO dimer) limit the importance of the chain. It has also been suggested that halogen oxide radicals might influence NO_x and HO_x chemistry in the troposphere, for example in the case of IO by forming $IONO_2$ and HOI (species to be compared with the analogous 'reservoir' compounds of the stratosphere, $ClONO_2$ and HOCl, Section 9.3.2). Thus, the halogens can perturb the O_3 budget not only by destruction of O_3 but also by halogen-induced shifts in the OH:HO_2 and NO:NO_2 ratios that lead to its reduced photochemical production.

Bursts of new particles have been observed at certain coastal sites that are ascribed to iodine chemistry. It now seems to be established that the exposure of seaweed at low tide leads to very strong emissions of both organic iodine and, especially, of molecular iodine (I_2), which has a very short photolytic lifetime of the order of 10–20 s. At night, I_2 mixing ratios of several tens of ppt have been observed at various locations. The reaction sequence

$$I_2 + h\nu \rightarrow I + I \qquad (8.67)$$

$$I + O_3 \rightarrow IO + O_2 \qquad (8.61)$$

first leads to the production of IO. Subsequent reaction to yield higher iodine oxides then most likely leads to the formation of clusters that grow to nanoparticles. IO concentrations are highest and particle bursts strongest when low tide and high Sun coincide. Many important questions remain, and the details of new particle formation from iodine oxides are not yet resolved. In terms of ozone destruction, the relevance of these fairly local events of very high iodine loadings might be limited. There are indications that the fine particles produced in these bursts can grow to CCN sizes, but the spatial scale on which such CCNs might be of relevance, especially when viewed in the context of competition with other continental particle sources, is not yet known.

Section 8.5 has introduced us to what is still quite a young area of tropospheric science. The polar phenomena seem to make a significant contribution to tropospheric chemistry, and the processes in the MBL similarly have large impacts in certain locations. If reactive halogen chemistry were indeed a widespread phenomenon, our understanding of photochemical processes in the clean MBL would have to be completely reassessed. Inorganic halogen chemistry, the unexpected component of tropospheric chemistry, provides fascinating new insights into the complex behaviour of our atmosphere.

8.6 HETEROGENEOUS CHEMISTRY

Although the Earth's atmosphere is made up mostly of gases, it has suspended in it liquid and solid particles that significantly influence the chemistry that occurs (Section 1.3). These particles also affect the way in which ultraviolet, visible, and infrared radiation is distributed within the atmosphere; in turn, the alterations to the radiation field can regulate both climate and the atmospheric chemical processes themselves. In the present section, we consider chemical changes occurring within liquid droplets, and on the surfaces of solid and liquid aerosol particles. *Heterogeneous chemical reactions* are usually thought of as those occurring at the interface between two phases. In some important atmospheric processes, however, initial transfer of reactants from the gaseous to a condensed phase is followed by homogeneous chemical change within that condensed phase. For our purposes, therefore, it is convenient to include in the category of heterogeneous reactions all multiphase processes. Atmospheric chemistry involving clouds, fogs, rain droplets, ice particles, and other solid and liquid aerosol particles is necessarily heterogeneous in the sense that it involves transfer of gas-phase molecules to the condensed-phase system across the particle interface. In fact, reactions occurring *inside* particles are really confined to the liquid phase (see later in this section). On the other hand, reactions *on* solid surfaces are thought to be of very considerable atmospheric significance. An added degree of complexity arises when the particle is liquid, as is the case for droplets in the troposphere, and possibly for stratospheric sulfate aerosol, which may be in the form of supercooled liquid sulfuric acid.

Our more detailed discussions of the troposphere (this chapter) and stratosphere (Chapter 9) demonstrate very clearly the range of influence of heterogeneous reactions. In the present chapter, there are several examples. Section 8.2.7 describes the formation of particles in the oxidation of biogenic organic compounds; Section 8.4 has as one of its themes the formation of cloud-condensation nuclei in the oxidation of natural sulfur compounds such as dimethyl sulfide (see also Section 7.2.4), and Section 8.4.2 includes the theme of oxidation of SO_2 on particles. Halogens may be released to the troposphere in processes involving seasalt particles, as described in Section 8.5. Sections in Chapter 9 that are particularly relevant include 9.3.4 (heterogeneous reactions in general, and their influence on reservoir species) and 9.5.5 (volcanoes), while Sections 11.2.2 and 11.4.5 include discussions of the impact of particles emitted by aircraft engines on the atmosphere. Almost the whole of Section 9.4, which deals with polar-ozone depletion, turns around the theme of chemical processing on particles, but Section 9.4.5 is particularly intimately involved with heterogeneous reactions. Our purpose here is to look at some of the more general aspects of heterogeneous chemistry in the troposphere, and to extend the discussion elsewhere with some additional examples.

Let us consider explicitly the steps involved in a heterogeneous reaction of a gas-phase species with either the bulk constituent of a droplet or with another species that is already dissolved in it.

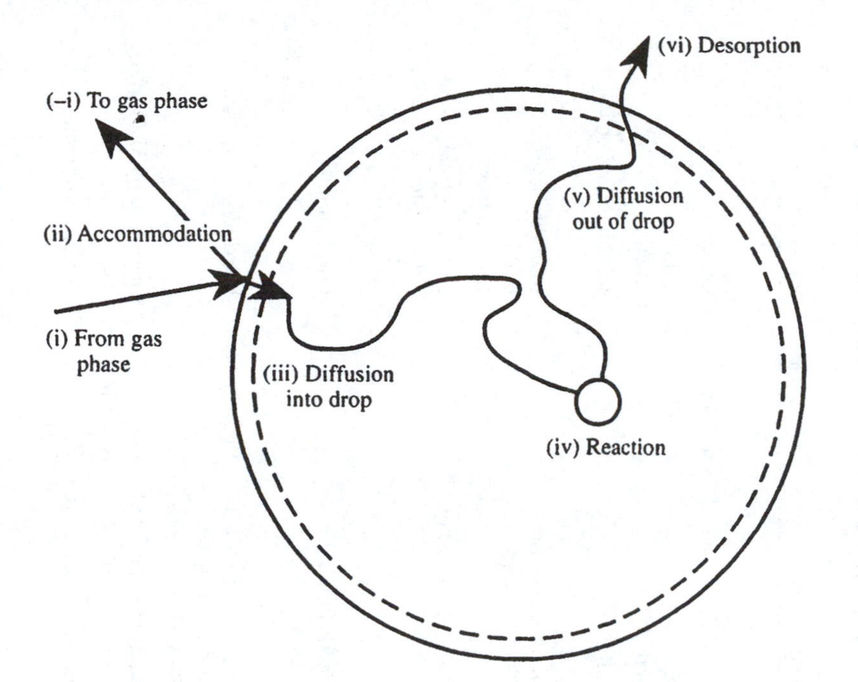

Figure 8.16 Processes involved in the uptake and chemical reaction of a gas-phase molecule by a liquid droplet, and chemical reaction in it. From A. R. Ravishankara and D. R. Hanson, in *Low-temperature Chemistry of the Atmosphere*, ed. G. K. Moortgat, A. J. Barnes, G. Le Bras and J. R. Sodeau, NATO ASI Series **21**, Springer-Verlag, Berlin, 1994.

Figure 8.16 illustrates the separate steps that together lead to reactive or non-reactive uptake. Chemical change corresponds to loss of the gas-phase molecule; uptake that is non-reactive can arise from physical dissolution or from reversible chemistry. The steps indicated are (i) gas-phase transport of the reactant to the surface of the droplet; (ii) accommodation at the surface; (iii) diffusion into the liquid; (iv) chemical reaction; (v) diffusion of unreacted molecules and products to the surface; and (vi) desorption of species from the interface. Characterization of each of these individual steps is obviously a formidable task, although one that may be simplified—as often happens in kinetics—by one of the steps being rate determining.

A useful starting point is provided by Table 8.2, in which an attempt is made to summarize our current state of knowledge about heterogeneous processes on different types of solid and liquid aerosol. The substrate materials are arranged so that they are increasingly hygroscopic, and then water containing, as the table is read from left to right. Although much has been discovered in recent years, it is striking how much uncertainty, or even complete lack of knowledge, remains.

As the table illustrates, the substrate surfaces on which atmospheric heterogeneous reactions occur are represented by a wide range of physical and chemical types. Condensed water is the predominant form of suspended matter in the troposphere, and clouds the most abundant form of condensed water. Liquid-water droplets, rather than ice, are favoured only in the lowest part of the troposphere. The cirrus clouds formed in the middle and upper troposphere are made up of ice crystals. More than 50 per cent of the surface of the Earth is usually covered by clouds at some altitude, and roughly seven per cent of the volume of the troposphere contains clouds. A fairly dense cloud will contain about 0.5 millionth of its volume as liquid water. If all the droplets possess a diameter of 10 μm, there will be almost 1000 droplets in every cm^3 of air, and the total surface area of water will be about $3 \times 10^{-3}\,cm^2$. The total surface area of water within the

Table 8.2 Heterogeneous processes of importance in the troposphere.[a]

	Increasingly hygroscopic or water containing →						
Substrate	*diesel or aircraft soot*	*mineral dust*	*organic matter (secondary)*	*sulfates bisulfates*	*seasalt (deliquescent)*	*sulfuric acid*	*cloud droplets ice particles*
Reactant							
OH	reactive loss[b]	unknown	reactive loss[b]	unknown	reactive loss[b]	uptake	uptake
HO_2	reactive loss[b]	unknown	unknown	unknown	uptake, reaction[b]		uptake reaction
RO_2	reactive loss[b]	unknown	uptake, reaction[b]		uptake		solubility limited
O_3	reactive loss surface ageing synergisms[b]	unknown	unknown		importance of direct uptake to be found		solubility limited uptake reactive loss
NO_2	chemisorption reduction to NO HONO formed	HONO formed[b]	nitration (more likely by N_2O_5/NO_2^+)		ClNO formation on dry NaCl		HONO formation on liquid water[b] or on ice[b]
NO_3	reactive loss[b]	unknown	reaction (*e.g.* with aromatic species)	unknown	limited solubility reaction with I^- *etc.*		low solubility
N_2O_5	hydrolysis or reaction as NO_2^+, depending on substrate reaction probabilities are very variable formation of particulate nitrate or gaseous HNO_3				formation of $ClNO_2$ and other halogen compounds	hydrolysis	hydrolysis
SO_2	catalytic oxidation	catalytic oxidation[b]			oxidation in polluted marine atmosphere	not important	oxidation by H_2O_2, O_3, *etc.*
Other processes	adsorption synergisms[b]	solubilization of catalytic metals	aerosols formed from biogenic VOC	pH-dependent uptake of acidic or alkaline gases		uptake of NH_3	aerosol processing

[a]From a figure presented by U. Schurath at *EUROTRAC Symposium, 1998*, Garmisch-Partenkirchen, March 1998, and reproduced by his kind permission.
[b]uncertain

troposphere, and on which surface reactions can occur, is thus enormous. As to the other substrates, they are much less abundant, but nevertheless significant. Sulfuric acid and sulfates are quite abundant, and have both natural and anthropogenic sources. Seasalt is widespread, especially in marine areas. The organic component remains rather poorly characterized at the time of writing. Some, at least, is of biogenic origin, presumably resulting from the oxidation of terpenes and other biogenic VOCs (Section 7.2.3). This biogenic source is strongly implicated in heavy organic aerosol loading in the Amazon basin during the wet season, when there is little biomass burning. One of the puzzles is the presence of 'black carbon' outside the burning season, since this material is often thought to be soot carbon from combustion. It seems possible that some of this material may be *primary* biogenic aerosol, such as particles of bacteria and fungi, plant debris (waxes and leaf fragments), and humic matter.

Four simple categories can be recognized for the chemistry that creates aerosols and droplets, and occurs on and in them:

(i) condensation of a single component;
(ii) reaction of more than one gas to form a new particle;
(iii) reaction of gases on a pre-existing particle; and
(iv) reactions within the particles themselves.

Category (i), the condensation of a single gaseous component to form a new suspended particle is *homogeneous, homomolecular* nucleation, and it is obviously central to cloud formation. The obstacles to such nucleation are touched on at the end of Section 3.3.1. The most obvious example is the aggregation of sufficient H_2O molecules from the gas phase to produce a droplet of liquid, or of solid, water. Category (ii) is the analogous process for reacting species involving two or more gases to form a condensable product species in a *homogeneous, heteromolecular* process. A typical example is the reaction

$$NH_{3\,(g)} + HNO_{3\,(g)} \rightarrow NH_4NO_{3\,(s)}, \tag{8.68}$$

which can form NH_4NO_3 that ultimately produces particles. Reaction (8.68) occurring on a particle that already exists—that is to say, a *heterogeneous, heteromolecular* reaction of type (iii)—may be a more important route to aerosol NH_4NO_3 than the direct formation of a new solid particle. Another example of the heterogeneous, heteromolecular process is afforded by the reaction of HNO_3 with seasalt particles (NaCl)

$$NaCl_{(cond)} + HNO_{3\,(g)} \rightarrow HCl_{(g)} + NaNO_{3\,(cond)}, \tag{8.69}$$

to form $NaNO_3$. Heteromolecular reactive condensation of gas-phase molecules on pre-existing particles is sometimes called *aerosol scavenging*. It can have an impact on bulk tropospheric chemistry by providing a sink for nitrogen and hydrogen species such as HNO_3, NO_3, N_2O_5, H_2O_2, and HO_2, as well as organic nitrates and peroxides. Clouds and raindrops have a major effect on gas-phase species through the scavenging mechanism. Rainout (removal of gases by cloud droplets) is believed to be more important than washout (removal of gases by raindrops) because of the longer lifetime and greater surface area of cloud droplets compared with raindrops. Water-soluble species, such as the acids, acid anhydrides, and peroxides are obviously particularly susceptible to removal by these mechanisms.

Finally, category (iv) includes chemical reactions that occur within the aerosol itself to form particles of changed composition, as in the oxidation of SO_2 to sulfate ions in clouds. These are multiphase processes, since they involve transfer from (and perhaps back to) the gas phase, and the condensed phase is almost always liquid, since diffusion rates in solids are too slow to permit significant extents of reaction. The particular case of the heterogeneous oxidation of SO_2 was presented in Section 8.4.2. As well as providing a liquid phase in which soluble gases can be dissolved, clouds also offer an active chemical medium for aqueous-phase reactions that can affect the distribution of active species in the troposphere. Despite the relatively small fractional volume of the troposphere occupied by cloud droplets or particles (see earlier in this section), the direct chemical influence of clouds on the tropospheric reactants may be more important in atmospheric photochemistry than the scattering and reflection of actinic ultraviolet radiation. It is therefore appropriate to consider how that influence is exerted.

The general principles that give rise to the differences between gas-phase and condensed-phase reactions should now be considered briefly. The solvent obviously has the potential to exert a considerable influence on the course of chemistry in the liquid phase. In air at one atmosphere pressure, and at ambient temperature, the molecules themselves occupy only roughly 0.2 per cent of the total volume; in liquids, the molecules can make up half the volume. At pressures of one atmosphere and below, we have been able to assume that the reactant molecules undergo essentially unhindered motion, and that assumption lies behind the various formulations of kinetics that we have discussed in previous sections. In distinction, in liquids, the reactive molecules must squeeze past the solvent molecules (or each other, if one species is also the bulk liquid) if they are to reach each other and undergo reaction. Reactants, activated complexes or intermediates, and products can also all interact with the solvent.

One manifestation of the interaction with intermediates is that energy removal in association reactions is virtually instantaneous. In the gas phase the energy that is liberated must be removed, and it is often necessary to explicitly write a chemical equation that includes the third-body M as, for example, in the reaction between OH and NO_2

$$OH + NO_2 + M \rightarrow HNO_3 + M. \tag{8.19}$$

As a result, reactions of this kind may display kinetics that are [M]-dependent and lie between third-order and second-order in the gas phase, whereas the systems always display pure second-order kinetics in the liquid phase.

Interactions of the reactants and the solvent (especially water) may make the formation of ions energetically more favourable than in the gas phase. New reaction channels may thus become accessible, and the kinetics of the processes can be influenced by the attractive or repulsive electrostatic interactions between the reactants, amongst many other factors.

Simple treatments of liquid-phase kinetics often start from the concept of the *encounter pair* of reactants that find themselves together within a *solvent cage*. Two extreme cases can be envisaged. In the first, the two species are very highly reactive towards each other, and undergo chemical transformation within a very few 'collisions' within the cage. The rate-determining process is then the diffusion of the reactants through the solvent to form the encounter pair, and the process is a *diffusion-controlled reaction*. At the other extreme, the activation energy for reaction may require the partners to pick up appreciable amounts of energy as they shake against each other within the cage, so that the kinetics are controlled by the rate of reaction within the cage, rather than by the rate at which they reach it. *Activation-controlled reaction* kinetics then result.

For many of the liquid-phase reactions of interest in atmospheric chemistry, the intrinsic reactivity of the partners is, indeed, very high, leading to diffusion-controlled kinetic behaviour. A very elementary treatment suggests a maximum rate coefficient for a diffusion-controlled reaction of, say, Na^+ ions in water, of about $5 \times 10^9\,dm^3\,mol^{-1}\,s^{-1}$, which is very roughly 40 times lower than the limiting collisional rate coefficient in the gas phase. In general, a rate coefficient of $> 10^9\,dm^3\,mol^{-1}\,s^{-1}$ for an aqueous-phase reaction is taken to be indicative of a diffusion-controlled mechanism.

One of the largest known rate coefficients for a condensed-phase process is that for the very important reaction

$$H^+ + OH^- \rightarrow H_2O \tag{8.70}$$

($1.4 \times 10^{10}\,dm^3\,mol^{-1}\,s^{-1}$ at 298 K). The magnitude mainly reflects the large diffusion coefficients in water of OH^- and, especially, of H^+; the rapid diffusion is itself a consequence of the special mechanisms by which these ions migrate in liquid H_2O.

Although the diffusion coefficient is most important in making reaction (8.70) so fast, there is another factor operating that may be dominant in other reactions. The positive and negative ions attract each other, so that the effective encounter distance can be much greater than the gas-kinetic collision distance. The result is that oppositely charged ions will react more rapidly (perhaps twenty times faster) than their neutral analogues, under similar conditions, while similarly charged ions can be assumed not to react at all.

While our discussion has so far centred on the behaviour of the atmospherically dominant class of diffusion-controlled reactions, some processes of interest are activation controlled. One characteristic of such reactions is that the activation energy may be smaller than for the equivalent gas-phase reaction, because the reactant pair undergoes many individual 'collisions' at each encounter, whereas in the gas phase the collision and the encounter are the same thing. A particularly interesting property shown by activation-controlled ionic reactions is that of the *kinetic salt effect*. Rate coefficients are affected by the presence of other ionic species present in the solution that do not themselves participate in the reaction. Interactions between oppositely charged partners are slowed down by the presence of such salts. In the atmosphere such effects may be of significance, since water droplets may contain substantial amounts of seasalt or other similar solutes.

A special feature of the tropospheric aqueous-phase reactions is the high solubility of certain key compounds such as HO_2 or N_2O_5; the reactions of these species within the droplets ensure that dissolution is irreversible. The partitioning into the aqueous phase is thus strongly favoured, and partially offsets the small relative volume of the water droplets. Because NO is relatively insoluble, HO_2 and NO are separated and prevented from participating in the reaction

$$HO_2 + NO \rightarrow OH + NO_2 \tag{8.12}$$

that converts NO to NO_2 (and thus promotes ozone formation) in the gas phase. The inhibition of this reaction in the liquid phase has a most pronounced effect on the chemistry and oxidizing capacity of the atmosphere in the vicinity of clouds. Photochemical processes may also be different in condensed phases, because of changes in absorption spectra, the opening of new photolytic channels, and alterations of actinic flux.

Some of the most important atmospheric reactions exemplify the principles very nicely. One case is the formation of HNO_3 from N_2O_5

$$[N_2O_5 + H_2O \rightarrow HNO_3 + HNO_3]_{aq}, \tag{8.71}$$

the square brackets written here being intended to demonstrate that the reaction occurs *within* a water droplet. The gas-to-aqueous phase transfer of N_2O_5 is limited by gas-phase diffusion and transfer through the interface, while reaction (8.71) within the aqueous phase is so fast as to be essentially 'instantaneous', so that the dissolution of N_2O_5 is irreversible. The reaction between N_2O_5 and H_2O is slow in the gas phase, and the difference between the gas-phase and liquid-phase reactivity is a consequence of interactions with the solvent opening up lower-energy ionic reaction pathways in solution.

The gas-phase processes involved in the formation of N_2O_5 are the steps (8.36)–(8.38) presented in Section 8.3.1. The NO_3 radical is the essential intermediate in these processes, but NO_3 is rapidly photolysed during the day, so that N_2O_5 is formed only at night. Reaction (8.71) should thus have a particularly noticeable effect during the dark winter period of the year, and at high latitudes. N_2O_5, NO_2, and NO_3 are essentially in thermal equilibrium, which is maintained by the opposing reactions (8.37) and (8.38). Since reaction (8.71) removes N_2O_5, it prevents regeneration of NO_2 in reaction (8.37). High altitudes accentuate the effect, because N_2O_5 is more stable at lower temperatures when the decomposition reaction (8.37) becomes slower. Observations confirm large losses in tropospheric NO_x (by up to 90 per cent) under such conditions.

Cloud reactions further influence the oxides of nitrogen by removal of NO_3 radicals. Because the radicals have a long lifetime against removal in the gas phase, they can be incorporated into cloud water as $(NO_3)_{aq}$. Although NO_3 is not particularly soluble in pure water, the dissolved radical reacts rapidly with the chloride ion

$$[NO_3 + Cl^- \rightarrow NO_3^- + Cl]_{aq} \tag{8.72}$$

present in the seasalt aerosols of the marine boundary layer; NO_3 is thus removed permanently from the gas phase, and conversion times are as little as a few tens of seconds. At the relatively higher temperatures and lower NO_x concentrations experienced in the marine boundary layer, the reaction may be more efficient in removing NO_x from ('denitrifying') the atmosphere than is the route involving hydrolysis of N_2O_5, since reaction (8.38), which generates N_2O_5, requires NO_2, and its rate *decreases* with increase of temperature.

The effect of reaction (8.72) is not only to remove oxides of nitrogen from the gas phase, but also both to enhance the concentration of NO_3^- dissolved in the cloud water and to generate atomic chlorine. Further oxidation may thus be initiated in the liquid phase (indeed OH can be formed from NO_3^-: see later in this section), and Cl can be exchanged to the gas phase, so that free halogen atoms can be formed without the high degree of acidification implied by reaction (8.62) discussed in Section 8.5.

Heterogeneous (multiphase) loss of N_2O_5 in reaction (8.71) is essentially irreversible; the equilibrium implied by reactions (8.37) and (8.38) is disturbed, and NO_2 and NO_3—neither of which is very soluble in water—are removed indirectly by the presence of cloud droplets. A similar sequence is responsible for the removal of PAN (Section 8.2.6), another important component of the NO_y system (see, again, the final paragraph of Section 4.5 for an explanation of NO_y). PAN is another compound that is not highly soluble in water, and clouds do not affect its concentration directly either by dissolving it or acting as a medium for chemical reaction. PAN and its precursors, it will be remembered, are connected by the equilibrium

$$CH_3CO.O_2 + NO_2 \rightleftharpoons CH_3CO.O_2NO_2. \tag{8.34}$$

Peroxy radicals, such as $CH_3CO.O_2$, *are* highly soluble, so that the equilibrium is shifted to the left in the presence of clouds, and the atmosphere is 'renoxified': that is, NO_y is converted to NO_x.

The discussion of the last few paragraphs has indicated how heterogeneous chemistry can bring about changes in concentration of gas-phase NO_x. These changes can in turn have a considerable impact on both O_3 and OH in the troposphere. One set of model predictions, for example, suggests a depletion, as far towards the equator as 25°N, of O_3 by 25 per cent and of OH by 20 per cent as a result of the occurrence of the heterogeneous reaction (8.71). The effects were even more pronounced at higher latitudes, for the reasons explained earlier. The models show that O_3 production is suppressed because of the lowered [NO_2], and that [OH] is diminished in two ways. First, the loss of ozone itself results in lowered OH generation in steps (8.1) and (8.2) (Section 8.2.1). Secondly, the all-important reaction

$$HO_2 + NO \rightarrow OH + NO_2 \tag{8.12}$$

is less able to convert HO_2 to OH both because there is less NO available, and because of the highly favoured partitioning of HO_2, but not NO, into the droplets, as explained earlier.

One very interesting result of the model simulations just described was that sulfate aerosol turned out to be the main agent in the reduction of atmospheric oxidizing efficiency, and that seasalt particles played a relatively small part in this aspect of heterogeneous tropospheric chemistry. One reason for the predominance of processing on sulfate particles is that anthropogenic emissions of sulfur oxides and nitrogen oxides tend to be concentrated in the same geographical locations. Much less NO_x is found in the marine boundary layer (MBL), which is where the seasalt particles are largely found, so that seasalt aerosol processing is limited to coastal regions immediately downwind of urban and industrial areas. It goes without saying, of course, that these considerations apply just to the one multiphase process, reaction (8.71), and that chemistry involving the heterogeneous chemistry of halogen atoms or of NO_3 radicals may be most important of all within the MBL.

Most of the examples discussed in the present section have referred to particles belonging to the categories in the three right-hand columns of Table 8.2: that is to say, particles that are aqueous or hygroscopic. Important new goals will be to understand better the origins and behaviour of primary and secondary organic aerosol, and of mineral particles. Biogenic organic aerosol may have a far-reaching influence on both atmospheric chemistry and climate. Such species act as effective cloud-condensation nuclei, and tropical continental clouds—which play a major role in planetary heat redistribution—are strongly influenced by the presence of the particles. Furthermore, there is some speculation that organic particles might be the true cloud-condensation nuclei in the marine troposphere as well, but that, since they can also act as surfaces for the deposition of H_2SO_4 from the oxidation of DMS (Section 8.4.1), they might be mistaken for cloud-condensation nuclei of marine biogenic origin. Mineral aerosols, too, are expected to have a part to play in tropospheric chemistry, although quantitative information is sparse. Experiments have demonstrated, for example, that minerals can increase the rate of photochemical degradation of adsorbed organic compounds, and that the reaction products are often different from those evolving from gas-phase reactions. There could be significant atmospheric consequences, especially if the products turn out to be more toxic than the reactants. The presence of water on minerals associated with atmospheric particles can have a significant influence on rates of reaction, especially in relation to the transfer of transition metal ions as a result of the formation of organic complexes. We should note that transition metals such as iron, manganese, and cobalt are now known to be very important catalysts in the atmospheric oxidation of sulfur (IV) compounds (Section 8.4.2).

A quantitative interpretation of heterogeneous chemistry requires a method for introducing the coupling of the chemistry that occurs in the two phases, as well as a knowledge of the solubilities, interfacial mass-transport kinetics, and the reaction kinetics of the species in the condensed phase. The theoretical background to the coupling needs further development, and the whole study of heterogeneous reactions in particles and droplets is in its infancy. However, in view of the likely involvement of reactions in and on surfaces in both the troposphere and the stratosphere, this field is one that merits the most detailed attention. The past few decades have seen great progress, with very considerable effort expended in both experimental and theoretical research, but enormous uncertainties remain. Understanding of heterogeneous chemistry has evidently reached the stage where its complexity and probable importance in tropospheric chemistry are apparent, but where basic information on composition and reaction channels and mechanisms is still sorely needed.

CHAPTER 9

The Stratosphere

9.1 PURPOSE OF THE CHAPTER

From the point of view of atmospheric chemistry, the most important trace species in the stratosphere is ozone, and the most important chemistry concerns the production and loss of ozone. Several of the basic components of that chemistry have been set out in earlier chapters of the book. The purpose of the present chapter is to look at various aspects in more detail, and to draw together the several threads that have appeared. One of the objectives will be to determine what factors might disturb the balance of production and loss, in preparation for trying to assess the impact of Man's activities in Chapter 11.

We have already covered a great deal of information about the stratosphere and about ozone in this book, and it will be a useful beginning here to summarize the key facts and indicate where they are to be found.

In Section 4.1, we learned of methods by which stratospheric ozone concentrations may be measured, and we subsequently examined a typical concentration–altitude profile (Section 5.1, Figure 5.1). The universally employed atmospheric measure of column ozone, the Dobson unit (DU), was explained there: 1 DU corresponds to 2.69×10^{16} molecule cm^{-2}.

Ozone is formed in the stratosphere following the photolysis of O_2 to form O atoms, which then add to a further O_2 molecule (Section 5.2). The basic scheme for the formation and loss of ozone was proposed by Sydney Chapman in about 1930. The reactions in his scheme were the suite

Change in odd oxygen

$$O_2 + h\nu \rightarrow O + O \qquad +2 \tag{9.1}$$

$$O + O_2 + M \rightarrow O_3 + M \qquad 0 \tag{9.2}$$

$$O_3 + h\nu \rightarrow O + O_2 \qquad 0 \tag{9.3}$$

$$O + O_3 \rightarrow 2O_2 \qquad -2 \tag{9.4}$$

$$[O + O + M \rightarrow O_2 + M \qquad -2]. \tag{9.5}$$

Reaction (9.5) is now known to be too slow for it to play a part in stratospheric chemistry. Both photolytic reactions (9.1) and (9.3) can yield excited fragments, but collisional deactivation to $O(^3P)$

Atmospheric Chemistry
By Ann M. Holloway and Richard P. Wayne

Published by the Royal Society of Chemistry, www.rsc.org

is the (almost) exclusive fate of any $O(^1D)$ formed. For the purposes of discussing oxygen-only chemistry, no account need be taken of excitation.

This simple *oxygen-only* scheme predicts that a layer of ozone should be formed in the stratosphere, and the form of a *Chapman layer* was shown in Figure 5.2. In the reactions, odd oxygen (meaning O_n with $n = 1$ and 3, odd numbers, as discussed in Section 4.5) is formed only in reaction (9.1) and lost only in reaction (9.4), so that these two processes are critical in determining the balance between rates of production and loss of O_3, and consequently of the ozone concentration at any altitude.

Measurements of ozone distributions and concentrations in the atmosphere show that the real situation is more complex than the simple ideas suggest. First, the predicted absolute concentrations are several times higher than measured ones. Secondly, there is a mismatch between predictions and observations of the latitudes and altitudes at which ozone ought to be found in highest concentrations.

The first problem has its origins in the rate of reaction (9.4) at stratospheric temperatures. Once accurate rate coefficients had been obtained in the laboratory, it became clear that at stratospheric temperatures the rate of loss of odd-O was too small to balance the rate of production of odd-O in reaction (9.1). Model calculations performed with oxygen-only chemistry show global rates of production of ozone in spring several times larger than the losses in reaction (9.4). A lack of balance of this magnitude, if it were real, would lead to a doubling of stratospheric ozone concentrations in just a few weeks. As we said near the end of Section 5.2, since ozone concentrations are not rapidly increasing, we conclude that something other than the Chapman reactions is very important in destroying ozone in the natural stratosphere. Catalytic cycles (Section 6.1) turn out to provide pathways that have the overall chemical effect of reaction (9.4), but that proceed with lower activation energies and thus faster at the relevant temperatures. In Section 7.2.2, we returned briefly to the theme of catalytic ozone destruction, and explained that the catalysts are atomic and free-radical intermediates. These intermediates originate in precursors that may be formed by non-living forces (such as volcanic eruptions), but are mainly produced by living organisms—including Man, who has shown himself capable of severely perturbing the stratospheric ozone layer, a matter to be explored in detail in Sections 11.2 and 11.3.

We heard in Section 5.2 that the second major area of mismatch between predictions of the simple theories and atmospheric measurements, that of the location of the highest ozone concentrations, was a consequence of the horizontal and vertical transport of ozone. Figure 5.3 showed the rate of ozone formation predicted by a model without any motions of the atmosphere: it reflects the photolysis rate, P, of O_2 in reaction (9.1) at different latitudes and altitudes on a particular day (22 March). Ozone concentrations ought, on the basis of the simple O-only chemistry, to be proportional to $P^{\frac{1}{2}}$ (see Section 9.2). But the spatial distribution of ozone looks like Figure 5.4, and not at all like Figure 5.3. The way in which atmospheric motions bring about the redistribution is explained in Section 5.2. These are properly matters of atmospheric physics, and we must take leave of them here to return to our current task of exploring the chemical processes in more detail, and in particular the chemistry associated with the catalysis of reaction (9.4).

9.2 INTERCONVERSION OF O AND O_3: VALIDITY OF THE ODD-OXYGEN CONCEPT

Reactions (9.2) and (9.3) interconvert O and O_3. Even at the top of the stratosphere, where pressures are lowest, reaction (9.2) has a half-life of as little as ~100 s. Ozone likewise has a very short photolytic lifetime in reaction (9.3) during the day. It is this rapid interconversion of O and O_3 that

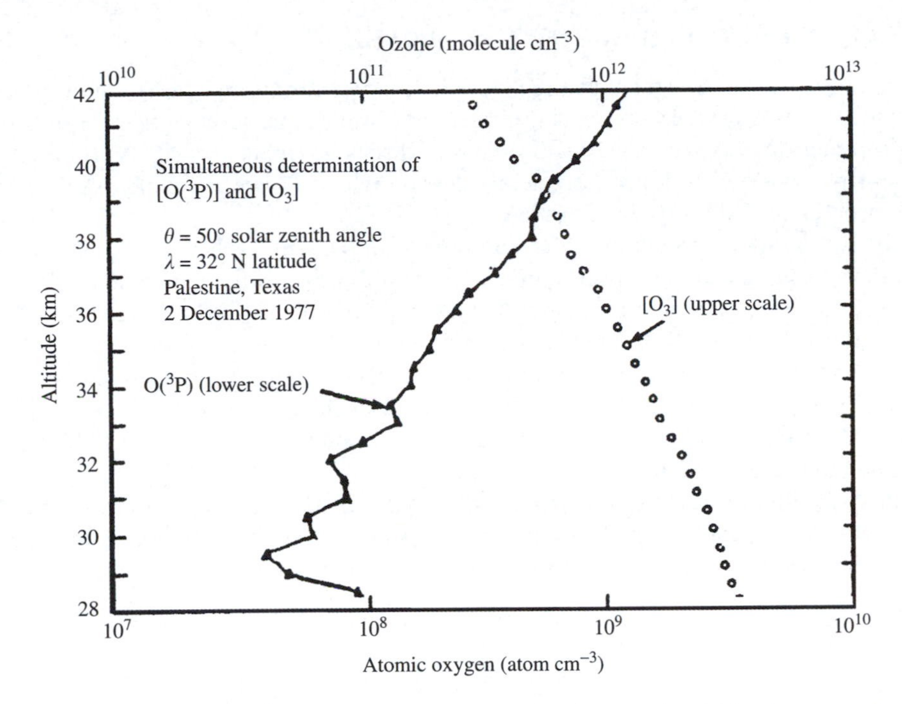

Figure 9.1 Concentrations of O(^{3}P) and of O_3 measured simultaneously within the same element of the stratosphere. From *The Stratosphere 1981*, editor-in-chief R. Hudson, World Meteorological Organization, Geneva, 1981.

provides the rationale for the concept of 'odd oxygen' (see preceding section). Reactions (9.2) and (9.3) 'do nothing' so far as odd oxygen is concerned, but merely determine the ratio [O]/[O_3]. Changes in the number of odd-oxygen species are shown alongside the chemical reactions in equations (9.1)–(9.5). Reaction (9.2) becomes slower with increasing altitude, while reaction (9.3) becomes faster. Atomic oxygen is thus favoured at high altitudes and ozone at lower ones. Simultaneous measurements of [O(^{3}P)] and [O_3], plotted in Figure 9.1, show that these expectations are borne out experimentally. Ozone is the dominant form of odd oxygen below ~60 km, and in the stratosphere (< 50 km) constitutes more than 99 per cent of odd oxygen (note that the scale for [O_3] in Figure 9.1 is shifted by a factor of 10^3 relative to that for [O(^{3}P)]). The rate of production of ozone at any altitude in the stratosphere may thus be equated with the rate of O-atom formation, which is twice the rate of photolysis, P, of O_2. The rate of loss, L, of O_3 is equal to $k[O][O_3]$, where k is the rate coefficient for reaction (9.4). O-atom concentrations reach a steady state that adjusts rapidly to the changes of solar intensity during the course of the day, and the ratio $\alpha = [O]/[O_3]$ is determined by the relative rates of reactions (9.2) and (9.3), so that $L = \alpha k[O_3]^2$. Finally, for a period over which [O_3] is constant, losses balance production, so that $L = 2P$, and $[O_3] = (2P/\alpha k)^{\frac{1}{2}}$, the square-root proportionality being the one stated earlier. The parameter α depends on the local solar intensity and spectral distribution, pressure and temperature, so that it varies with altitude, latitude, season and time of day.

After sunset, atomic-oxygen concentrations fall very rapidly at relatively low altitudes ($\lesssim$40 km) where the sink reactions (9.2) and (9.4) remain, but the sources (9.1) and (9.3) are cut off. With little atomic oxygen present, ozone is no longer destroyed in reaction (9.4), even though none is formed. Diurnal variations in stratospheric O_3 are therefore expected—and found—to be small. Above about 55 km, diurnal changes in concentration become increasingly pronounced, as daytime O_3 photolysis becomes faster, and conversion of O back to O_3 becomes slower.

9.3 CATALYTIC CYCLES IN THE GAS PHASE

Until about 1964, it was thought that the Chapman, oxygen-only, reactions could explain atmospheric ozone abundances. However, improved laboratory measurements of rate coefficients, especially for reaction (9.4), revealed the discrepancy discussed in Section 9.1. As pointed out there, reaction (9.4) is too slow to destroy ozone at the rate it is produced globally. An additional, or faster, loss process for odd oxygen is needed.

At first sight, it would seem that no trace component in the stratosphere could be responsible for loss of odd oxygen, since the species involved would be rapidly consumed. This objection disappears if the trace constituent participates in a catalytic process that removes O or O_3. The idea of catalytic atmospheric loss processes had originated with a suggestion of Bates and Nicolet in 1950 about the influence of H and OH from water photolysis on ozone concentrations in the upper mesosphere. Serious consideration of the influence of such processes on ozone concentrations followed the discovery that the laboratory rate measurements could also be affected by trace impurities reacting in a cyclic manner.

The essence of catalytic schemes for loss of odd oxygen is the provision of a more efficient route than the direct one for reaction (9.4). A chain mechanism that achieves the same result as reaction (9.4) can be represented in simplified form by the pair of reactions

$$X + O_3 \rightarrow XO + O_2 \tag{9.6}$$

$$XO + O \rightarrow X + O_2 \tag{9.7}$$

$$\text{Net} \qquad O + O_3 \rightarrow O_2 + O_2.$$

The reactive species X is regenerated in the second reaction of the pair, so that its participation in odd-oxygen removal does not lead to a change in its abundance. Since the overall reaction consists of two separate steps, each one must be exothermic to be efficient. That requirement places a constraint on the X–O bond energy: $107\,\text{kJ mol}^{-1} < D(\text{X–O}) < 498\,\text{kJ mol}^{-1}$. Rates of the catalytic destruction of ozone may then exceed those of the elementary step (9.4) either because [X] exceeds [O], or because the rate coefficient for reaction (9.6) exceeds that for (9.4). The rather large activation energy for reaction (9.4) of $17.1\,\text{kJ mol}^{-1}$ does, as we shall see shortly, often favour the indirect route at low stratospheric temperatures.

Several species have been suggested for the catalytic 'X' in the atmosphere. The most important of these for the natural stratosphere are X = H and OH, X = NO, X = Cl, X = Br, and possibly X = I. Reactions (9.6) and (9.7) are then replaced by specific reactions of these X and XO species. For the catalysts just mentioned, the reactions become

Cycle 1:

$$H + O_3 \rightarrow OH + O_2 \tag{9.8}$$

$$OH + O \rightarrow H + O_2 \tag{9.9}$$

$$\text{Net} \qquad O + O_3 \rightarrow O_2 + O_2.$$

Cycle 2:

$$OH + O_3 \rightarrow HO_2 + O_2 \tag{9.10}$$

$$HO_2 + O \rightarrow OH + O_2 \tag{9.11}$$

$$\text{Net} \qquad O + O_3 \rightarrow O_2 + O_2.$$

Cycle 3:

$$NO + O_3 \rightarrow NO_2 + O_2 \qquad (9.12)$$

$$NO_2 + O \rightarrow NO + O_2 \qquad (9.13)$$

Net $$O + O_3 \rightarrow O_2 + O_2.$$

Cycle 4:

$$Cl + O_3 \rightarrow ClO + O_2 \qquad (9.14)$$

$$ClO + O \rightarrow Cl + O_2 \qquad (9.15)$$

Net $$O + O_3 \rightarrow O_2 + O_2.$$

The catalytic cycles are then said to involve HO_x, NO_x, and ClO_x species (see the final paragraph of Section 4.5). A bromine ('BrO_x') cycle can be written that is exactly analogous to the pair of reactions (9.14) and (9.15). An IO_x cycle has also been suggested, but its efficiency is limited because IO, the chain carrier, is rapidly photolysed in the atmosphere. Rate parameters for the key reactions are given in Table 9.1, and show how the catalytic cycles can be faster than the direct reaction (9.4). The entry for the reaction '$O_3 + O$' shows that the activation energy for reaction (9.4) with ozone is far higher than that for any of the other reactions of atomic oxygen. At a temperature of 220 K, typical of the stratosphere, the rate coefficient for the direct reaction is thus orders of magnitude smaller than the corresponding value for the catalytic reactions. Whether or not the catalytic reactions will be actually *faster* then depends on the relative concentrations of XO and of O_3.

Figure 9.2 shows the relative contribution to ozone destruction of several of the catalytic cycles as well as of the oxygen-only 'Chapman' reactions for a particular location (mid-latitude) and time of year (equinox). The model on which the figure is based contained a total of 438 reactions, including 43 heterogeneous ones. It is apparent that the catalytic cycles all make a major contribution to destruction of odd oxygen, but that different cycles dominate at different altitudes. The calculations demonstrate clearly that the catalytic cycles *can* explain atmospheric ozone concentrations if the trace constituents are present at levels consistent with atmospheric observations.

Table 9.1 Rate parameters for some reactions in catalytic cycles.

Reaction	*X + O₃*			*XO + O*		
	A	E_a/R	k_{220}	*A*	E_a/R	k_{220}
			$O_3 + O$:	8.0×10^{-12}	2060	6.9×10^{-16}
H	1.4×10^{-10}	480	1.6×10^{-11}	2.4×10^{-11}	–110	4.0×10^{-11}
OH	1.7×10^{-12}	940	2.4×10^{-14}	2.7×10^{-11}	–224	7.5×10^{-11}
HO_2	2.0×10^{-16}	–380	1.2×10^{-15}			
NO	1.4×10^{-12}	1310	3.6×10^{-15}	5.5×10^{-12}	–188	1.3×10^{-11}
Cl	2.8×10^{-11}	250	9.0×10^{-12}	2.5×10^{-11}	–110	4.1×10^{-11}
Br	1.7×10^{-11}	800	4.5×10^{-13}	1.9×10^{-11}	–230	5.4×10^{-11}

A and E_a are the Arrhenius parameters, and k_{220} is the rate coefficient calculated for a temperature of 220 K. Units for *A* and k_{220} are $cm^3\ molecule^{-1}\ s^{-1}$. For the reaction between HO_2 and O_3, the published recommendation is recalculated to give the table entry of E_a/R at $T = 220$ K. The apparent small negative activation energies for the XO + O reactions essentially mean that E_a is near zero for these atom–radical processes, but the values represent current recommendations. Source of data: *IUPAC Subcommittee for Gas Kinetic Data Evaluation, Summary Tables of Gas Kinetic Data: June 2006* , except for the reaction $H + O_3$, where the current NIST recommendation (*Standard Reference Database 17, Version 7.0 (Web Version), Release 1.4.2 Data Version 2009.01*) was used.

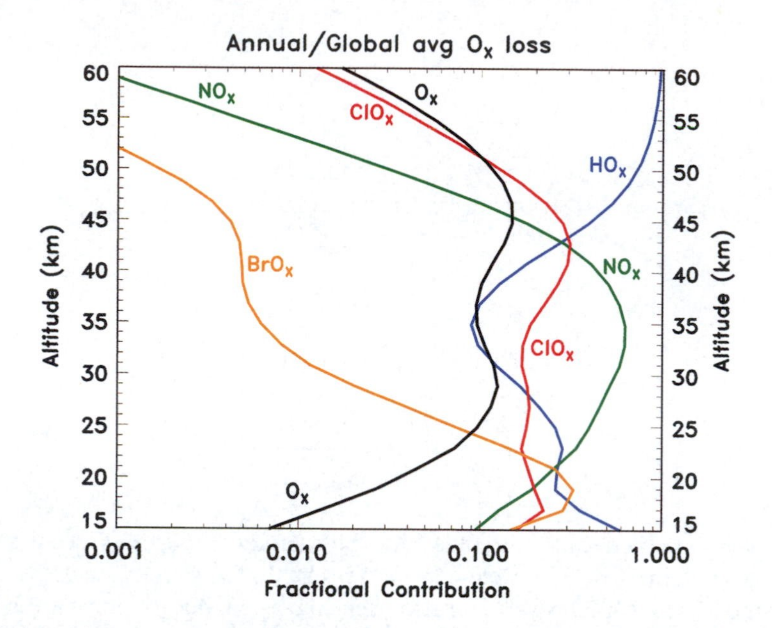

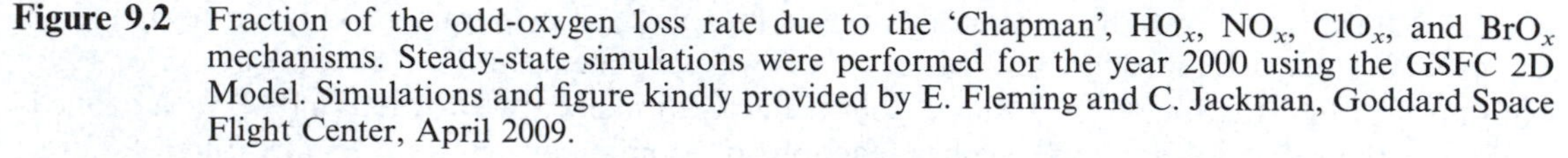
Figure 9.2 Fraction of the odd-oxygen loss rate due to the 'Chapman', HO_x, NO_x, ClO_x, and BrO_x mechanisms. Steady-state simulations were performed for the year 2000 using the GSFC 2D Model. Simulations and figure kindly provided by E. Fleming and C. Jackman, Goddard Space Flight Center, April 2009.

The set of catalytic cycles based on the X → XO → X processes (9.6) and (9.7) gives a first indication of how the natural ozone balance is maintained in the atmosphere, although the simple ideas have to be modified before a fully quantitative result emerges. First, even within a 'family' (*e.g.* HO_x), catalytic cycles can be identified that do not fit into the category so far discussed. For example, the sequence of

Cycle 5:

$$OH + O \rightarrow H + O_2 \quad (9.9)$$

$$H + O_2 + M \rightarrow HO_2 + M \quad (9.16)$$

$$HO_2 + O \rightarrow OH + O_2 \quad (9.11)$$

$$\text{Net} \quad O + O + M \rightarrow O_2 + M$$

has the same overall effect as reaction (9.5), which destroys two odd-oxygen species. Direct reaction was disregarded as being of no consequence in the atmosphere, but the catalytic cycle 5 is of major importance above 40 km. Similarly, at lower altitudes (< 30 km) the cycle

Cycle 6:

$$OH + O_3 \rightarrow HO_2 + O_2 \quad (9.10)$$

$$HO_2 + O_3 \rightarrow OH + O_2 + O_2 \quad (9.17)$$

$$\text{Net} \quad O_3 + O_3 \rightarrow O_2 + O_2 + O_2$$

becomes important because it does not utilize atomic oxygen, whose concentration is very low at those altitudes. Partitioning of the overall odd-hydrogen budget between the reactive species H, OH, and HO_2 thus determines which sets of reactions really lead to destruction of odd oxygen.

The identification of particular cycles and of where they are atmospherically important is somewhat artificial but does give an insight into the chemistry. One important conclusion is that the effects of the various catalytic families are definitely not additive. Indeed, adding NO_x to the calculation leads to *less* destruction of O_3 than is obtained with ClO_x or, especially, $HO_x + ClO_x$ alone. The explanation is that the various families are not isolated; rather, members of one family can react with members of another. Two reactions have shown themselves to be of particular importance in this context, and, as it turns out, to the overall balance of odd oxygen in the stratosphere. They are

$$HO_2 + NO \rightarrow OH + NO_2, \tag{9.18}$$

$$ClO + NO \rightarrow Cl + NO_2. \tag{9.19}$$

Reaction (9.18) is so central to stratospheric chemistry that each change in rate coefficient reported from different laboratory experiments has necessitated drastic revision of stratospheric models. The reactions do not appear to have any unusual characteristics, and we shall have to examine the chemistry of the catalytic cycles in more detail to discover why the two processes are so important. That theme is taken up again in Section 9.5, but first we ask where the catalytic species come from.

9.3.1 Natural Sources and Sinks of Catalytic Species

The catalytic families HO_x, NO_x, ClO_x, and BrO_x appear to be present in the 'natural' atmosphere unpolluted by Man's activities. Table 7.1, found in Section 7.2.2, showed in bare outline some biological precursors of the first three of these catalytic families. We shall now consider the *natural* sources of the catalyst precursors in more detail. It must be remembered, however, that in the contemporary atmosphere the background concentrations, especially of ClO_x and BrO_x, have already been supplemented by anthropogenic sources. This aspect of our survey will be held over until Section 11.2.

In the conversion of the precursors to the catalytic radicals themselves, excited atomic oxygen plays an important role. Ozone photolysis, reaction (9.3), yields O atoms in the ^1D state as explained in Section 5.1; for simplicity we shall represent $O(^1D)$ here as O*. Oxygen photolysis (reaction 9.1) also yields O*, but to a lesser extent and at higher altitudes. Because of the excess energy it carries, O* can react with N_2O, H_2O and CH_4 in a way that ground-state O cannot (see equations (5.2)–(5.4) and the associated discussion).

Most of the stratospheric NO_x originates with tropospheric N_2O, whose source is largely biological (see Section 7.2.2). The reaction with O* yields NO,

$$O^* + N_2O \rightarrow NO + NO, \tag{9.20}$$

and thus initiates the NO_x chemistry. Several other sources of NO_x exist, but they are of less importance than the N_2O source. Cosmic rays penetrate the atmosphere and contribute to NO production in the altitude range 10–30 km. Ionization reactions involving N_2 lead, after several steps, to NO. The source is probably the major one during the polar night, although it is not important on a global scale. Short-wavelength ultraviolet, and solar-particle, ionization in the mesosphere and thermosphere also lead to NO production. However, rapid photodissociation at levels above the stratopause seems to prevent this high-altitude source from reaching the stratosphere. Tropospheric NO or NO_2 (*e.g.* that produced by lightning) generally does not survive long enough to be transported in significant quantities to the stratosphere, although some lightning-derived NO_x in the tropics does seem to reach the lower stratosphere.

Reactions of O* with water vapour and methane,

$$O^* + H_2O \rightarrow OH + OH \tag{9.21}$$

$$O^* + CH_4 \rightarrow OH + CH_3, \tag{9.22}$$

are the main sources of OH radicals. The stratosphere is very dry, probably because tropospheric water vapour would have to pass through the very low temperatures at the tropopause, and is instead frozen out by this natural 'cold trap'. Much stratospheric H_2O is in fact a product of CH_4 oxidation. Nevertheless, CH_4 constitutes more than a third of the total hydrogen ($4\times[CH_4]+2\times[H_2O]$) in the lower stratosphere, so that reaction (9.22) is important. In fact, the reaction contributes more than one HO_x species, since subsequent oxidation of CH_3 to CO yields two or three more odd-hydrogen species (Section 8.2.2)

By far the most abundant *natural* precursor of ClO_x is methyl chloride (CH_3Cl). Tropospheric mixing ratios show little difference between Northern and Southern Hemispheres, suggesting a non-industrial source. The main contributor appears to be the world's oceans. Burning of vegetation also produces CH_3Cl, and forest fires must be regarded as natural events. Primitive agriculture employs 'slash and burn' techniques that lead to enhanced methyl chloride levels near the practising communities, but, properly speaking, that is an anthropogenic source. Volcanic emissions can also show high CH_3Cl levels. Tropospheric lifetimes for methyl chloride are estimated as $\sim$1 yr, and some CH_3Cl can be transported across the tropopause. Stratospheric concentrations show a decrease with altitude, as expected if CH_3Cl is reactive, but extrapolate well at lower altitudes to the tropospheric mixing ratios. Once in the stratosphere, CH_3Cl is primarily removed by reaction with OH,

$$CH_3Cl + OH \rightarrow CH_2Cl + H_2O, \tag{9.23}$$

although above $\sim$30 km about one-third is photolysed. In either case, atomic chlorine ultimately becomes available to enter the ClO_x cycle.

Alternative natural sources of ClO_x have been investigated. These include volcanic release of HCl, either slowly emitted to the troposphere followed by transport to the stratosphere, or, in major volcanic eruptions, directly to the stratosphere. Acidification (*e.g.* by H_2SO_4) of marine aerosols containing NaCl is another potential source of HCl. None of these processes seems capable, however, of providing enough ClO_x to have a marked influence on atmospheric ozone concentrations.

Natural bromine enters the stratosphere mainly as CH_3Br, which is produced by algae in the oceans, together with smaller amounts of species such as $CHBr_3$, CH_2Br_2, CH_2BrCl, and $CHBrCl_2$. Substantial additional amounts of bromine are attributable to Man's activities (see Section 11.2.3). There is still considerable uncertainty about the relative contributions, but one current estimate puts the natural mixing ratio of CH_3Br at about 10×10^{-12}, with Man responsible for adding perhaps a further 5×10^{-12} to the mixing ratio.

Sink processes for the catalytic families largely involve slow transport of the reservoir species across the tropopause. In the troposphere, the species then dissolve in water, and are subsequently 'rained out'. Reservoir compounds such as $HONO_2$, HOCl, HCl, HBr and probably $ClONO_2$, $BrONO_2$, and HO_2NO_2, are all highly soluble, and are thus specially good candidates for rainout. Radical–radical reactions such as

$$OH + HO_2 \rightarrow H_2O + O_2 \tag{9.24}$$

are important in determining the total concentrations of active reaction intermediates; in this case, of course, the products O_2 and H_2O are hardly conventional reservoir compounds that must be removed from the stratosphere.

9.3.2 Null Cycles, Holding Cycles, and Reservoirs

In competition with every catalytic cycle that destroys odd oxygen, cycles can be written that interconvert the species X and XO without odd-oxygen removal. Such processes are null cycles and can be illustrated by writing catalytic and null cycles alongside each other for NO_x

	Cycle 3 (catalytic):		Cycle 7 (null):	
	$NO + O_3 \rightarrow NO_2 + O_2$	(9.12)	$NO + O_3 \rightarrow NO_2 + O_2$	(9.12)
	$NO_2 + O \rightarrow NO + O_2$	(9.13)	$NO_2 + h\nu \rightarrow NO + O$	(9.25)
Net	$O + O_3 \rightarrow O_2 + O_2$		$O_3 + h\nu \rightarrow O_2 + O$	

Reaction (9.12) converts NO to NO_2 in both cases, but photolysis of NO_2 in cycle 7 leads to atomic-oxygen formation. Cycle 7 has the overall effect of ozone photolysis, reaction (9.3), formally photosensitized by NO_2: it is a 'do-nothing' cycle so far as odd oxygen is concerned. However, that fraction of the NO_x that is tied up in the null cycle is ineffective as a catalyst. During the day, when reaction (9.25) can occur, a given atmospheric NO_x concentration then destroys less odd oxygen than originally anticipated because of the null cycle.

Competitive processes abound in stratospheric chemistry. Reactions (9.13) and (9.25) are not, for example, the only reactions open to NO_2. The nitrate radical, NO_3, can be formed by reaction of NO_2 with ozone

$$NO_2 + O_3 \rightarrow NO_3 + O_2. \tag{9.26}$$

Most atmospheric NO_3 is removed by photolysis during daytime. The products of photo-decomposition are $O + NO_2$ (in which case there is no net loss of odd oxygen), and $O_2 + NO$ (in which case odd oxygen is consumed): the competition between these processes depends on the quantum yields for the different channels at the wavelengths available for photolysis. In addition, some NO_3 reacts in the three-body process

$$NO_3 + NO_2 + M \rightarrow N_2O_5 + M. \tag{9.27}$$

In the gas phase, N_2O_5 is rather unreactive in the stratosphere. Ultimately, it decomposes back to $NO_2 + NO_3$ thermally or photochemically, so that its formation does not constitute a permanent loss of odd nitrogen. On the other hand, it does behave as an unreactive reservoir of NO_x, containing typically 5 to 10 per cent of the total NO_x budget. The cycle involving its formation and destruction is then a holding cycle. However, in the presence of stratospheric aerosol, the situation is altered dramatically, since the process

$$N_2O_5 + H_2O \rightarrow 2HNO_3 \tag{9.28}$$

becomes efficient on surfaces (see Section 8.6), and the nitrogen is diverted into a new reservoir, nitric acid.

Holding cycles involving members of two families also turn out to be important routes for conversion to HNO_3. Three-body formation of nitric acid (HNO_3)

$$OH + NO_2 + M \rightarrow HNO_3 + M. \quad (9.29)$$

is an important storage step in the HO_x–NO_x system. Nitric acid is photolysed to regenerate $OH + NO_2$, but the process is relatively slow, and about half the stratospheric load of NO_x is stored in the nitric acid reservoir. For the ClO_x family, hydrochloric acid (HCl) is the main reservoir, reaction of Cl with stratospheric methane being the major source. The holding cycle for Cl involving HCl as reservoir can then be written

Cycle 8:

$$Cl + CH_4 \rightarrow CH_3 + HCl \quad (9.30)$$

$$OH + HCl \rightarrow H_2O + Cl \quad (9.31)$$

$$\text{Net} \quad CH_4 + OH \rightarrow CH_3 + H_2O.$$

About 70 per cent of stratospheric ClO_x is thought to be present as HCl (although in the lower stratosphere, the partitioning is strongly dependent on heterogeneous processes, as will be described in Section 9.4.5).

As the understanding of stratospheric ozone chemistry has become more complete, so yet more exotic molecules have been called into service as reservoir species. The list includes HOCl (hypochlorous acid), HO_2NO_2 (pernitric acid), and $ClONO_2$ (chlorine nitrate): the 'trivial' names of the compounds are almost always used in the published literature, and are given here for identification. Suggested routes to the compounds mentioned once again emphasize the interaction between families

$$ClO + HO_2 \rightarrow HOCl + O_2 \quad (9.32)$$

$$HO_2 + NO_2 + M \rightarrow HO_2NO_2 + M \quad (9.33)$$

$$ClO + NO_2 + M \rightarrow ClONO_2 + M. \quad (9.34)$$

Chlorine nitrate plays several important roles in stratospheric chemistry. Formation of $ClONO_2$ strongly couples the chlorine and nitrogen cycles, and at high concentrations of stratospheric chlorine could lead to non-linear responses of ozone to chlorine perturbations (see Section 9.4). Diurnal variations in [ClO] appear to be driven by $ClONO_2$ chemistry. HOCl has also been identified positively, and provides further information for testing current interpretations of stratospheric chlorine chemistry.

We defer most of our discussion of bromine chemistry in the stratosphere until Sections 9.4.5 and 11.2.3, but observe here that bromine reservoirs are worthy of note, in part because they are so inefficient! In the gas phase, conversion of Br to HBr in the analogue of reaction (9.30)

$$Br + CH_4 \rightarrow CH_3 + HBr \quad (9.35)$$

is extremely slow, because the reaction is endothermic for Br (while reaction (9.30) with Cl is exothermic). Furthermore, although the bromine analogue of reaction (9.34) can generate bromine nitrate, $BrONO_2$, this compound is readily photolysed

$$BrONO_2 + h\nu \rightarrow BrO + NO_2 \quad (9.36)$$

at longer wavelengths than $ClONO_2$, in the visible region where there is higher intensity. Reversal of the formation of $BrONO_2$ is thus favoured, and neither of the reservoirs HBr or $BrONO_2$ are anything like as important as the chlorine analogues.

Another feature of the bromine reservoirs is that *heterogeneous* hydrolysis of $BrONO_2$

$$BrONO_2 + H_2O \rightarrow HOBr + HNO_3 \tag{9.37}$$

is apparently efficient on the surface of stratospheric sulfate aerosol or of polar stratospheric clouds (see Sections 9.4.4 and 9.4.5). HOBr is another photochemically labile species

$$HOBr + h\nu \rightarrow OH + Br, \tag{9.38}$$

and this reaction therefore enhances the OH (and, indirectly, the HO_2) concentration. In turn, the enhanced HO_x concentration reduces the lifetime of HCl, and thus its efficiency as a reservoir. One consequence of each of these considerations is that the BrO_x chains are even more efficient than the ClO_x ones in catalysing the destruction of stratospheric ozone.

The reservoir compounds are evidently of very great importance in stratospheric chemistry. They act to divert potentially catalytic species from active to inactive forms, but the compounds remain available to liberate active catalysts. The assumed rates of production and destruction have a large influence on the predictions of stratospheric models. It has recently become increasingly apparent that heterogeneous chemistry occurring on aerosol particles present in the stratosphere plays a very important part in the partitioning between active and inactive species. We shall see in Section 9.4 that unexpected release of active catalytic species from reservoir compounds is probably responsible for the phenomenon of the Antarctic ozone hole, and that the surface reactions are central to the explanation.

We now return to the importance of reactions (9.18) and (9.19) in stratospheric chemistry. In chemistry involving O_x and NO_x reactions alone, there are null cycles (*e.g.* cycle 7) and holding cycles. For ClO_x on its own, there are no corresponding null cycles, and those for HO_x, which involve H_2O_2, are rather inefficient. Reactions (9.18) and (9.19) are thus critical in providing effective null cycles for the HO_x and ClO_x families. Cycles 9 and 10 show these null cycles

Cycle 9:

$$OH + O_3 \rightarrow HO_2 + O_2 \tag{9.10}$$

$$HO_2 + NO \rightarrow OH + NO_2 \tag{9.18}$$

$$NO_2 + h\nu \rightarrow NO + O \tag{9.25}$$

Net $$O_3 + h\nu \rightarrow O_2 + O.$$

Cycle 10:

$$Cl + O_3 \rightarrow ClO + O_2 \tag{9.14}$$

$$ClO + NO \rightarrow Cl + NO_2 \tag{9.19}$$

$$NO_2 + h\nu \rightarrow NO + O \tag{9.25}$$

Net $$O_3 + h\nu \rightarrow O_2 + O.$$

Not only are these cycles effective null paths for HO_x or ClO_x, but they also provide an additional null cycle for NO_x. It is the photolytic step (9.25) that completes the null cycle in each case.

In the real atmosphere, photolysis of NO_2 is, of course, in competition with reaction (9.13), which would lead to an overall loss of two odd oxygens.

The combination of OH and HO_2 as a direct step is a loss process for HO_x

$$OH + HO_2 \rightarrow H_2O + O_2 \tag{9.24}$$

However, catalysed by NO_x, and with the participation of reaction (9.23), a cyclic process with the effect

$$OH + HO_2 + 2h\nu \rightarrow H_2O + O + O \tag{9.39}$$

can be written. The importance of a knowledge of the rate coefficient for reaction (9.18) can now be appreciated. Sequences of steps that would destroy odd oxygen if the rate coefficient were small become diverted into null or even generating cycles for larger values.

9.3.3 Summary of Homogeneous Chemistry

The information presented in the previous sections can most conveniently be summarized in flow charts showing source, radical, and sink species, and their reactive interconnections: such diagrams are sometimes referred to as *Nicolet diagrams* to honour a pioneer in atmospheric chemistry. Figure 9.3 is a simplified flow chart for the HO_x and NO_x species, while Figure 9.4 shows similar information for the ClO_x and BrO_x species. The diagrams presented here are deliberately restricted to the homogeneous processes described so far. The reaction partner that effects a transformation is given within the arrows, photochemical change being represented by $h\nu$. Cycles are readily identifiable as closed loops of arrows, and the reservoirs are generally the non-radical species in the centre of the diagrams that are formed and destroyed by radical and photochemical processes. The coupling between cycles is particularly clearly seen for the case of ClO_x (left side of Figure 9.4), where HO_x, NO_x, and BrO_x all play a role in the interconversions.

The contribution made to ozone removal by the direct $O + O_3$ reaction and each of the cycles is a function of altitude, as already shown in Figure 9.2. Different pathways within the family cycles of Figures 9.3 and 9.4 also make altitude-dependent contributions. The relative efficiency of the cycles and pathways at different altitudes depends on the availability of O and O_3. Thus, in the lower

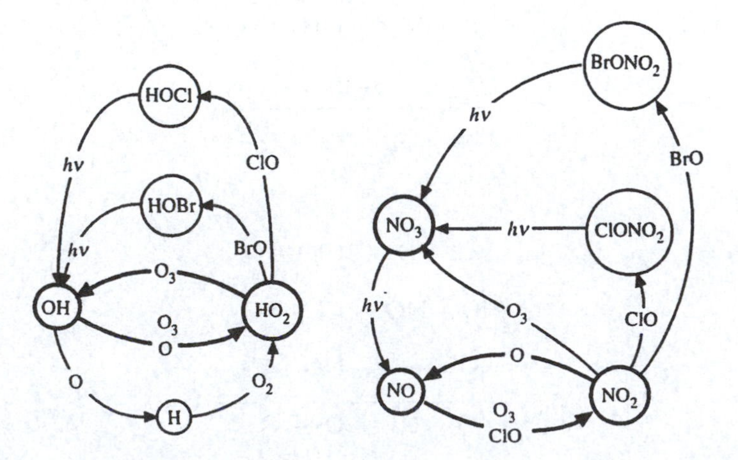

Figure 9.3 Chemical cycles for HO_x and NO_x trace species. Diagram modified from a figure of D.J. Lary, *J. Geophys. Res.* **102**, 21515 (1997).

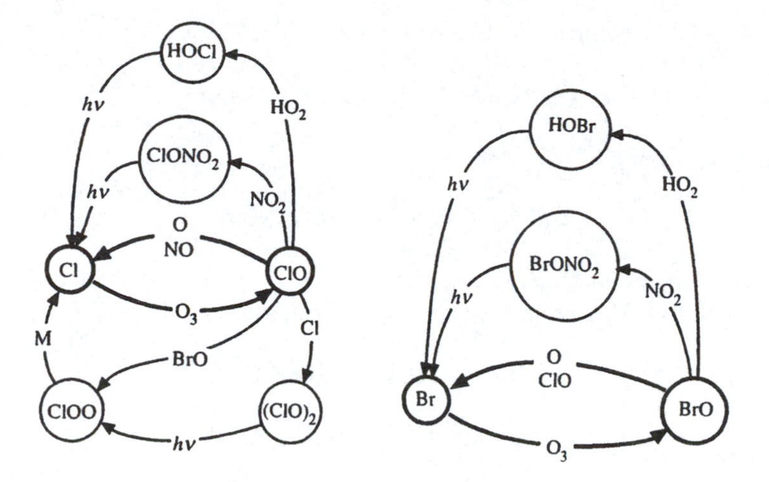

Figure 9.4 Chemical cycles for ClO_x and BrO_x species. Source as for Figure 9.3.

stratosphere, where $[O]/[O_3]$ is low (Figure 9.1), cycles having the net effect $2O_3 \rightarrow 3O_2$ are dominant, while in the upper stratosphere, where $[O]/[O_3]$ is high, the cycles producing the conversion $2O \rightarrow O_2$ are the most efficient. In the mid-stratosphere, the most effective cycles are those where the net reaction is $O + O_3 \rightarrow 2O_2$.

9.3.4 Heterogeneous Chemistry in the Stratosphere

In Section 9.4, we shall describe unusually large losses of ozone that have occurred in the Antarctic every spring for at least the past 30 years. The explanation for these unexpected depletions calls on several factors, one of which is unexpected chemistry. It has emerged that reactions on or in cloud particles provide a previously unknown route for the release of active catalysts from reservoir compounds. These heterogeneous reactions are now known to play a most important part in polar stratospheric chemistry in ways that will be explained in subsequent sections. However, it is now quite evident that heterogeneous reactions occur in regions of the stratosphere distant from the polar regions, and that particles of ice, soot, nitric acid hydrates, and sulfuric acid and sulfates can all be active. Volcanic eruptions that inject particles into the stratosphere are thus expected to have an effect on chemistry, and clear signatures are now recognized that demonstrate the influence of eruptions such as that of Mount Pinatubo.

Heterogeneous chemistry was discussed in relation to tropospheric chemistry in Section 8.6, and the general principles set out there apply equally to the stratosphere. Indeed, it was the discovery of heterogeneous processing in the stratosphere that led to a reawakening of interest in the contribution that multiphase chemistry might make in the atmosphere. That interest in turn stimulated research that has enlightened our perception of tropospheric chemistry.

Heterogeneous processing in the stratosphere is particularly significant in relation to the reservoir compounds, as just noted. Major alterations are brought about in the partitioning of species in the catalyst families between active chain carriers and temporarily inactive reservoirs, and in the partitioning between reservoirs of differing reactivity. Examples that we have already met include the hydrolysis reactions

$$N_2O_5 + H_2O \rightarrow 2HNO_3 \tag{9.28}$$

$$BrONO_2 + H_2O \rightarrow HOBr + HNO_3 \tag{9.37}$$

Reaction (9.37) has an analogue for $ClONO_2$

$$ClONO_2 + H_2O \rightarrow HOCl + HNO_3, \quad (9.40)$$

and a critical conversion of $ClONO_2$ and HCl

$$ClONO_2 + HCl \rightarrow Cl_2 + HNO_3 \quad (9.41)$$

occurs on particles in the polar stratosphere (Section 9.4.5). Note that the reaction influences two reservoirs at the same time. Other surface reactions proposed include

$$HOCl + HCl \rightarrow H_2O + Cl_2 \quad (9.42)$$

$$HOBr + HCl \rightarrow H_2O + BrCl \quad (9.43)$$

$$N_2O_5 + HCl \rightarrow ClNO_2 + HNO_3 \quad (9.44)$$

$$HNO_4 \rightarrow HONO + O_2. \quad (9.45)$$

9.4 POLAR OZONE CHEMISTRY

9.4.1 Abnormal Polar Ozone Depletion

The Antarctic ozone depletions with which we introduced Section 9.3.4 are what have become known as *ozone holes*. Very substantial reductions of total column ozone can occur over Antarctica as the Sun first rises in the (Austral) spring, and at some altitudes there can be a nearly complete loss of ozone just where it should be highest. To exemplify the huge ozone losses now experienced, Figure 9.5 shows some data for ozone abundances as a function of altitude over the South Pole. Abundances are shown for one particular day in 2001, and for comparison the averages for the period 1962–71 are displayed. The ozone layer that was centred on 15 km in the earlier years had essentially vanished in October 2001. Similar depletions occur over the Arctic, but they are less intense, persistent and regular.

The ultimate cause of the depletions is Man's release of halogen-containing compounds in increasingly large quantities in the second half of the 20th century. Ozone holes are thus formed anthropogenically, and the main part of this story unfolds in Section 11.3, where the subject is examined as part of Man's impact on the atmosphere. However, it makes sense to discuss the physical and chemical aspects of the polar phenomena in the present chapter as a continuation of our study of stratospheric chemistry. For the time being, we shall just accept that there is a surprisingly large amount of chlorine and bromine in the contemporary stratosphere, and not ask until Chapter 11 how those species reached such high concentrations.

9.4.2 Special Features of Polar Meteorology

There are two features of polar stratospheric meteorology and dynamics that appear to have a close bearing on the interpretation of polar ozone loss. Figure 9.6 shows that temperatures in the lower stratosphere drop to very low values during winter. In the Antarctic, temperatures below –90 °C are typical during July and August. As pointed out in Section 9.3.1, the stratosphere is very dry because H_2O from the troposphere is condensed out by the 'cold trap' of the tropopause, and what H_2O there is in the stratosphere is largely formed by CH_4 oxidation. The low water-vapour pressures

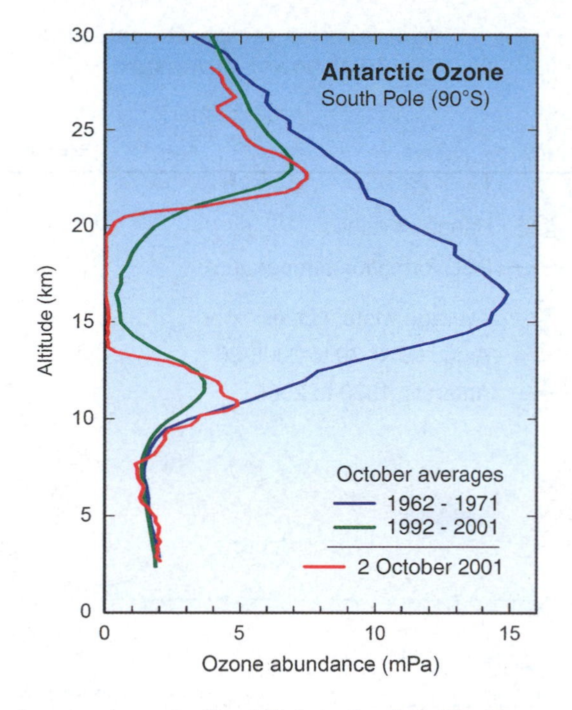

Figure 9.5 Distribution of ozone above the South Pole in October. The blue trace is the average over a ten-year data series obtained before there was significant halogen-driven ozone loss. Thirty years later (green trace) ozone has been enormously depleted at the altitudes around 17 km where ozone was formerly most abundant: this depletion is the 'ozone hole'. On one particular day in 2001, the ozone had completely vanished, and there is no sign of an ozone layer. Source: *Scientific Assessment of Ozone Depletion: 2006*, World Meteorological Organization, Global Ozone Research and Monitoring Project, Report No. 50, WMO, 2007.

mean that water clouds do not usually form in the stratosphere. However, at the very low temperatures of the polar winter, *polar stratospheric clouds* (PSCs) can be created. The critical temperature for formation of the clouds is −78 °C, temperatures experienced on average for 1–2 months over the Arctic and 5–6 months over Antarctica. The different behaviour arises, in part, because there are significant meteorological differences between the hemispheres that result from the differences in the distributions of land, ocean, and mountains at middle and high latitudes.

Winter temperatures are low enough for PSCs to form for nearly the entire Antarctic winter but usually only for part of every Arctic winter (and sometimes not at all). There is clear evidence that the polar stratospheric clouds are involved in polar ozone destruction, and it is significant that years when stratospheric temperatures are particularly low are also those in which ozone depletions are greatest. For example, lower-stratospheric temperatures were unusually depressed (by up to 5 °C) in southern middle and polar latitudes in 1998, and this was a year in which the hole was exceptionally wide and deep. In the years when temperatures in the Arctic are not low enough for PSC formation, significant ozone depletion is not seen. We return to a further discussion of the composition and formation of PSCs in Section 9.4.4.

The second, and related, feature of polar meteorology is that a vortex forms as air cools and descends during the winter. A westerly circulation is set up, as illustrated in Figure 9.7, with very high wind speeds of perhaps 100 m s^{-1} or more by spring. The vortex develops a core of very cold air, and it is these low temperatures that allow the polar stratospheric clouds to form in the lower stratosphere. When the Sun returns in September, temperatures rise, the winds weaken, and the vortex breaks down in November. But in the winter and early spring, the stability of the vortex is so

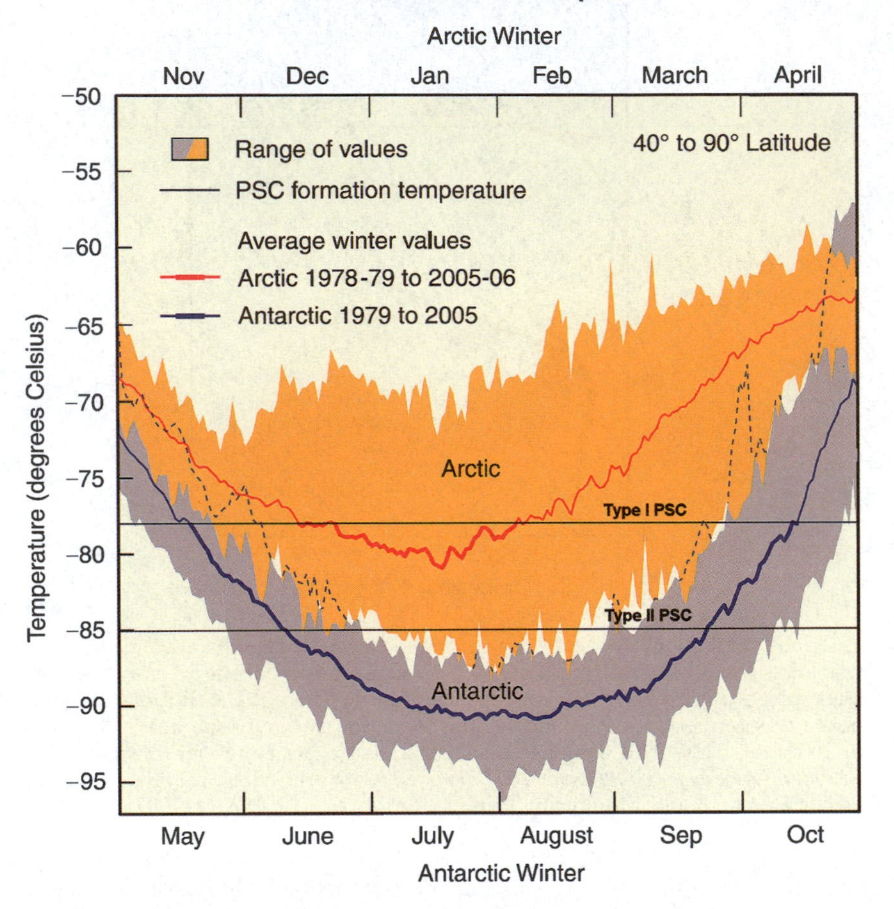

Figure 9.6 Arctic and Antarctic temperatures. Stratospheric air temperatures reach their minimum values in the winter. Average minimum values over Antarctica fall as low as –90 °C in a typical year. Over the Arctic, average minimum temperatures are near –80 °C in January and February. Polar stratospheric clouds (PSCs) are formed only below a threshold temperature of about –78 °C. Source: *Scientific Assessment of Ozone Depletion: 2006*, World Meteorological Organization, Global Ozone Research and Monitoring Project, Report No. 50, WMO, 2007.

great that air at polar latitudes is *almost* sealed off from that at lower latitudes. One simple view is that the air over the pole is more or less confined to what is effectively a giant reaction vessel. There is a slow downward circulation that drives the polar air through the cold core of the vortex that contains the polar stratospheric clouds. As we shall see shortly, these clouds seem to be involved in unusual chemistry, with the downward circulation allowing the core region to act as a 'chemical processor'. The concept is thus of the vortex as a 'containment vessel'. However, in reality material is slowly lost from the vortex, perhaps at about one per cent a day during the period of early August to late October. In that case, the vortex would be best characterized as a 'flow reactor' with a continual slow supply into it of reactants and a flow out of products. A further modification to the simple view is also necessary because the Antarctic vortex is sufficiently extensive that its edges are exposed to sunlight even at the winter solstice. There is evidence for ozone depletion starting in mid-winter that probably originates from such regions. Whatever the details of the processes, the low temperatures, the presence of polar stratospheric clouds, and the unusual dynamics and

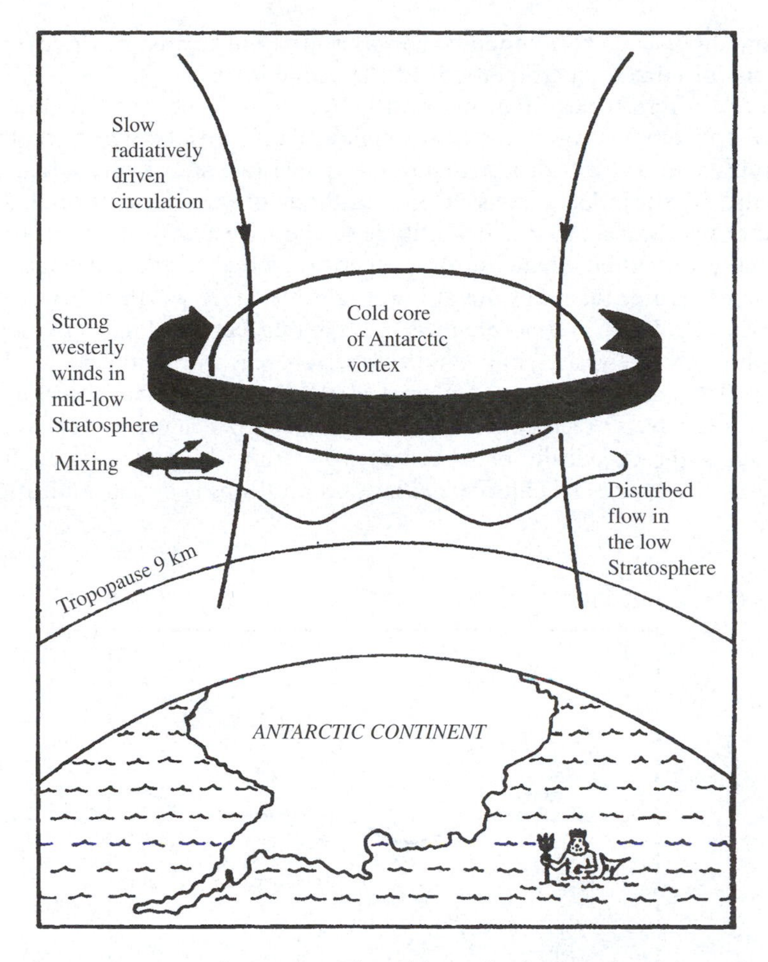

Figure 9.7 The winter vortex over Antarctica. The cold core is almost isolated from the rest of the atmosphere, and acts as a reaction vessel in which the constituents may become chemically 'preconditioned' during the long polar night. Source: R.P. Wayne, *Chemistry of Atmospheres,* 3rd edn, OUP, 2000, by permission of Oxford University Press.

meteorology of the vortex most assuredly provide the backdrop to some unexpected and surprising chemistry.

9.4.3 Anomalous Chemical Composition

The discovery of anomalous ozone depletion naturally generated a great deal of scientific activity. An 'ozone expedition' in 1987 involved not only ground- and satellite-based measurements and balloons, but two 'airborne laboratories' that flew into the depleted region several times to obtain detailed information about the extent of the perturbed region, and about the chemistry in it. The instrumented aircraft were a DC8 and the stratospheric ER–2 (see Section 4.2). The airborne laboratories were able to measure concentrations of ozone; particulate matter, aerosols, and condensation nuclei; the oxides of nitrogen and of the halogens; water-vapour concentrations; and temperatures, pressures, and other meteorological parameters. An enormous collaborative effort was involved, with some 150 scientists and support staff from 19 organizations and four nations. The wealth of data obtained has permitted much interpretation of the ozone-hole phenomenon.

Several more campaigns have been mounted subsequently, but it was the discoveries in 1987 that provided the basis for current interpretations of the ozone hole.

The stratospheric aircraft measurements of 1987 confirmed, beyond all doubt, that there is anomalous chemistry going on within the vortex region. First, consider the concentrations of ozone and of catalytically active ClO radicals. Figure 9.8, panel (a), shows measurements obtained at altitudes of around 18 km in late August. Concentrations of ozone are normal. But the chlorine oxide concentrations show a sharp rise at a latitude of about 65°S. The concentration goes up by a factor of 10 over a few hundred kilometres. In the perturbed region, the concentrations are, in fact, more than 100 times greater than they are at lower latitudes. It is worth remembering that almost all the chlorine in the Earth's atmosphere is a consequence of Man's release of compounds containing it! Only a few weeks later, in mid-September, there is a dramatic change in the behaviour of ozone (Figure 9.8, panel (b)). This is a time at which the hole has developed fully. Ozone concentrations decrease over exactly the latitude range where the chlorine monoxide concentration increases on entering the chemically perturbed region within the polar vortex. The strong anti-correlation between the ozone and chlorine oxide concentrations is a clear indication that chlorine

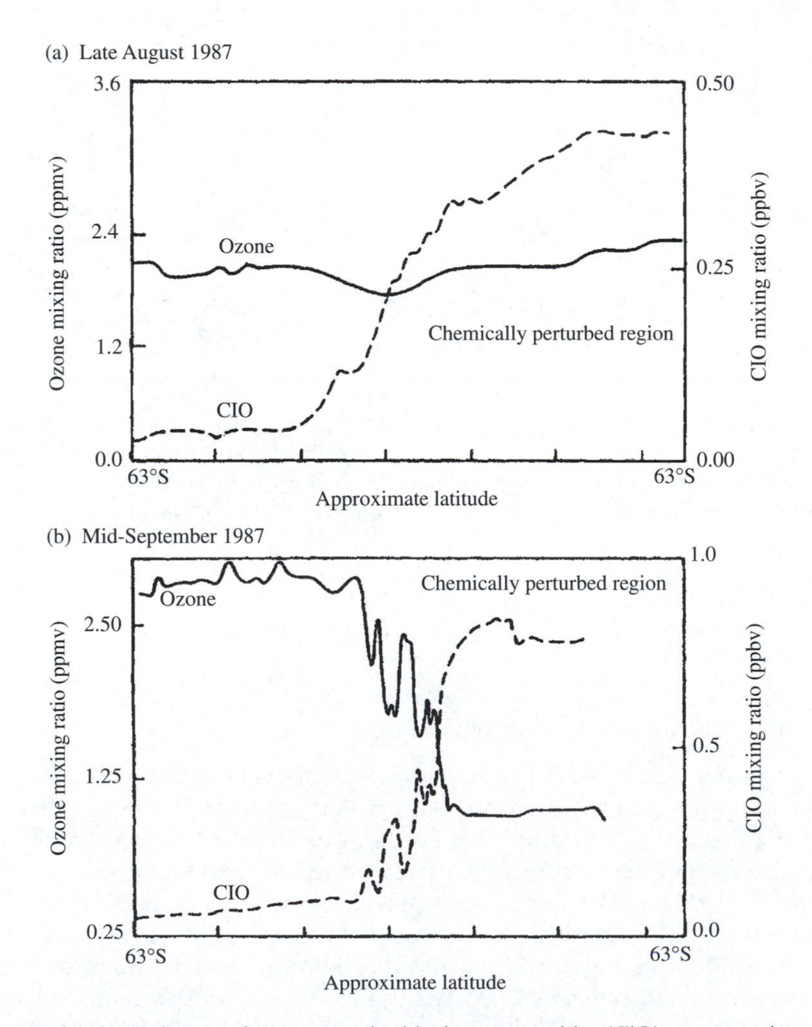

Figure 9.8 Latitude dependence of ozone and chlorine monoxide (ClO) on entering the chemically perturbed region: (a) late August; (b) mid-September 1987.

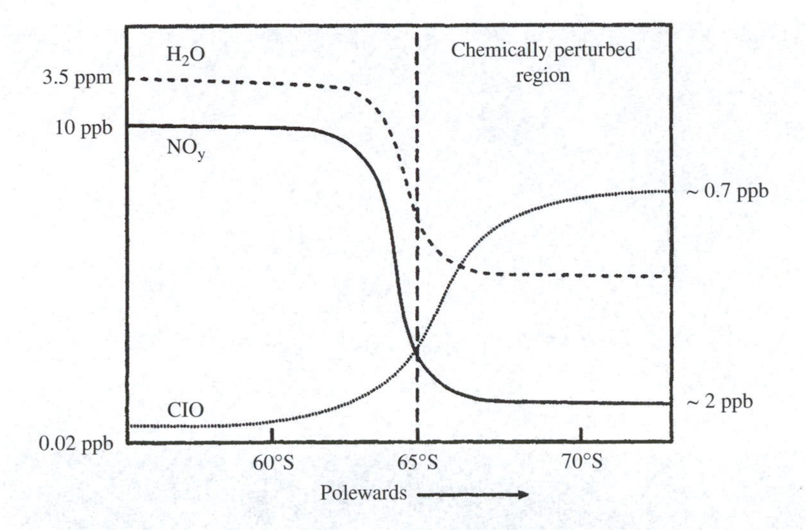

Figure 9.9 Schematic representation of the changes in concentration of some of the species discussed in the text on entering the chemically perturbed region.

chemistry is somehow responsible for the ozone depletion. This anticorrelation is exhibited not only by the large-scale changes but also by the smaller fluctuations in concentration seen on traversing the critical latitude region.

Other measurements, summarized in Figure 9.9, show that the stratosphere within the disturbed region is abnormally dry, and that it is highly deficient in the NO_y compounds (NO_x *plus* N_2O_5, HNO_3, HNO_4 and $ClONO_2$). The dehydration and denitrification are explained by the condensation of water and the conversion of the oxides of nitrogen to nitric acid in the polar stratospheric clouds, the particles of which may become large enough to undergo sedimentation over appreciable distances. If the PSCs contain HNO_3 (see next section), NO_y may be removed irreversibly: up to 90 per cent of the available reactive nitrogen has been observed to be lost in this way in the Antarctic (and the Arctic) vortex. It is this removal of the oxides of nitrogen which leads to the anomalous chlorine chemistry. Only the core of the vortex is cold enough for the larger particles to form, so we see again that it is the combination of low temperature and special dynamics in the atmosphere that set up the conditions needed for perturbed chemistry.

Dehydration may be brought about by a similar sedimentation of PSC particles containing large fractions of water-ice. As discussed in Sections 9.4.2 and 9.4.4, the formation of such particles requires very low temperatures; for this reason dehydration is observed in the Antarctic, but not in the Arctic where temperatures are higher. Although intense dehydration has not been observed without intense denitrification, the processes are quite possibly independent.

9.4.4 Polar Stratospheric Clouds

An idea of the visual appearance of PSCs is revealed by Figure 9.10. The optical properties of PSCs suggest that there are two main classes. Type I PSCs are small ($<1\ \mu m$ diameter) HNO_3-rich particles, and have a *mass* mixing ratio of about 10^{-8}. Type II PSCs are larger (from 10 μm to perhaps more than 1 mm diameter), are composed primarily of H_2O-ice together with minor amounts of HNO_3 as hydrates, and can constitute up to 10^{-6} of the stratosphere when they are present. Several hydrates of the acids may be present in the PSC particles. The hydrates include nitric acid dihydrate (NAD: $HNO_3 \cdot H_2O$), nitric acid trihydrate (NAT: $HNO_3 \cdot 3H_2O$), and sulfuric acid tetrahydrate (SAT: $H_2SO_4 \cdot 4H_2O$). Type I PSCs often appear to belong to one of two

Figure 9.10 Polar stratospheric clouds. This photograph shows polar stratospheric clouds lit from below near Kiruna, Sweden. Source: NASA, http://earthobservatory.nasa.gov/images/imagerecords/0/622/PSCbest.jpg

subcategories, Type Ia solid particles consisting of nearly pure NAT and Type Ib particles, which are supercooled *liquid* ternary solutions of HNO_3–H_2SO_4–H_2O.

Figure 9.11 shows that Type-I PSCs are formed at temperatures 5–10 K higher than the Type-II PSCs. Both physical size and chemical composition are thus temperature dependent. The extent of denitrification (and dehydration) will be affected by the rate of sedimentation, itself obviously more rapid for larger particles. It is worth noting here a difference between 'denoxification' (removal of NO_x) and 'denitrification' (removal of NO_y). NO_x removed by formation of HNO_3 that is incorporated as NAT or as a ternary mixture in Type-I PSCs may ultimately be available for release again. The efficacy of the HNO_3 reservoir has been enhanced: NO_x is (temporarily) lost, but NO_y is not, and the atmosphere is denoxified. Subsidence of NAT–ice mixtures to the troposphere in Type-II PSCs, on the other hand, removes NO_y permanently, and the atmosphere is denitrified. The chemical consequences of denoxification and denitrification are different, a point to which we shall return in the section that follows.

Some of the heterogeneous processes introduced in Section 9.3.4, such as reactions (9.28) and (9.40)–(9.42), are known to proceed with efficiencies that depend on the nature of the surface. For example, the hydrolysis reactions (9.28) and (9.40) are fast on Type-II PSCs, but slow on Type-I PSCs. On the other hand, reactions (9.41) and (9.42) are probably fast on both Type-I and Type-II PSCs, but the rates may depend on the relative humidity since that will determine the solubility of the HCl reagent in the aerosol particles. Once again, then, reaction probabilities will show an unusually strong dependence on temperature that may be more related to surface composition than to the ordinary considerations of reaction kinetics.

9.4.5 Perturbed Chemistry

Figure 9.12 provides, in diagrammatic form, a summary of the chemistry that explains the appearance of the Antarctic ozone holes, and it should be used to map the description that now

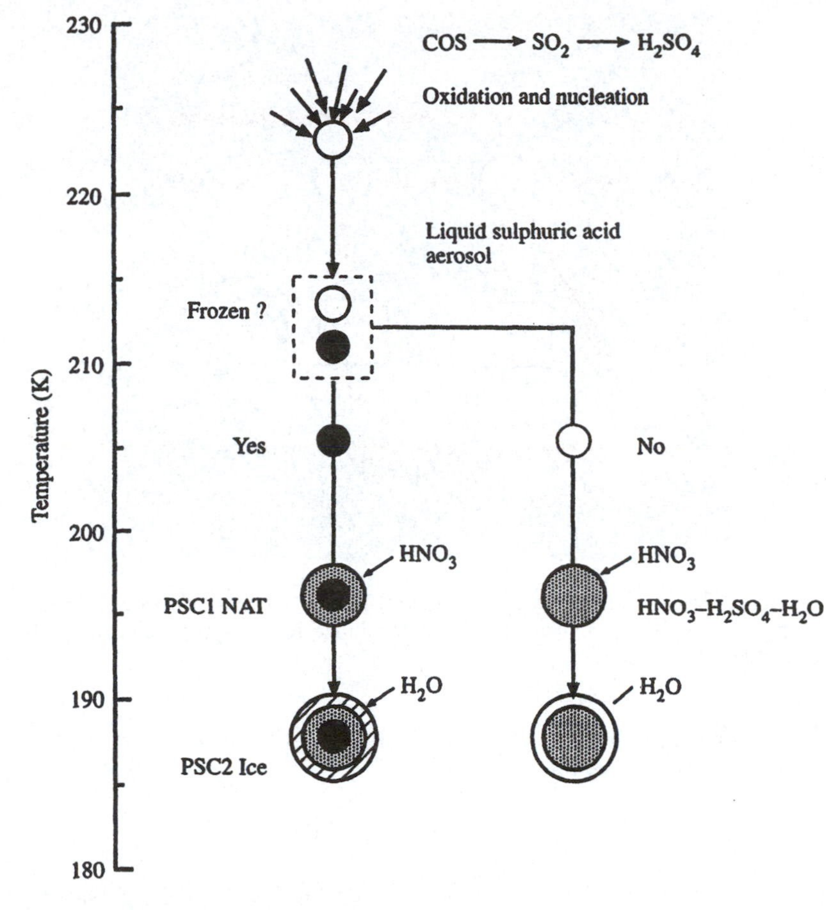

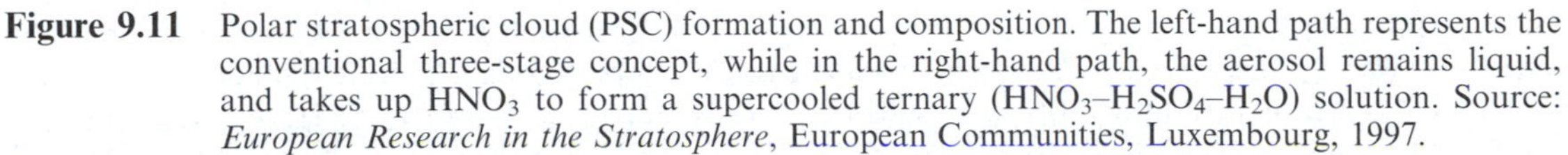

Figure 9.11 Polar stratospheric cloud (PSC) formation and composition. The left-hand path represents the conventional three-stage concept, while in the right-hand path, the aerosol remains liquid, and takes up HNO_3 to form a supercooled ternary (HNO_3–H_2SO_4–H_2O) solution. Source: *European Research in the Stratosphere*, European Communities, Luxembourg, 1997.

follows. The central feature of the perturbed chemistry of the polar stratosphere is the conversion of reservoir compounds (Section 9.3.2) to catalytically active species (or their precursors) on the surface of the polar stratospheric clouds (PSCs). Most of the chlorine in the stratosphere is usually bound up in the reservoir molecules hydrogen chloride and chlorine nitrate, as a result of the reactions

$$Cl + CH_4 \rightarrow CH_3 + HCl \qquad (9.30)$$

$$ClO + NO_2 + M \rightarrow ClONO_2 + M. \qquad (9.34)$$

Liberation of the active chlorine from the reservoirs is normally rather slow. But the two reservoir molecules can react together on PSC particles,

$$ClONO_2 + HCl \rightarrow Cl_2 + HNO_3 \qquad (9.41)$$

as explained in Section 9.3.4. The outcome is that molecular chlorine is released as a gas, and the nitric acid remains in the ice particles (as hydrates such as NAT), which can ultimately transport

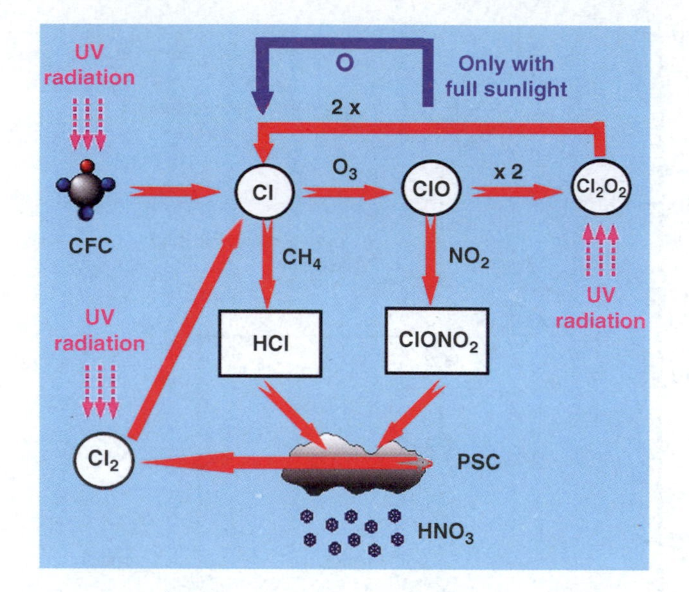

Figure 9.12 Schematic diagram of chemical conversions in the ClO_x-catalysed decomposition of ozone in the presence of PSCs. Chemical species inside circles are active chlorine, while those inside rectangles are reservoirs.

water and nitric acid out of the vortex, and perhaps even to the troposphere if the temperature is low enough. The Cl_2 is photodissociated to atoms

$$Cl_2 + h\nu \rightarrow Cl + Cl \tag{9.46}$$

if sunlight is present, even at very low intensities. The surface reaction (9.41) has now been characterized in many laboratory studies. The PSCs disturb the balance between active and reservoir chlorine in two related ways. They provide surfaces on which unusual chemical change can occur, and they also transport active nitrogen out of the stratosphere, in the form of HNO_3, reducing the amount of $ClONO_2$ reservoir that can be formed in the first place. Denitrification (permanent loss of NO_y) requires lower temperatures than denoxification (loss of NO_x, possibly temporarily), as explained in Section 9.4.4. Low temperatures thus particularly favour reduction of [$ClONO_2$].

Surface reactions such as reaction (9.41) can obviously play an important role in atmospheric chemistry whenever particles are present, but their involvement in atmospheric chemical transformations was frequently neglected before 1985. Ignorance of the existence of reaction (9.41) is a major reason why the ozone hole was not predicted at that time by any atmospheric model. Other surface reactions that have subsequently been demonstrated to occur in the laboratory, and that may be involved in polar chemistry, were introduced in Section 9.3.4, and mentioned again in Section 9.4.4. As examples, the reactions

$$N_2O_5 + H_2O \rightarrow 2HNO_3 \tag{9.28}$$

$$ClONO_2 + H_2O \rightarrow HOCl + HNO_3 \tag{9.40}$$

also both produce HNO_3, and reaction (9.28) removes another important reservoir molecule, N_2O_5. The reactions occur on ice particles (with dissolved HCl in the second case), and HNO_3 remains in the ice particle after reaction. HOCl, the gas-phase product of reaction (9.40), is readily photolysed by near-ultraviolet and visible light to yield ClO_x radicals.

Because the conversion can occur on the surface of the polar stratospheric clouds, release of molecular chlorine from the major reservoir molecules can continue in the chemically perturbed region throughout the polar winter and early spring. The vortex largely isolates the air within it, so that this part of the stratosphere can become chemically altered, or 'preconditioned', over the long polar night.

Consider, as an example, the molecular chlorine released by reaction (9.41). As we have noted already, when the Sun finally returns again in the spring, the Cl_2 generated from the reservoir gases as a result of the preconditioning is rapidly split into chlorine atoms that can destroy ozone, and at the same time liberate chlorine monoxide

$$Cl_2 + h\nu \rightarrow Cl + Cl \quad (9.46)$$

$$Cl + O_3 \rightarrow ClO + O_2 \quad (9.14)$$

The beginnings of an explanation for enhanced concentrations of chlorine monoxide accompanying ozone depletions are already apparent. However, on their own, reactions (9.46) followed by (9.14) cannot lead to much ozone loss. It will be recalled from Section 9.3 that, in the mid-stratosphere, a *chain* process (Cycle 4)

$$Cl + O_3 \rightarrow ClO + O_2 \quad (9.14)$$

$$ClO + O \rightarrow Cl + O_2 \quad (9.15)$$

destroys large numbers of ozone molecules for each chlorine atom made available. Some cycling like this must be going on in the perturbed Antarctic stratosphere, but it cannot be this chain, because the concentration of oxygen atoms in the lower stratosphere is far too small.

Alternative catalytic cycles are required to explain substantial ozone depletion in the polar stratosphere in early spring. The most important of these cycles involves the ClO dimer, $(ClO)_2$, formed in the self-reaction of ClO. These dimers have been shown experimentally to be readily photolysed to yield, by an indirect route, two free chlorine atoms. The cycle is thus

Cycle 11:

$$ClO + ClO + M \rightarrow (ClO)_2 + M \quad (9.47)$$

$$(ClO)_2 + h\nu \rightarrow Cl + ClOO \quad (9.48)$$

$$ClOO + M \rightarrow Cl + O_2 + M \quad (9.49)$$

$$\underline{2(Cl + O_3 \rightarrow ClO + O_2)} \quad (9.14)$$

Net $$2O_3 + h\nu \rightarrow 3O_2.$$

Dimers such as $(ClO)_2$ are only formed at low temperatures, so that, once again, the low Antarctic polar temperatures are an essential component of another part of the perturbed chemistry. High concentrations of chlorine monoxide also favour the formation of the dimer, and such high concentrations are a feature of the chemically perturbed region of the Antarctic vortex. The photolysis of $(ClO)_2$ in reaction (9.48) was another process completely unknown when the ozone hole was first found. Experimental proof that the reaction liberates atomic chlorine gives some confidence in the validity of the explanation for the Antarctic ozone hole.

The influence of temperature on the chemistry of PSCs was discussed in Section 9.4.4. Here, we return briefly to the subject. The general conclusion is that all the chlorine-activation reactions proceed more rapidly on Type-II PSCs than they do on Type-I PSCs. The timescales for the

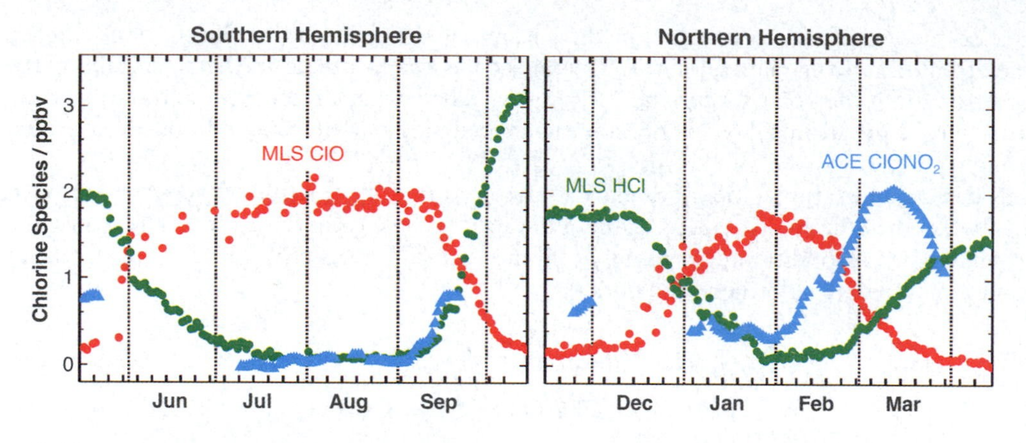

Figure 9.13 Daily averages of ClO (red dots), HCl (green dots) and $ClONO_2$ (cyan triangles) observed at an altitude of about 20 km and at 70–75° equivalent latitude. (a) 2005 Antarctic winter–spring; (b) 2004–2005 Arctic winter–spring. In the Arctic, the magnitude and duration of chlorine activation are much smaller than in the Antarctic, even in a relatively cold winter. MLS = Aura Microwave Limb Sounder; ACE-FTS = Atmospheric Chemistry Experiment Fourier-Transform Spectrometer. Source: as for Figure 9.6.

activation processes are thought to range from about ten days at $T = 198$ K on Type-I PSCs to one day on Type-II PSCs at their threshold temperature (~ 188 K), to just a few hours at temperatures of 180 K and below. Bearing in mind, as well, the much greater extent of denitrification at the lower temperatures, it becomes evident why stratospheric temperatures are so intimately connected with the depth and extent of the Antarctic ozone holes.

Subsequent developments of *in-situ* and remote-sensing measurements have both confirmed the underlying ideas and provided further information for more detailed interpretations. Figure 9.13 shows some measurements of key species made in the Antarctic winter of 2005 (left-hand panel) and the preceding Arctic winter (right-hand panel) for an altitude of about 20 km. Obtaining such data is a great achievement, and the results happily confirm the ideas that had been pieced together earlier from sparser evidence. Figure 9.14 combines some of the data obtained, in smoothed form for clarity, in order to summarize pictorially the photochemical and dynamical features of polar-ozone depletion. The upper panel represents the conversion of chlorine from inactive to active forms during winter, and the formation of the inactive reservoirs again in the spring. Evolution of the polar vortex is indicated in the lower panel, where the temperature scale is meant to indicate changes in minimum temperatures in the lower polar stratosphere. The surface conversion of the reservoirs $ClONO_2$ and HCl to active chlorine occurs rapidly, and virtually to completion, in early winter; the reservoirs begin to reappear in late winter as temperatures become too high for further chemical processing on PSC surfaces. It is interesting that, during the 'normal' part of the year (late spring, summer, and early autumn), $ClONO_2$ is a slightly more abundant reservoir than HCl. Even more interesting is the observation that, on recovery, $ClONO_2$ initially overshoots its 'normal' level by a factor of up to two. It seems that the chlorine rapidly returns to the reservoirs during the recovery period, but that it takes time to re-establish the steady-state partitioning *between* the reservoirs. This behaviour arises because reaction (9.34) forming $ClONO_2$ is much more rapid than reaction (9.30) that yields HCl. While the overshoot shown in the figure is temporal, a similar spatial effect is seen for $ClONO_2$, and for the same reasons. A *collar region* of enhanced $ClONO_2$ concentration is found surrounding the vortex in late winter; in this region, PSC processing is limited or infrequent, and sunlight is available to produce NO_2 from the photolysis of HNO_3.

Other cycles may supplement cycle 11 in the catalytic destruction of ozone in the polar spring. In explaining catalysis of ozone by halogen-containing species, we have considered mainly ClO_x

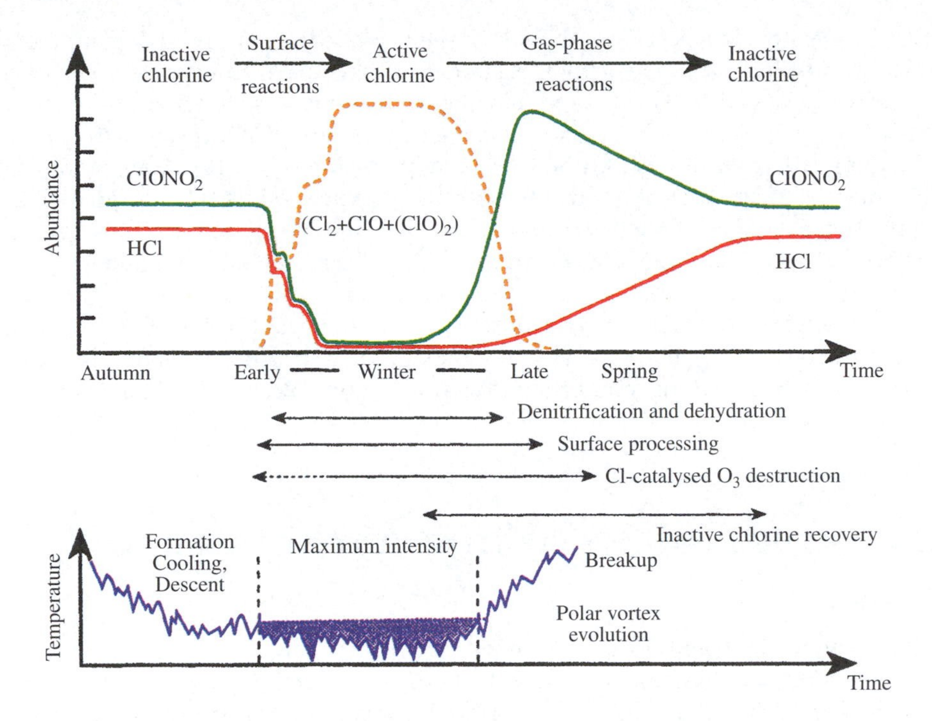

Figure 9.14 Photochemistry and dynamics in the polar stratosphere. From *Scientific Assessment of Ozone Depletion: 1998*, World Meteorological Organization, Geneva, 1999.

chemistry. In fact, atom-for-atom, Br is an even more effective catalyst than Cl, one estimate suggesting a factor of 58 increased depletion power. We pointed out in Section 9.3.2 that the reservoirs for Br and BrO are much 'shallower' than those for Cl and ClO, which means relatively more of the Br is available for catalysis. Thus, the bromine analogue of cycle 4 can more effectively destroy ozone than does the chlorine-based cycle. There is also significant synergism between the Br- and Cl-catalysed processes. As a result, increases in Br loading in the atmosphere may potentiate Cl, and *vice versa.*

The chlorine–bromine synergism with respect to ozone depletion arises because of a coupling reaction between BrO and ClO. The reaction possesses three branches

$$BrO + ClO \rightarrow Br + Cl + O_2 \tag{9.50a}$$

$$BrO + ClO \rightarrow Br + OClO \tag{9.50b}$$

$$BrO + ClO \rightarrow BrCl + O_2, \tag{9.50c}$$

of which the first produces Br and Cl atoms that react with ozone to complete a catalytic cycle

Cycle 12:

$$BrO + ClO \rightarrow Br + Cl + O_2 \tag{9.50a}$$

$$Br + O_3 \rightarrow BrO + O_2 \tag{9.51}$$

$$Cl + O_3 \rightarrow ClO + O_2 \tag{9.14}$$

Net $$O_3 + O_3 \rightarrow O_2 + O_2 + O_2.$$

It has been argued that this cycle could make a significant contribution to depletion of ozone in the Antarctic stratosphere. A noteworthy aspect of reaction (9.50a), and cycle 12 in general, is that halogen oxides are recycled to the corresponding halogen atoms *without* the need for sunlight. The other branches of the interaction between BrO and ClO produce OClO (reaction (9.50b)) and BrCl (reaction (9.50c)). BrCl regenerates Br and Cl, but only on photolysis. Reaction (9.50b) is the only confirmed stratospheric source of OClO, so that the presence of OClO is a good indicator of chlorine activation as well as of the presence of active bromine compounds in the stratosphere. The species has been observed in both polar vortices, with the largest column abundances being found in the Antarctic, and the quantities are broadly consistent with the expectations from model simulations. Enhancements in OClO were observed after the eruption of Mount Pinatubo. The increases occurred before temperatures were low enough for PSC processing, and imply additional ozone destruction due to chlorine (and bromine) activation on volcanic sulfate aerosol (see Section 9.5.5).

9.5 PERTURBATIONS OF THE STRATOSPHERE BY NATURE

9.5.1 Overview

Chapter 11, Sections 11.2 and 11.3, will show that Man's activities have led to a decline in stratospheric ozone concentrations over the past 30–40 years, primarily as a result of the release of halogen-containing compounds that are long-lived enough in the troposphere that they can exchange into the stratosphere. However, a variety of natural phenomena can also affect the stratosphere, and those are the processes that we shall examine in the present section. It is important to understand the mechanisms and extent of ozone depletions brought about in this way, so that the signatures of the natural and anthropogenic disturbances can each be read and distinguished clearly. In general, the ozone depletions brought about by nature seem not to be very persistent, lasting only for a few years at most. Longer-term trends in ozone concentrations appear to be associated with the disturbance by humans.

Some of the natural perturbations result from impulsive phenomena such as solar-particle storms (Section 9.5.4) and volcanic eruptions (Section 9.5.5). A study of the response of the atmosphere, and especially its ozone content, to these events is potentially very instructive. It can provide an alternative way of validating our view of stratospheric chemistry that complements the measurements of concentrations of the participating chemical species. Other influences involve more gradual variations of factors such as solar UV output or galactic cosmic-ray intensity. Various periodic ozone changes have been established as well, including annual oscillations and the *semiannual oscillation* (SAO), *quasibiennial oscillation* (QBO), and the *El Niño–Southern Oscillation* (ENSO). In terms of total ozone, the amplitude of the annual cycle ranges from about six per cent in the tropics to 30 per cent at 60°N or S. The amplitude of the SAO is about four per cent in the Northern Hemisphere and two per cent in the Southern Hemisphere, while that of the QBO ranges from four to seven or eight per cent (Section 9.5.2). In those regions of the globe where ENSO affects ozone, the variations can reach five per cent (Section 9.5.2). Changes in solar activity over the solar cycle (Section 9.5.3) can bring about changes in total ozone between one and two per cent.

The effects arise because of changes in chemistry (for example, altered catalyst concentrations or changed surface areas) or in temperature and dynamics, which can redistribute atmospheric ozone. It is not always possible to disentangle the processes responsible for ozone perturbations, as we shall see shortly. Meteorological behaviour, including circulation and vertical motions, is often affected by the primary driver, such as changes in solar intensity or volcanic eruptions. Another complication is that, although an event like a volcanic eruption cannot be ascribed to Man, the

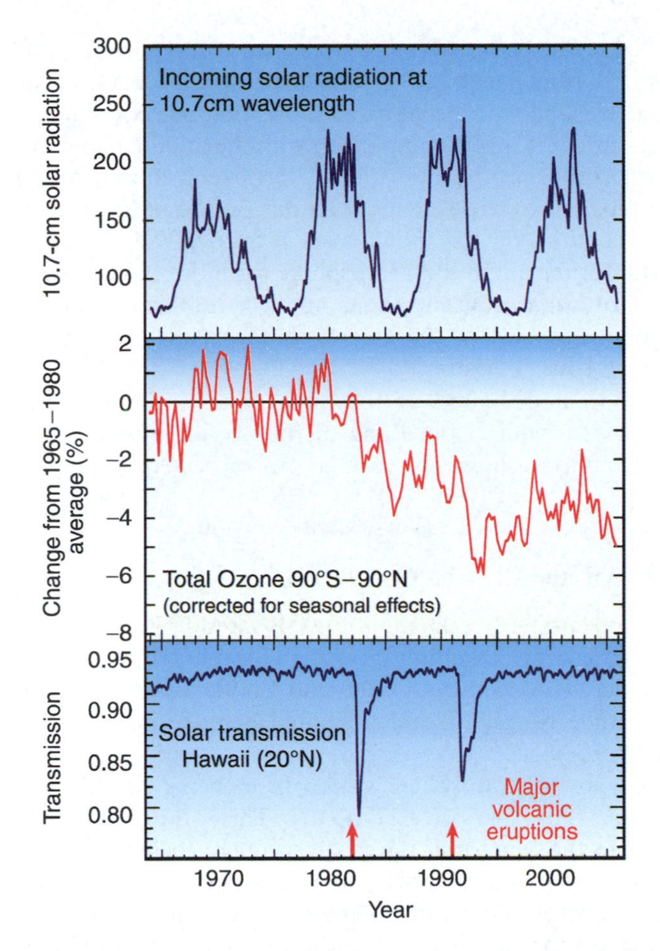

Figure 9.15 The solar cycle, global ozone and volcanic eruptions. The top panel shows the solar intensity measured at 10.7 cm, which tracks the 11-year solar cycle. Ozone levels shown in the middle panel are 3-month averages corrected for seasonal effects, and have decreased since the early 1980s. Optical transmission through the atmosphere (bottom panel) gives an indication of dust and aerosol present, and the major volcanic eruptions of El Chichón (1982) and Mt. Pinatubo (1991) made considerable impacts. Source as for Figure 9.6.

perturbed chemistry that follows an eruption may be partially or even principally a consequence of the presence of chemical species in the stratosphere that are associated with human activities.

Figure 9.15 serves to introduce the sections that follow by displaying experimental data that back up the statements made in this overview. The middle panel shows ozone concentrations over the period 1964 to 2006. Up until about 1980, concentration fluctuations were centred on the mean shown here as a horizontal line. Since the early 1980s, total ozone values have clearly been dropping below this long-term mean, and apparently showing a persistent decline. Looking more closely at the ozone variations, other patterns can be seen. Examine now the top panel, which shows solar radiation intensities over the same period (the particular very long wavelength of 10.7 cm is often used as a diagnostic for total solar intensity). The well-known 11-year (sunspot) cycle stands out clearly. A comparison of the top and middle panels indicates that the cyclic changes in solar output correlate with in-phase, but 'noisier' signals from the ozone. But it is also evident that the changes in solar intensity cannot account for the long-term decreases in ozone. Solar intensity changes are discussed further in Section 9.5.3. Volcanic eruptions (Section 9.5.5) occurred frequently in the period from 1965 to 2005. The largest recent eruptions are El Chichón (1982) and Mt. Pinatubo (1991) (see red arrows in bottom panel). Large volcanic eruptions can lead to the formation of new particles in the stratosphere, and the bottom panel shows that one consequence is decreases in solar transmission to Earth's surface. Careful comparison with the middle panel shows that the eruptions of 1982 and 1991 were accompanied by anomalously low ozone concentrations, experienced with a time delay after the explosion. Heterogeneous reactions on the particles deplete ozone in an analogous manner to the reactions on PSCs discussed in Section 9.4. The effect is only temporary because the particles do not remain in the stratosphere for more than a few years. The recovery of solar transmission and of ozone concentrations confirm this expectation, and show that large volcanic eruptions cannot account for the long-term decreases found in global total ozone any more than can changes in solar intensity.

There is a further point that Figure 9.15 illustrates. The correlations just described might be thought rather weak and to require something of the eye of faith to be discerned. But that is deceptive: mathematical analysis shows the correlations to be very strong indeed.

9.5.2 The Quasibiennial and El Niño Oscillations

At first sight, it might seem curious to talk about the QBO and ENSO phenomena before discussing the impact on stratospheric ozone resulting from the volcanic eruptions, solar-proton events and changes in solar irradiance. However, both QBO and ENSO are closely linked with stratospheric perturbations, and the underlying ideas are presented here in order to simplify the subsequent discussions in Sections 9.5.3–9.5.5.

Tropical winds in the lower stratosphere switch from being easterly to westerly in a rather irregular cycle that has an average period of 27 months. This is the *quasibiennial oscillation* or QBO. It dominates variability in the equatorial lower stratosphere (below an altitude of about 35 km). Not only are meteorological parameters such as winds and thermal structure affected, but the distributions of ozone and other minor constituents are altered at all latitudes. A QBO modulation is observed in the concentrations of NO_2 and O_3 in the middle stratosphere. The NO_2 signal is a consequence of variations in NO_y concentrations linked to the altered meteorology, and the O_3 signal is an indirect chemical response to the modulation in NO_2, at least at altitudes above 30 km.

Yet another periodic phenomenon, the *semiannual oscillation* (SAO), dominates the seasonal variation of winds in the tropics at higher altitudes (above about 35 km). Evidence of the influence of the SAO has been adduced from satellite measurements of stratospheric O_3.

ENSO describes a phenomenon in which there is a significant shift in the location of the warmest water from its 'average' location in the western equatorial Pacific across to the central and eastern area. A particularly strong El Niño was observed in 1997–98, with an extensive warm anomaly east of the date-line (a classic El Niño signature). Temperatures up to 5 °C above average were observed. Such unusually warm water produces very heavy rain showers and flooding locally (in Peru, for example) where such conditions do not normally occur. Incidentally, temperatures in the western Pacific were average or lower than average during this period. El Niño influences not only the equatorial Pacific region. Because the jet-streams in the atmosphere are strengthened during these warm phases, there are characteristic warm–cool and wet–dry anomalies outside the Pacific that are forced by El Niño. Amongst the consequences are changes in stratospheric ozone concentrations.

The ENSO phenomenon appears to be aperiodic, and its effect on total ozone does not show zonal symmetry. Rather, a wave-train propagation is followed that is known over the Northern Hemisphere as the *Pacific–North American teleconnection pattern*. Major ENSO warm events have been linked to observed decreases in ozone columns, while cold ENSO events have their teleconnection pattern in a global map of ozone departures from the mean that mirrors the pattern produced by the warm events. The changes in ozone column are closely associated with anomalies in the circulation of upper air known as *centres of action*, and can be traced back to a wave-train propagating into the winter hemisphere. At the centres of action of ENSO, total deficiencies of ozone of as much as four per cent can be expected. These expectations seem to be matched by observations of satellite measurements for the 1997–98 event. Components from seasonal, QBO, and solar influences must be filtered out, illustrating the difficulty in pin-pointing any one influence on ozone amounts. The vertical structure of the ENSO effect on ozone has also been determined. For the 1997–98 event, the amplitude of the ENSO signal in ozone was as much as 18–19 per cent at altitudes between 13 and 20 km.

Regionally, then, the ENSO-induced variations in ozone concentrations can be large, but they are located longitudinally in the teleconnection patterns. Zonal averages of ozone changes may thus remain small during ENSO events.

9.5.3 Solar Ultraviolet Irradiance

The solar ultraviolet irradiance determines the overall amount of stratospheric ozone, principally through the rate of odd-oxygen production in photodissociation processes, but also by affecting atmospheric temperature and dynamics. Some time-variations exist in the solar UV irradiance as a result of solar rotation, and variations in solar activity (coupled to the sunspot cycle). Natural changes in ozone concentrations due to solar UV variability must be taken into account in any attempt to isolate long-term anthropogenic effects on the ozone layer.

The Earth's orbit changes over time in ways that influence the amount of energy received at the surface. The *eccentricity* of the Earth's orbit changes with a period of 100 000 years. At the moment the Earth's orbit is fairly circular but in 50 000 years it will be more eccentric with the difference between *aphelion* (farthest) and *perihelion* (nearest) points in the orbit becoming larger. Over time, the day of year when the Earth reaches perihelion also changes. This *precession of the equinox* combined with changing eccentricity results in alterations in the intensity of solar radiation. In addition, there are variations of *obliquity* over a 41 000-year period: the Earth wobbles on its axis of rotation, changing the tilt of the Earth, and hence its seasonality. Changes of solar radiance on these timescales undoubtedly had a major influence on ozone-production rates, and thus on ozone concentrations. However, our measurements of stratospheric ozone concentrations that extend over less than one century are hardly likely to show much sign of the orbital effects.

On a (much) shorter timescale, the most prominent solar UV variations occur with periods of about 11 years for the solar cycle and of 27 days for solar rotation. Over the 11-year cycle, the intensity of Lyman-α ($\lambda = 121.6$ nm) radiation varies by a factor of about two. In the mid-ultraviolet ($\lambda = 200$–300 nm), measured intensities increase by around 10 per cent from the solar minimum to the maximum. There is a close correlation between observed ozone levels and solar ultraviolet intensity, as indicated in Figure 9.15. Globally, ozone-column amounts are expected to vary by 1.2 per cent over the cycle, and this prediction is in good agreement with the observations. According to satellite data, the ozone response depends on latitude and altitude, as shown in Figure 9.16 for the period 1979–2005 by the SAGE (Stratospheric Aerosol and Gas Experiment) I and II studies. The results of the SAGE and other satellite experiments show (i) a statistically significant response (shaded areas) in the upper stratosphere ($\sim$2 to 4 per cent) where solar UV variations directly affect ozone-production rates; (ii) a statistically insignificant response in the tropical middle stratosphere; and (iii) a statistically significant response in the lower stratosphere with amplitude $\sim$1 to 3 per cent.

Despite the excellent accord for the total ozone columns, prediction and measurement of vertical profiles of ozone fail to match. Most simulations for low latitudes using both 2D and 3D models underpredict the changes observed in the upper stratosphere (say at altitudes of 50 km), but overpredict them severely in the mid stratosphere, at altitudes around 30 km.

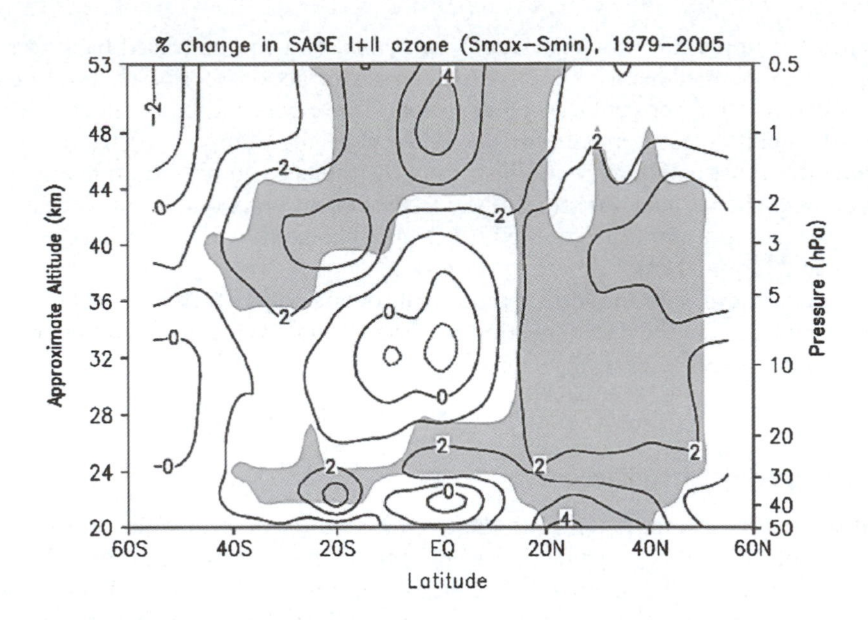

Figure 9.16 Ozone change (per cent) from solar minimum to maximum based on data from the SAGE I and II satellite instruments over the period 1979–2005. Shaded areas are significant at the 95 per cent confidence level. Adapted from L. L. Hood, *Vertical Profile of the Stratospheric Solar Cycle Ozone Variation at Low Latitudes: Observations and Mechanisms*, 14th Conference on Middle Atmosphere, Portland, OR, 20–24 August 2007.

The lower stratospheric ozone response appears to be mainly responsible for the solar-cycle variation of total column ozone at low latitudes. The minimum equatorial response centred on about 30–32 km may be caused, at least in part, by a solar-modulated QBO. One hypothesis that needs further study is that reduced tropical upwelling under solar-maximum conditions increases the descent rate of QBO easterlies, making the east phase dominant. Relative equatorial downwelling near 30 km would increase [NO_x] and decrease the ozone concentration. Increased [NO_x] further decreases [O_3] chemically through its catalytic activity (Section 9.3). As a result, the expected solar-UV induced increases in ozone at solar maximum may be at least partially cancelled. In the upper stratosphere, the response to solar variations occurs in the region where direct photochemical responses of ozone-production rate to solar UV variations are expected theoretically. The behaviour therefore undoubtedly has a direct, dominantly photochemical, origin. The differences between observation and model, if real, are likely to result from solar-induced variations of active ozone-destroying catalysts (Section 9.3). Charged particles (mainly electrons and protons) from the Sun are continually entering the Earth's atmosphere, the protons generally at high latitudes because the moving charges are guided by the Earth's magnetic field lines (see also Section 10.2). Bombardment of the atmosphere by the particles is a source of odd hydrogen and nitrogen. The rate at which the energetic particles arrive is linked to the solar cycle, and the particles are more plentiful as the solar minimum is approached. Increased catalyst concentrations would enhance losses of O_3, so that the effect of the reduced production rates of [O_3] would be amplified. Odd nitrogen (NO_x) is one possible catalyst whose concentration could be altered. However, satellite measurements show no significant evidence for a solar-cycle variation of [NO_x] throughout most of the stratosphere, except at high latitudes. Enhanced catalytic loss of O_3 through the HO_x mechanism remains a possible explanation. Long-term measurements of odd hydrogen near the tropical stratopause are needed to investigate this issue further. For the time being, we note that the response of stratospheric ozone to changes in UV intensity nicely illustrates some points made earlier: that the changes in [O_3] may have a chemical or a dynamical origin, or both; that the dynamics may alter the chemistry; and that the QBO is an important part of the story!

9.5.4 Solar Proton Events

Every few years, a burst of solar activity leads to great enhancements of the particle flux known as a *solar proton event* (SPE). During an SPE, NO_x and HO_x are produced very rapidly, and only above about 60° latitude. Any resulting changes in ozone concentration are relatively easy to identify because of this SPE 'signature', and SPEs thus act as natural 'experiments' to check elements of the stratospheric chemistry problem.

Figure 9.17 shows proton fluxes recorded by a satellite during an SPE of October–November 2003. At their peak around 30 October, the fluxes reached almost five orders of magnitude more than their normal value. Figure 9.18 demonstrates that exceptionally low ozone concentrations were found in high-latitude regions of the Northern Hemisphere after the SPE, while Figure 9.19 compares altitude profiles for O_3 before and after the SPE that show clearly a significant depletion of ozone. As particles propagate down into the atmosphere they lose their energy in collisions with atmospheric gases. With enough energy, the atmospheric molecule is ionized, and an ion–electron pair is created. In addition to the primary protons, the secondary electrons produced in ionization may have enough energy to further ionize and dissociate atmospheric gases. Approximately 36 eV of energy is required in the production of one ion pair, so that a proton with 10 MeV initial energy is able to ionize about 280 000 molecules along its path before all the energy is lost. The atmospheric penetration depth is dependent on the particle energy. Solar protons with an energy of 1–500 MeV deposit their energy in the mesosphere and stratosphere but the most energetic protons, with $E > 1$ GeV, are able to reach ground level, although at these energies galactic cosmic rays generally predominate.

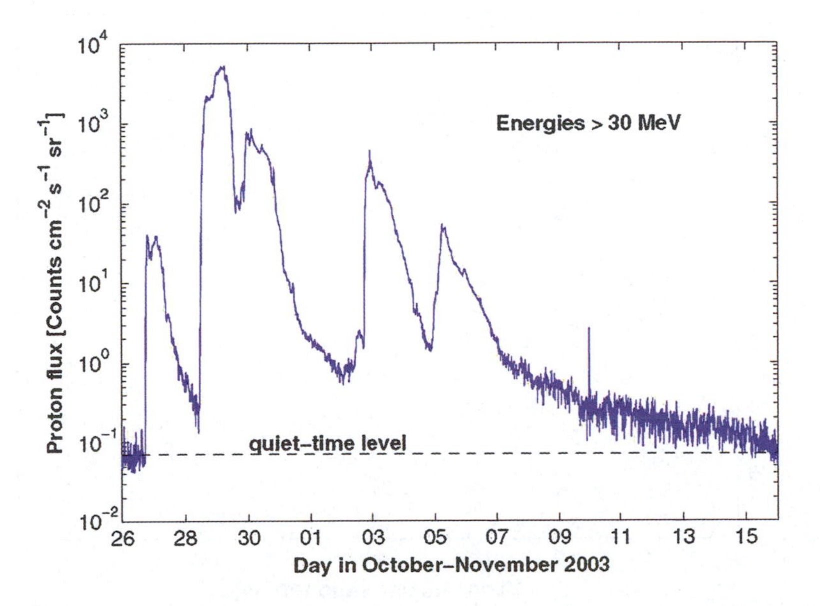

Figure 9.17 High-energy proton flux measurements by GOES satellite in late October 2003, showing the increase by several orders of magnitude. The series of SPEs begins on October 26 and the flux values return to the quiet level in mid-November. From P.T. Verronen, *Ionosphere–Atmosphere Interaction During Solar Proton Events*, Finnish Meteorological Institute Contributions, No. 55 (2006).

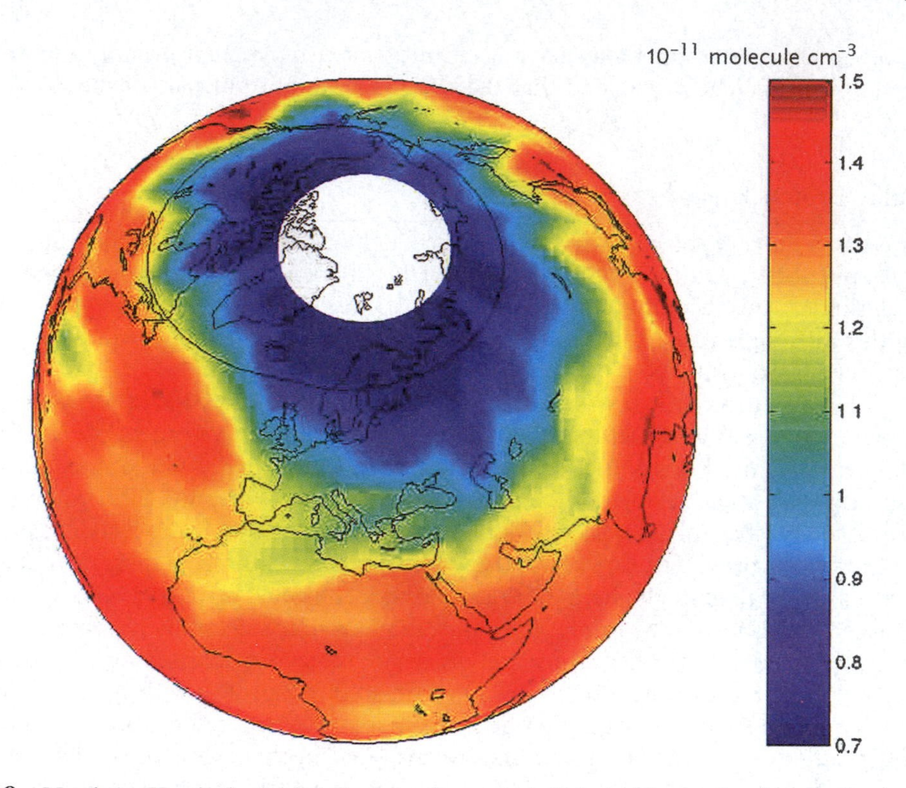

Figure 9.18 Northern Hemispheric distribution of ozone at 46 km altitude showing the low amounts of ozone in the polar region after the SPE of October–November 2003. The map is constructed using GOMOS measurements. Source as for Figure 9.17.

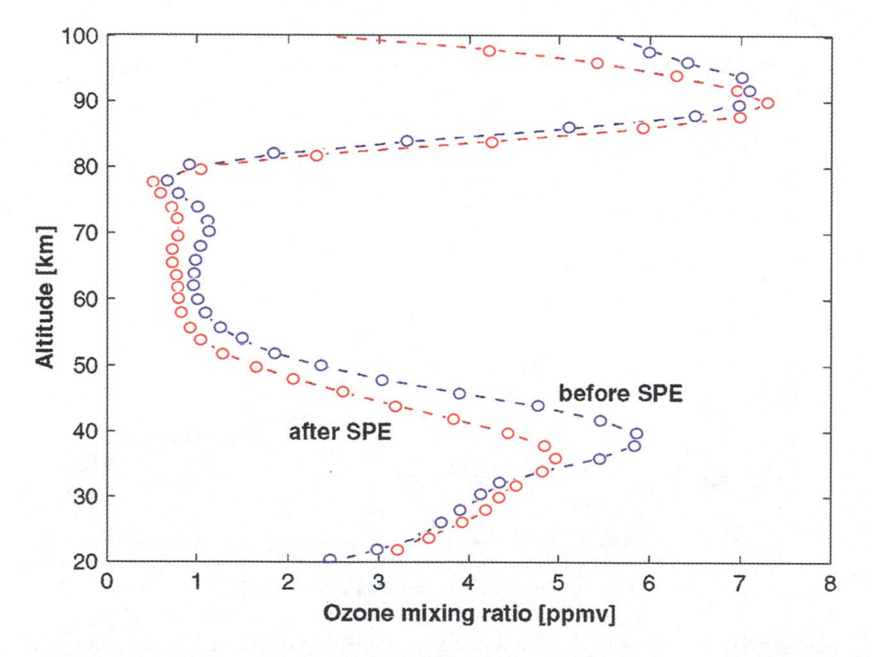

Figure 9.19 Ozone depletion due to the SPE of October–November 2003 as observed by the GOMOS instrument. The profiles are zonal averages for latitudes 70°N–75°N. The blue points are night-time measurements for October 18–24, while the red points are for November 3–6. Source as for Figure 17.

NO_x is produced following dissociation of molecular nitrogen by the primary and secondary solar particles and, to a lesser extent, in ionic chemical reactions following the ion-pair production. Production of HO_x is solely due to ion chemistry, involving a rather complex scheme of water-cluster ion reactions. A depletion of ozone is caused by the increase of NO_x and HO_x, which accelerates the catalytic ozone loss cycles involving these species. The magnitude and duration of depletion depends on the particle flux, altitude, season (solar illumination level and atmospheric dynamics), and the chemical state of the atmosphere. Rocket measurements during an earlier SPE (1989) had shown greatly enhanced NO_x concentrations, and a satellite instrument was able to measure [O_3], [NO] and [NO_2] simultaneously for the first time in an SPE of July 2000. In both cases, the experimental results were essentially consistent with model simulations if due allowance was made for transport effects. Chlorine and HO_x chemistry have to be included in the reaction scheme, and the models showed the importance of including feedback mechanisms involving atmospheric temperature (see the last paragraph of this section).

The short-term depletion of O_3 due to the increase in HO_x lasts for some hours and can be greater than 90 per cent in the middle mesosphere, while the long-term decrease, several tens of per cent, is typically seen in the upper stratosphere and is due to the increase in NO_x. Because of the long chemical lifetime of NO_x, the effects on ozone can last for months or years, and the NO_x generated can be transported from the location of the proton precipitation, so that lower altitudes and latitudes may also be affected. The effect on global, total ozone is probably moderate, of the order of few per cent at the maximum.

Depletion of ozone will lead to a decrease in stratospheric temperature, since solar heating by ozone is a major contributor to the radiative balance. Lower temperatures reduce the rate of odd-oxygen destruction, and there is thus a tendency for ozone depletion to be damped out. That is, there is a negative-feedback mechanism that opposes changes in ozone concentrations. On the other hand, heating caused by collisions between electrons, ions and neutral molecules moving at different velocities during proton precipitation could increase the temperature in the mesosphere temporarily. These opposing cooling and heating effects have been estimated to change the temperature by about -3 K and $+10$ K, respectively, and could play a role in SPE-related wind changes.

9.5.5 Volcanoes

Volcanic eruptions are powerful events, and they are capable of injecting H_2O, HCl and sulfur compounds into the atmosphere. However, only the most powerful eruptions are able to inject much material into the stratosphere. During the latter part of the 20th century, the explosive eruptions of El Chichón (1982), and especially Mount Pinatubo (1991) were of this kind. Following the eruption of Mount Pinatubo, low values of ozone were observed in the lower stratosphere, particularly in winter and spring. Reductions in the total ozone column ranged from 2 per cent in the tropics to 7 per cent in mid-latitudes. Ozone depletion in the aerosol cloud was higher, reaching around 20 per cent. These depletions are unlikely to have their origin in chlorine compounds injected from the volcano. Enhancements in chlorine compounds after the El Chichón eruption did occur (discussed towards the end of this section), but the eruption of Mount Pinatubo apparently made a negligible direct contribution to stratospheric chlorine loading. It is evident that the true cause of the ozone depletions involves not increased concentrations of catalysts, but rather the aerosols produced by the eruptions.

Stratospheric aerosols of natural and artificial origin can affect the chemistry and climatology of the Earth. Sulfur compounds (SO_2, OCS, CS_2) from the volcano are likely to increase the loading of stratospheric aerosol. As we have seen (Sections 9.3.4 and 9.4.5), aerosol can have a direct chemical effect on stratospheric chemistry and lead to ozone depletion. In addition, the aerosols alter, by scattering and absorption, the ultraviolet flux available for odd-oxygen production as well as changing stratospheric temperatures and dynamics.

The eruption of Mount Pinatubo produced the largest loading of stratospheric sulfate aerosol in the 20th century. It injected up to 20 Mtonne of SO_2 into the lower stratosphere, and the SO_2 cloud encircled the Earth in tropical regions in 22 days. Within two months, virtually all the SO_2 had been converted into liquid H_2SO_4 aerosol, and the resultant aerosol surface area reached 100 times that of the background aerosol present before the eruption, with consequential changes in stratospheric circulation and heterogeneous chemistry. The surface reactions locked NO_x into inactive HNO_3, thus liberating active chlorine from its reservoirs (see later and Section 9.4.5).

Temperature increases in the lower and middle stratosphere were significant, with peak increases at an altitude of 23 km of around 3.5 K. This *positive anomaly* gradually decreased over the course of 1992 as the tropical aerosol was dispersed. The heating of the stratosphere also caused increased tropical up-welling, raising air from regions with low ozone, reducing the ozone column. The associated subsidence at high latitudes brought ozone down, potentially masking some of the destruction by chemical processes. The ozone column in the Pinatubo period was also influenced by the phase of the QBO and the phase of the ENSO (a recurrent theme: see the end of Section 9.5.3, for example). The heating of the lower stratosphere by the aerosols produced by Pinatubo affected the QBO, locking the phase of the QBO for several months. The years 1991 and 1992 had pronounced El Niño conditions, which makes deconvolution of the chemical and dynamical contributions to the ozone depletion yet more difficult. Figure 9.20 shows the results of an attempt to separate the contributions for the latitude range 60°N–60°S. Note the large depletion due to enhanced heterogeneous chemistry (blue line) that persists for several years, and the distortion of the mid-1989–mid-1991 dynamical cycle (black line) after the eruption so that the next peak occurs in mid-1994 rather than mid-1993 as expected.

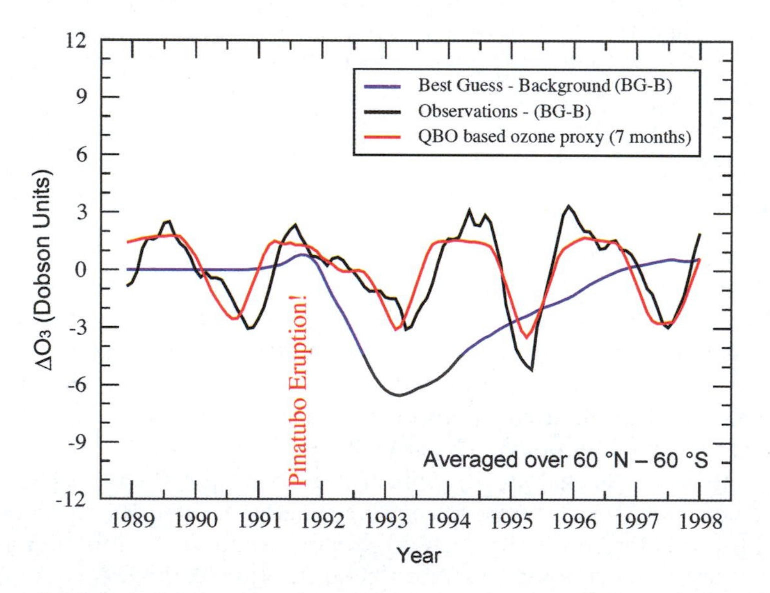

Figure 9.20 Contribution to global ozone variability from chemical (ΔO_3^{chem}, blue line) and from dynamical effects (ΔO_3^{dyn}, black line). The chemical effects are evaluated from the differences between the best-guess run (BG) and that with background surface-aerosol density (B). The dynamical effects are evaluated by subtracting this difference (BG–B) from the observations. A regression of ΔO_3^{dyn} to the QBO proxy, with a lag of 7 months, is superimposed for comparison (red line). From P. Telford, P. Braesicke, O. Morgenstern and J. Pyle, *Atmos. Chem. Phys. Discuss.*, 2009, **9**, 5423.

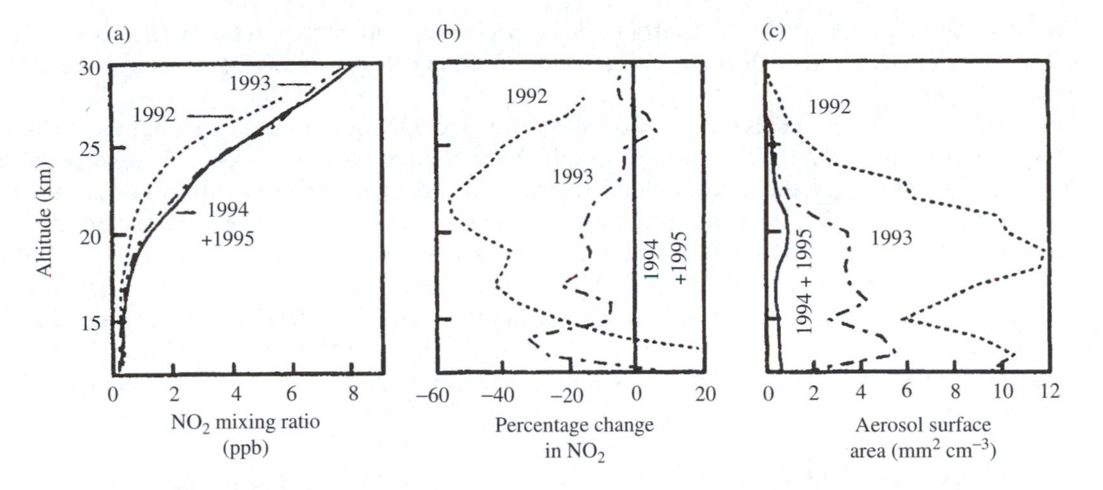

Figure 9.21 NO_2 and volcanic aerosol: observations made over southern France in June of 1992–95. (a) Sonde profiles of NO_2; (b) percentage reductions in NO_2; (c) aerosol surface area. From G. Mégie *et al., European Research in the Stratosphere*, EC Publications, Luxembourg, 1997.

In terms of chemistry, the eruption rapidly led to a strong modification of the partitioning between nitrogen compounds in the stratosphere. In both the Arctic and at middle latitudes, the total NO_2 column was reduced by up to 35 per cent, and full recovery was not seen until at least 1995 (Figure 9.20). Figure 9.21 shows the results of some profile measurements: reductions of NO_2 concentrations by as much as 60 per cent were found at altitudes of 22 km one year after the eruption, and recovery took two to three years. The anticorrelation between NO_2 reduction and aerosol surface area is most striking. Entirely similar behaviour was observed for NO, and the stratosphere was said to have been 'denoxified' as a consequence of the eruption. Concentrations of some other compounds *increased* after the eruption, with both local concentrations and total column amounts of HNO_3 showing an enhancement. The obvious interpretation of the observations taken together is that reactions (9.26) and (9.27) had converted much of the NO_x into N_2O_5, and that the heterogeneous reaction

$$N_2O_5 + H_2O \rightarrow 2HNO_3 \tag{9.28}$$

was acting as a route for the conversion of NO_x to HNO_3.

It will be recalled that the reservoir $ClONO_2$ is formed from ClO and NO_2 in reaction (9.40). Strong denoxification as a consequence of the volcanic aerosol loading thus also has a major impact in the partitioning between chlorine-containing species, because the latter are denied access to one of their most important reservoirs. Furthermore, there is substantial evidence that $ClONO_2$ is hydrolysed on volcanic aerosol surfaces in reaction (9.40). Observations of the amount of active chlorine showed the expected enhancement after the eruption, thus confirming the influence of the heterogeneous reactions. The oxide OClO was observed in 1992 in larger quantities than in previous years, again suggesting the activation of chlorine on sulfate aerosols.

Since reservoirs for HO_x also depend in part on the availability of NO_2 (see, for example, reactions (9.29) and (9.33)) they, too, are influenced by the reductions of NO_x following the eruption, with the HO_x radical species showing a general increase. Figure 9.2 indicates that reactions involving HO_x are responsible for over half the photochemical destruction of ozone in the lower stratosphere, with halogen chemistry accounting for a further third. Although NO_x catalytic cycles themselves are relatively unimportant, NO_x species are critical in regulating the other cycles. The effect of increased aerosol loading is thus to enhance the halogen and HO_x cycles at the expense of the NO_x cycles.

One of the curious anomalies highlighted by the eruption of Mount Pinatubo is that ozone losses were apparently smaller in the Southern Hemisphere than in the Northern Hemisphere, despite the more rapid movement of the aerosol cloud to the south. A possible explanation centres on the different altitudes to which the cloud penetrated in the two hemispheres. The peak concentrations of aerosol in the south were found at or above 22 km, where NO_x catalytic cycles are important in ozone loss (see Figure 9.2). Reductions in NO_x would thus be expected to *increase* ozone concentrations so far as these cycles are concerned, and the overall effect for the higher altitude injections might be a compensation between enhanced halogen and HO_x losses and decreased NO_x losses. In the north, however, the injections were lower, where NO_x cycles are relatively much less important. The compensatory mechanism would not apply; substantial ozone depletions would be expected and are, indeed, observed.

The strong evidence linking heterogeneous chemistry with ozone losses seen after the eruption of Mount Pinatubo suggests that smaller, but similar, depletions seen after the eruption of El Chichón should be attributed to the same cause. The loading and nature of the stratospheric injection from El Chichón were somewhat different from those of Mount Pinatubo. So far as homogeneous chemistry is concerned, the HCl injected might also be expected to provide enough chlorine to add to the background depletion of ozone in the ClO_x cycle. In the six months following the El Chichón eruptions, column HCl under the cloud of volcanic debris was measured to be enhanced by 40 per cent. Ozone depletions of as much as 10 per cent were found at altitudes between 18 and 25 km that could be attributed, through their timing and properties, to the El Chichón event. However, the period involved in 1982–1983 was also a period of tropospheric climatic anomalies, making it difficult to ascribe the stratospheric changes to the volcanic injections alone.

Before leaving the topic of the effects of volcanoes on the stratosphere, it is worth restating explicitly two features that have emerged in our discussion. The first is that the anomalous chemistry that occurred after the eruption of Mount Pinatubo, and probably after that of El Chichón as well, involves the processing of halogen-containing species whose origins lie largely (probably > 83 per cent) with Man. Thus, although the volcanic eruption is itself a natural event, its impact on stratospheric ozone is a consequence of human activities. The second feature is that stratospheric warming due to the injected aerosol could produce climatic changes sufficient to have an indirect effect on ozone concentrations. The 25th anniversary of the El Chichón eruption was an occasion for reanalysis of the available data to try to find a general link between volcanic activity and climate. It has been noted that during the winters following the three biggest eruptions in the last decades (Agung, El Chichón and Mount Pinatubo) El Niños took place. It is currently uncertain to what degree ENSO influences the atmospheric response to volcanic forcing, or in what way a volcanic eruption influences the strength and the timing of an El Niño event. Current models suggest that only eruptions larger than that of Mount Pinatubo can shift the likelihood and amplitude of an El Niño event above the level of the model's internal variability. This conclusion reconciles, on one hand, the demonstration of a relationship between explosive volcanism and El Niño; and, on the other hand, the ability to predict El Niño events of the last 148 years without knowledge of volcanic forcing. The strongest eruption of the previous millennium (1258 AD) is likely to have triggered a moderate-to-strong El Niño event in the midst of prevailing *La Niña* (the cold phase of El Niño) conditions induced by increased solar activity during the *medieval climate anomaly*.

Volcanic eruptions in the future may offer scope for investigating the response of the stratosphere to sudden disturbances that will inevitably occur, and those of the past as well as the future may complicate predictions of how the stratosphere might respond to the decrease (or increase) of Man's own contribution to the chemical inventory of the stratosphere.

9.6 VARIATIONS AND TRENDS IN STRATOSPHERIC OZONE

Beyond the intrinsic interest of the phenomena described in Section 9.5, one of the main objectives in carrying out the many investigations is to see if, having assessed the contributions to changes in stratospheric ozone resulting from solar variations, volcanoes, SPEs, QBOs, ENSOs, and all the other acronym-bearing phenomena, there remain any short-term variations or long-term trends *not* accounted for by all known natural causes. That is, can Man's hand be seen in the ozone record? We have already made enough statements in this book for the answer to this question to be obvious, but it is instructive now to carry out the exercise in the way just formulated.

Look again at Figure 9.15, which served to introduce Section 9.5. The middle panel of the figure showed global ozone concentrations over more than four decades in the form of deviations from the average of the first 15 years of the period. Above and below the ozone record are panels showing a measure of the incoming solar radiation and a measure of the volcanic dust in the atmosphere. Even to the eye unaided by mathematical trickery, a response in ozone concentration to fluctuations in solar intensity and to certain volcanic eruptions can be picked out. Now consider Figure 9.22, where the studies summarized in Section 9.5 have been used to deconvolute the contributions of each of the phenomena we have described. The top left-hand panel here (black trace) provides the absolute ozone column densities in Dobson units from which the middle panel of Figure 9.15 was calculated. Below that, the blue trace shows how the annual cycle affects ozone. In the right-hand panel, we see the contributions to ozone change from the solar cycle (red), the QBO (cyan) and from volcanoes (green). Subtract these all from the top trace, and something significant is indeed left behind: the yellow line of the final trace. This yellow line must therefore be Man's contribution to ozone changes, and it is evident that the change reflects a long-term trend, and not just a quick variation. The curve has been labelled *EESC*, which is an acronym for *equivalent effective stratospheric chlorine*: in other words, the amount of halogen-containing material present in the stratosphere scaled for the ozone-depleting power of each individual compound. The existence of this EESC-related trend is the reason why Sections 11.2 and 11.3 of this book had to be written!

Elaborate analyses of the available data have been undertaken several times by the NASA/WMO International Ozone Trends Panel. The first report (1988) exemplifies the methods used for the analysis. Satellite data and ground-based measurements were used to provide cross-checks. The conclusions of the report published in 2007 were that global-mean total column-ozone values for

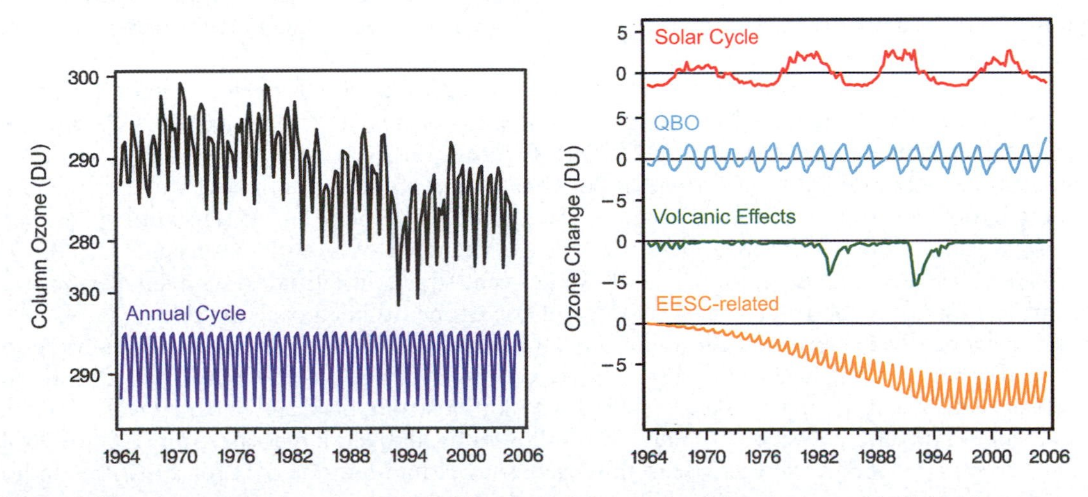

Figure 9.22 Ozone variations for 60°S–60°N estimated from ground-based data and individual components that comprise ozone variations. Source as for Figure 9.6.

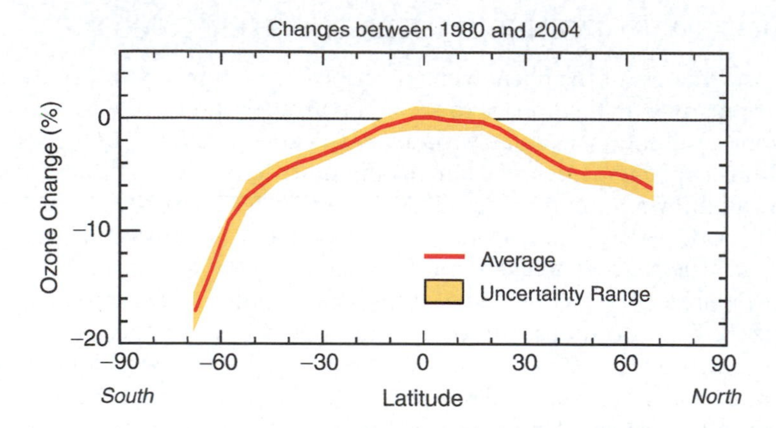

Figure 9.23 Ozone changes between 1980 and 2004 for different latitudes. The largest decreases occur at the highest latitudes in both hemispheres because of the seasonal depletion in polar regions. The losses in the Southern Hemisphere are greater than those in the Northern Hemisphere because Antarctic depletion is much larger than the Arctic losses. Long-term changes in the tropics are much smaller because of the relatively small abundance of reactive halogen gases in the tropical lower stratosphere. Source as for Figure 9.6.

2002–2005 were approximately 3.5 per cent below the 1964–1980 average, but were similar to the 1998–2001 values, suggesting that ozone is no longer decreasing. Ozone depletion shows a strong latitudinal dependence, and there are marked differences between the two hemispheres, as indicated in Figure 9.23 (the data have been smoothed for clarity). The largest decreases have occurred at the highest latitudes in both hemispheres because of the large winter/spring depletion in polar regions. Averaged for the period 2002–2005, losses in the Southern Hemisphere (SH: 5.5 per cent) are greater than those in the Northern Hemisphere (NH: 3 per cent) because of the Antarctic ozone hole. Long-term changes in the tropics are much smaller because reactive halogen gases are less abundant in the tropical lower stratosphere. The behaviour of column ozone in the two hemispheres during the 1990s was different. Ozone in the NH reached a minimum around 1993, followed by an increase. In the SH the decrease continued through the late 1990s, followed by the recent levelling off. The discussion of trends detected at middle and relatively high latitudes must not, of course, allow the reader to forget the large declines in ozone seen in individual months over polar regions. Taken together with the trends at mid-latitudes, it is evident that Earth's ozone shield has taken a considerable blow.

Once a change in total ozone has been found, it is important to discover at what altitude the change has occurred. Figure 9.24 gives an overview of the results from one series of satellites (SAGE I and II, as described in connection with Figure 9.16). However, since the contours show differences between 2005 and 1979, it is not possible to discern from this figure an interesting, and possibly important, result. That is, the largest changes occurred before 1995. Up until 1995, there were significant decreases in upper stratospheric O_3 averaged over 60°N–60°S and altitudes of 35–50 km. The net ozone decrease was ~10–15 per cent over mid-latitudes, with smaller but significant changes over the tropics. Lower stratospheric ozone declined over the period 1979–1995, but has been relatively constant with significant variability over the last decade. At mid-latitudes of both hemispheres, ozone declined by up to 10 per cent by 1995 between 20 and 25 km altitude. In the lowermost NH stratosphere, between 12 and 15 km, a strong decrease in ozone was observed from ozonesonde data between 1979 and 1995, followed by an overall increase from 1996 to 2004. The net result is no long-term decrease at this level. The SH mid-latitude data for 20–25 km do not show a similar compensating increase. Changes in the lowermost stratosphere have a substantial influence on the total ozone column because of the high proportion of the ozone present in this

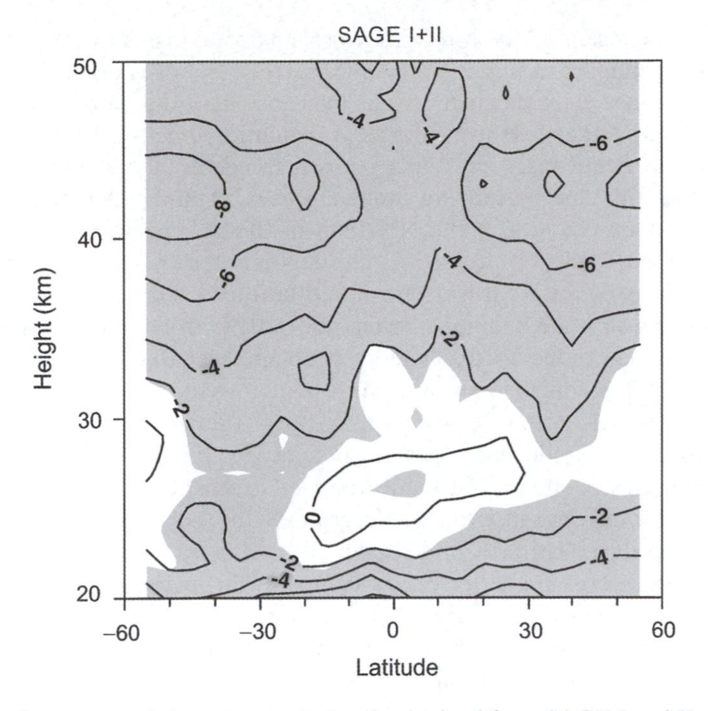

Figure 9.24 Annual ozone trends in per cent per decade obtained from SAGE I and II satellite data, plotted as a function of latitude and altitude for the period 1979–2004. The trends were estimated using regression to an EESC curve (see text) and converted to per cent per decade using the variation of EESC with time in the 1980s. Shadings indicate that the changes are statistically significant at the 2σ level. Source as for Figure 9.6.

region. Figure 9.24 suggests significant ozone decreases between the tropopause and 25 km during 1979–2004 even at 25°S–25°N. Since no change is found in total ozone over the tropics, this decrease in the lower stratosphere may be offset by significant increases in tropospheric ozone in the tropics.

We now turn to the consensus explanation of the quite complex trends just discussed. Halogen increases have been the principal driver of ozone depletion over the past few decades. The link is supported by the statistical fit of globally averaged measurements of ozone decay with equivalent effective stratospheric chlorine (EESC). There is good overall agreement between observed long-term changes in extrapolar ozone and model simulations that include the effects of increasing halogens. Models generally reproduce the observed ozone changes as a function of altitude, latitude, and season, confirming our understanding that halogen changes are the main driver of global ozone changes. Empirical and model studies have shown that changes in tropospheric and stratospheric dynamics have been partially responsible for the observed NH mid-latitude winter ozone decline from 1979 to the mid-1990s and the increase thereafter. Whether this behaviour is due to dynamical variability or results from a long-term trend in stratospheric circulation is not yet clear. The dynamical and chemical impact of aerosols on mid-latitude ozone was greatest in the early 1990s after the eruption of Mount Pinatubo in 1991. The observed decrease in NH column ozone in 1993 agrees with chemical–dynamical models that include these effects. An outstanding issue is the absence of an influence of the Mount Pinatubo eruption on the observed SH ozone column, which contrasts with model predictions.

Several independent modelling studies covering periods in the 1990s confirm that dilution of ozone-depleted polar air makes a substantial contribution to mid-latitude ozone depletion, especially in the SH as a result of the much larger polar ozone losses in the Antarctic than in the Arctic. The vertical dependence of upper stratospheric ozone trends is generally consistent with our understanding of gas-phase chlorine chemistry as the cause, modulated by changes in temperature and other gases such as methane. However, global dynamical–chemical models have not demonstrated that they can simultaneously reproduce realistic trends in all relevant parameters, although observations over the full time period are limited. Both 2D and 3D models perform better in reproducing observed past changes in the NH than in the SH, perhaps because of the stronger influence of polar-ozone depletion in the Antarctic than the Arctic. Polar processing in the subvortex region can directly lead to low ozone at mid-latitudes. Break-up of the polar vortices can lead to a dilution of mid-latitude ozone by ozone-poor air. Unfortunately, the models do not yet accurately simulate filamentation, stratosphere–troposphere exchange, or the break-up of the polar vortices, and there remains controversy over the extent to which the vortices behave as strictly isolated containment vessels or as 'flowing processors' (Section 9.4.2). Further, it has been suggested that cirrus clouds could heterogeneously activate chlorine compounds in stratospheric air, and thus contribute to ozone loss in the lowermost stratosphere. However, the role of cirrus remains unclear, and certainly not quantified. Nevertheless, at the *qualitative* level, the variations of ozone depletion with altitude, latitude, and season all hold clues to the causes of changing ozone losses. The space–time patterns are entirely consistent with the increasing ozone losses being linked through chemistry to increases in chlorine and bromine loading at mid-latitudes, and not to natural changes such as those in solar irradiance or dynamics. This conclusion is reinforced by the impact of the eruption of Mount Pinatubo. The short-term rapid decreases in ozone abundances over a few years can be connected demonstrably to increased halogen loading. It seems fitting to conclude this chapter with the observation that one of the best indicators of Man's perturbation of the stratosphere was provided by one of nature's most dramatic exhibitions of her power.

CHAPTER 10

Airglow, Aurora and Ions

10.1 BEYOND THE STRATOSPHERE

As we progress higher in the atmosphere above the ozone layer, the ozone that is responsible for heating the stratosphere becomes too dilute to win the battle against the cooling effects of greater altitudes. We have reached the *stratopause*, and the atmospheric region that lies above it is the *mesosphere* in which temperatures decrease with increasing altitude. Figure 1.1 shows the turning-point of temperature as lying at roughly 55 km, and the mesosphere as extending to somewhat above 80 km, when atmospheric temperatures begin to increase yet again in the *thermosphere*. The *mesopause*, envisaged as dividing the upper and lower regions, is where the lowest temperatures in the entire atmosphere are encountered. The mesopause of Figure 1.1 is intended to possess a 'standard atmosphere' temperature of 180 K, but in polar regions temperatures can be as low as 110 K, during the summer (a surprise, perhaps). Further examination of Figure 1.1 shows that there are some interesting phenomena to be discovered at these altitudes, and it is the purpose of this chapter to discuss some of them briefly in order to round off our examination of chemistry typical of the atmospheric regions.

There are some general characteristics of the chemistry that it is useful to bear in mind when reading the rest of this chapter, and they all are ultimately a result of the low pressures (densities) of the local air and that which lies above, the low temperatures, and the distance from the Earth's surface.

- Only rather simple chemical species persist to these altitudes. Most of the reactive compounds originating from geological emissions, the natural biosphere and Man have been oxidized in the stratosphere or below. What remains is principally the permanent gases N_2, O_2, CO_2 and H_2O, and atoms (H, N, O) and radicals (OH, NO) derived from them, along, as ever, with a little O_3. In addition, as some types of meteor enter the atmosphere and burn up (*ablate*), they deposit sodium and other metals into the mesosphere.
- Because the atoms and molecules lying yet higher are relatively sparse, shorter wavelengths of solar radiation reach the thermosphere and mesosphere than are encountered lower down. Similarly, energetic particles from the Sun and elsewhere in the universe are substantially more abundant than they are even in the stratosphere.
- The low densities mean that collisions between potential reaction partners are infrequent. Figures 1.1 and 2.1 show that at 80 km, for example, the mean free path is of the order of centimetres, whereas at ground level it was 10^5 times less. As a consequence, rates of many

Atmospheric Chemistry
By Ann M. Holloway and Richard P. Wayne

Published by the Royal Society of Chemistry, www.rsc.org

chemical reactions become very slow. But there are exceptions. Some reactions (such as three-body combinations) are favoured by low temperatures, while others (such as those involving ions) are accelerated by long-range attractive forces.

These features taken together mean that, although only chemically simple species are involved, the higher energies can *excite* these species from the *ground state* in which they are normally found, and can even *ionize* them. The infrequency of collisions also allows the excited or ionized atoms or molecules to live for longer than they would were they formed lower down in the atmosphere. Excitation in most cases means a rearrangement of the electrons of the species, and the formation of ions means the complete removal of one or more electrons (positive ions) or the addition of new electrons (negative ions).

One of the possible fates of an electronically excited species is that it can emit radiation, often in the 'optical' spectral region (visible, ultraviolet and near-infrared), but the operation of this pathway depends on it competing effectively with *deactivation* (*quenching*), which is why low pressures are so important. So far as ions are concerned, there are types of chemical reaction open to them that just do not exist for neutral atoms or molecules. Thus, although the upper atmosphere has fewer types of 'chemicals' in it than, say, the troposphere, there are potentially many excited states and ionic reactions that can bring a great richness to the 'chemistry'. It is this chemistry and the atmospheric phenomena that accompany it that we now propose to survey.

10.2 AIRGLOW

10.2.1 Optical Emission from Planetary Atmospheres

Sources located outside our atmosphere and within it illuminate the night sky. The Moon, stars, and surface lights all contribute, but if the light from these sources were eliminated, the sky would not be completely black. A faint glow would remain that has its origins in atmospheric photochemical processes, and to which the name *airglow* is given. Astronauts see this airglow as an envelope that sheaths the night side of the Earth, but Earth-bound observers are also able to perceive it. The visible radiation is rather feeble, and has been equated in intensity to the light of a candle at a distance of 100 m. In fact, the night sky would appear far brighter were it not that the strongest airglow emission features lie in the near-infrared region, just beyond the response of the human eye. By day, the airglow intensities are orders of magnitude larger, but are not detectable by eye because they are completely dominated by atmospherically scattered sunlight. According to the time of day at which observations are made, the airglow is described as *nightglow* or *dayglow*. When the Sun is below the horizon at ground level, but in view from the upper atmosphere, the emission is termed the *twilight glow*. Several of the lines and bands are identical to those seen in the *aurora* (Section 10.3.1), so that we should now distinguish between the two phenomena of airglow and aurora. Airglow occurs continuously, is weak, and is observed at all latitudes. In contrast, auroras are much more intense, but irregular in form and occurrence, and are restricted to a region near the geographical poles. The differences arise from the excitation mechanisms. Airglow is driven by photons from the Sun, whereas auroras are excited by the impact of energetic solar particles.

Airglow emissions are a feature of most planetary atmospheres. The light consists of atomic and molecular line, band, and continuum systems of atmospheric constituents, both neutral and ionized. Because many of the emission features are measurable at the Earth's surface, airglow investigations provided a valuable source of information about the composition of our own as well as other planetary atmospheres long before the era of direct investigations by rocket probes and satellites. Other emission features are absorbed by the terrestrial atmosphere, and thus require instrumentation on space vehicles for their study. Airglow investigations continue to assist interpretations of the photochemistry and dynamics of planetary atmospheres.

Most of the spectral features of the airglow have their origins in transitions from *electronically excited* species (although very important components of the Earth's airglow are due to *vibrationally* excited OH,[i] as described in Section 10.2.6). For many of the atoms and molecules of the atmosphere, the first allowed optical transition to and from the ground state (the *resonance* transition) lies in the 'vacuum' ultraviolet, and so by definition does not penetrate the Earth's atmosphere. Resonance emissions from helium ($\lambda = 58.4$ nm), nitrogen (120.0 nm), hydrogen (121.6 nm), and oxygen (130.2, 130.4, 130.6 nm) that might otherwise be expected to be strong, are not observed at the ground, although they are major features of the Earth's airglow as seen from space. Satellite measurements have an important part to play in airglow studies. For example, the Upper Atmosphere Research Satellite (UARS, 1991–2005) carried two very interesting experiments designed to examine different aspects of airglow phenomena. These experiments were the *Wind Imaging Interferometer* (WINDII) and the *High-Resolution Doppler Imager* (HRDI): in general, the Doppler shifts of the airglow lines allow information about winds and other motions to be derived, while the line shapes and strengths yield information about atmospheric composition and temperature.

Emissions at wavelengths longer than the atmospheric cut-off may sometimes be attenuated or removed by specific absorption processes. For example, the strongest feature of the terrestrial dayglow is due to a transition from the first excited state of O_2, the $^1\Delta_g$, to the ground state, $^3\Sigma_g^-$, and is known as the *Infrared Atmospheric Band*. The (0,0) band[ii] of this system is strong and lies at $\lambda = 1270$ nm, just into the infrared region, and in a part of the spectrum generally free of atmospheric absorptions. But O_2 in the atmosphere below the regions of emission obviously absorbs exactly this wavelength, since it is almost all in the $\upsilon'' = 0$ vibrational level. Only experiments carried on aircraft, balloons or satellites can detect this band. However, almost no atmospheric O_2 is populated in the $\upsilon'' = 1$ level, so that the (0,1) band at $\lambda = 1580$ nm can be detected at the ground, although it is much weaker than the (0,0) band in the absence of interfering absorption.

10.2.2 Excitation Mechanisms

During the day, sunlight can be absorbed directly by atmospheric constituents, to yield electronically excited products. Radiative decay then contributes to the airglow, and the absorption–radiation sequence is *resonance fluorescence* or *resonance scattering*. Atoms such as H have particular roles to play in resonance scattering, since the hydrogen resonance line is a major component of the solar spectrum and there is thus a match that efficiently couples energy transfer from Sun to Earth. The upper and lower electronic levels involved are the ^{2}P and ^{2}S states of H, so that the excitation–emission sequence may be written

$$H(^2S) + h\nu(\text{solar}) \rightarrow H(^2P) \qquad (10.1)$$

$$H(^2P) \rightarrow H(^2S) + h\nu(\text{airglow}), \qquad (10.2)$$

with $h\nu$ in both cases corresponding to $\lambda = 121.59$ nm.

[i] Electronic states of atoms and molecules are identified by *term symbols*, a combination of numbers, letters and sometimes + and − signs. This book is decidedly not the place to explain term symbols in any detail whatever, and they should be regarded by readers who are not familiar with them merely as convenient, if somewhat bizarre, labels. Typical examples are P, D and S for the ground and first two excited states of atomic O, and Σ_g, Δ_g and Σ_g for the ground and first two excited states of molecular O_2. A little more explanation is given in footnote 1 to section 5.1.

[ii] The nomenclature is that a vibrational band in an electronic transition is written in the form (υ',υ''), υ' being the vibrational quantum number in the upper electronic state and υ'' in the lower. The upper state is always referenced first, regardless of whether the transition is upwards (absorption) or downwards (emission). A (0,0) transition thus means a transition to or from the level 0 in the upper state from or to level 0 in the lower state, while a (0,1) transition to or from level 1 in the lower state.

Photodissociation of molecules quite often results in the formation of electronically excited fragments. One of the most important excitation reactions of this kind involves ozone photolysis

$$O_3 + h\nu(\lambda < 310\,\text{nm}) \rightarrow O(^1D) + O_2(^1\Delta_g). \tag{10.3}$$

At $\lambda < 310$ nm *both* the atomic O *and* the molecular O_2 fragments are excited (see Section 5.1), and contribute to the oxygen airglow, as will be discussed in Section 10.2.4.

At night, the source of photochemical excitation is removed, yet the existence of the nightglow shows that excited species persist. In general, this result must mean that solar energy has been stored during the day, and that reactions are releasing the energy at night. Neutral atoms (especially oxygen) are a very important energy reservoir at altitudes below ~100 km, while ions are very important at higher levels. We shall see in Section 10.2.4 how several atomic and molecular transitions in the nightglow owe their existence to atomic oxygen. Reaction exothermicity is used to populate preferentially certain states of the species, which are thus out of thermal equilibrium with their surroundings. Emission of *chemiluminescence* then contributes to the nightglow. A typical example from the laboratory is the reaction

$$NO + O_3 \rightarrow NO_2^* + O_2, \tag{10.4}$$

where NO_2^* represents a molecule of NO_2 possessing both electronic and vibrational excitation. Emission is seen from NO_2^* populated virtually up to the exothermicity (205 kJ mol^{-1}) of the reaction. For many years, it was not clear if reaction (10.4) is of significance in the airglow, but an airglow layer in the stratosphere (altitude 40–60 km) thought to arise from the process has now been observed by the WINDII instrument on UARS.

Positive ions are neutralized by electrons in the process of *dissociative recombination* (see Section 10.4). The molecular ionization energy is liberated in the fragments, and electronic excitation may result. The first electronically excited state of N, the ^2D, is probably populated in the ionosphere by dissociative recombination of NO^+

$$NO^+ + e \rightarrow N(^2D) + O. \tag{10.5}$$

Nightglow excitation mechanisms can obviously operate during the day, but they are *usually* dominated while the Sun is present by processes more directly utilizing solar energy. Some exceptions do exist, a notable one being the chemical excitation of *vibrationally* excited OH radicals described in Section 10.2.6.

10.2.3 Airglow Intensities and Altitude Profiles

Intensities of airglow features are related to atmospheric concentrations of excited species A* through the rate law

$$I = k_r[A^*], \tag{10.6}$$

where I here is a total emission rate in quanta (or photons) per unit volume per unit time.[iii] Radiative decay and non-radiative loss processes, of which physical quenching is often the most

[iii] The rate coefficient, k_r, is equivalent to the transition probability, or Einstein *'A' coefficient for spontaneous emission*; a *radiative lifetime*, τ, is often defined by the relation $\tau = \ln 2/k_r$. To a first approximation, k_r is independent of concentration or pressure (although for certain forbidden transitions, collision-induced processes do endow k_r with some pressure dependence).

important, compete with excitation. If we write the quencher species as M and the rate coefficient as k_q, then the generalized excitation–de-excitation mechanism becomes

$$\text{Source} \rightarrow \text{A}^* \qquad \text{Excitation;} \quad \text{Rate} = P \tag{10.7}$$

$$\text{A}^* \rightarrow \text{A} + h\nu \qquad \text{Emission;} \quad \text{Rate} = k_r[\text{A}^*] \tag{10.8}$$

$$\text{A}^* + \text{M} \rightarrow \text{A} + \text{M} \qquad \text{Quenching;} \quad \text{Rate} = k_q[\text{A}^*]\,[\text{M}]. \tag{10.9}$$

Production and loss of A* must ultimately balance, and so long as the excitation rate, P, does not change rapidly, then a *steady state* for [A*] may be set up, and it may readily be shown that

$$I = k_r[\text{A}^*] = P(1 + k_q[\text{M}]/k_r)^{-1}. \tag{10.10}$$

In circumstances where P changes rapidly with time, the steady state may not be maintained. Such a situation may arise at twilight or dawn, or during an eclipse, if the excitation process is dependent on the presence of sunlight. For the extreme case, where P goes to zero instantaneously, airglow intensity will decay with a first-order rate determined by the composite coefficient ($k_r + k_q$ [M]). More realistic cases can be treated numerically, and allowance made for physical transport. Observations of the time dependence of the airglow intensity during periods of change can clearly provide information about [M], and hence emission altitude, if k_q and k_r are known, or conversely about the rate constants if the emission altitude can be estimated.

Radiation is an isotropic process, so that each volume element of the airglow emits equally in all directions. A human observer or an instrument does not integrate all this radiation, but rather perceives a 'brightness' that depends on the flux of photons per unit area per unit time. For this reason, airglow brightnesses are usually measured in units of the *Rayleigh* (R). One Rayleigh is the brightness of a source emitting 10^6 photons $\text{cm}^{-2}\,\text{s}^{-1}$ in all directions. The convenience of the unit lies in the direct relationship between it and the emission rate of equation (10.6), since (for an optically thin medium) I need only be multiplied by the depth of the emitting layer to give the Rayleigh brightness. For correct scaling of the 10^6 factor in the definition, the depth has to be measured in units of 10^6 cm (10 km), which is, in fact, a typical thickness for airglow emission layers. The name of the unit honours Robert John Strutt, Fourth Lord Rayleigh, who performed much pioneering work on the airglow. He is sometimes called 'the airglow Rayleigh' to distinguish him from his father, the Third Lord Rayleigh, or 'scattering Rayleigh'.

10.2.4 Atomic and Molecular Oxygen

Excited states of O and O_2 make an extremely important series of contributions to the airglow of Earth and other planets. Figures 10.1 and 10.2 introduce this topic by showing some transitions in O and O_2 that can be observed at ground level. All the relatively long wavelength ($\lambda > 300$ nm) transitions are 'forbidden' by selection rules, and the excited states are therefore long-lived in the absence of quenching collisions.

The O(^{1}S $\rightarrow$ ^{1}D) transition at $\lambda = 557.7$ nm was the first component of the airglow to be associated with a specific atomic or molecular event. Because the radiation was already known in the aurora, this well-known line is often called the *auroral green line*, even when it originates in the airglow. The green line and a red doublet at $\lambda = 630.0$ and 636.4 nm that is emitted by O(^{1}D) both seem to originate in two different altitude regions of the atmosphere. The high-altitude excitation

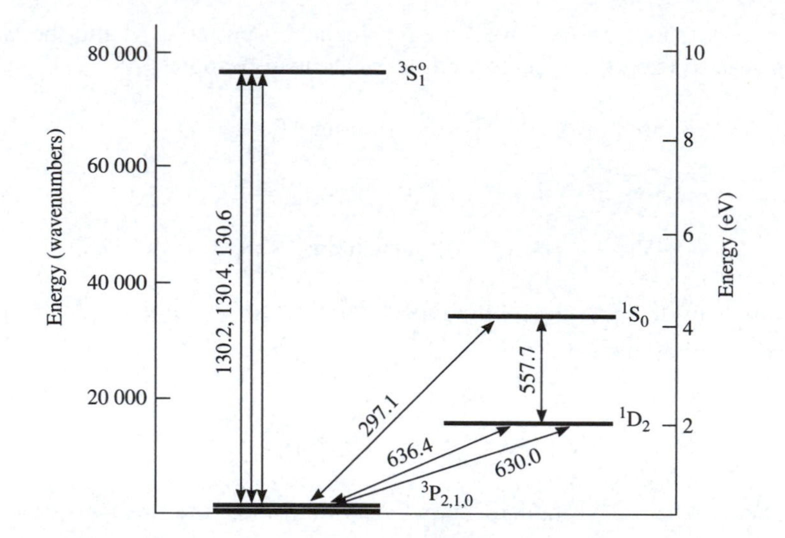

Figure 10.1 Some low-lying energy levels of atomic oxygen. Atmospherically important optical transitions are shown by the arrows, and the wavelengths of the lines are given in nanometres. Data from C.E. Moore, *Atomic Energy Levels*, Vol. 1, NSRDS–NBS35, Washington DC, 1971.

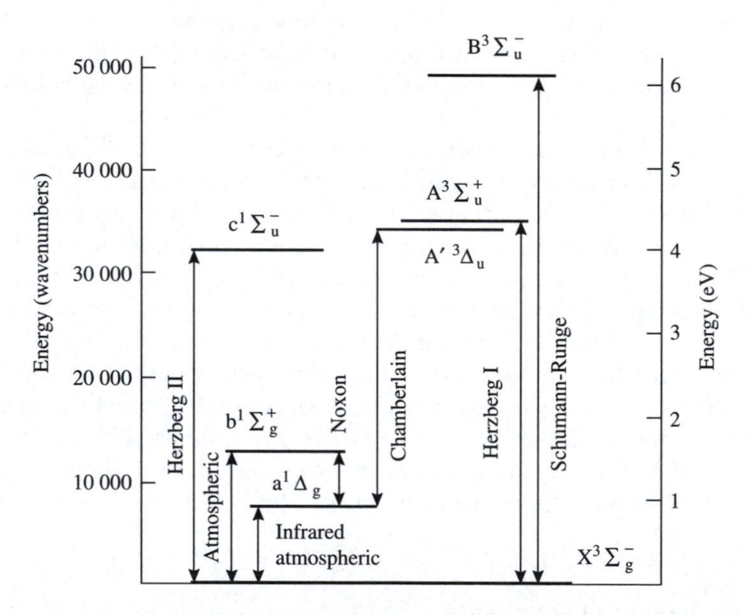

Figure 10.2 Nomenclature for some optical transitions in molecular oxygen. With the exception of the B $\leftrightarrow$ X (Schumann–Runge) system, all these transitions are forbidden by electric-dipole selection rules. Data from P.H. Krupenie, *J. Phys. Chem. Ref. Data*, 1972, **1**, 423.

processes include dissociative recombination of O_2^+ analogous to reaction (10.5). There are two channels

$$O_2^+ + e \rightarrow O(^1S) + O(^3P) + 2.78 \text{ eV} \tag{10.11}$$

$$O_2^+ + e \rightarrow O(^1D) + O(^3P) + 4.99 \text{ eV} \tag{10.12}$$

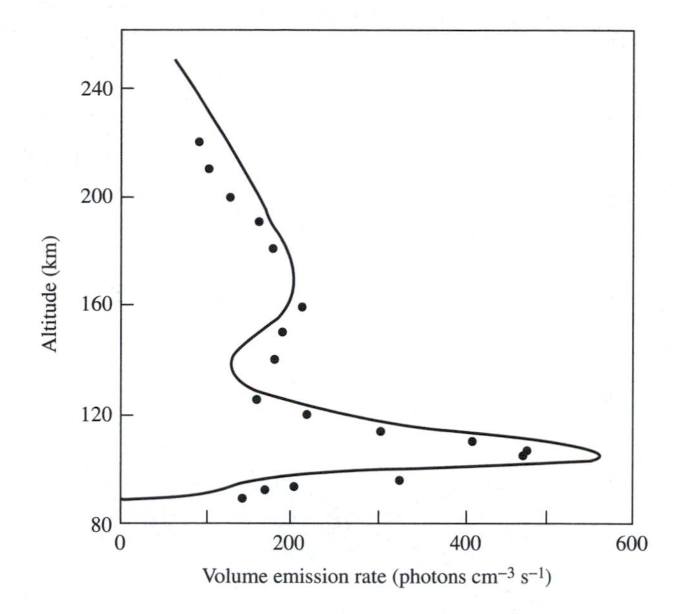

Figure 10.3 Measured (WINDII) and modelled $O(^1S)$ green-line dayglow emission profiles for 53°S, 126°E. From S. Tyagi and V. Singh, *Ann. Geophysicae*, 1998, **16**, 1599.

capable of exciting either $O(^1S)$ or $O(^1D)$. This two-channel dissociative recombination process is, of course, one of the two most important charge-neutralization steps in the ionosphere (see Section 10.4.2). The processes are significant sources of $O(^1S)$ and $O(^1D)$ in the aurora (Section 10.3.1).

Reactions of neutral species are responsible for the lower altitude (90–100 km) O-atom airglow. The main emitting layer of $O(^1S)$ is in this region, and the excited atoms are formed by day and by night in a process that we shall describe in detail later. Figure 10.3 shows measurements of the $O(^1S)$ green-line emission made with the WINDII satellite instrument. The two emitting layers are clearly evident in the observations. The solid line shows the predictions of one model in which several excitation processes are included. For the particular location and time to which the figure refers, the agreement between model and observation is quite good throughout the two emitting regions.

The $^1\Delta_g$ molecular fragment produced in the photolysis of ozone (reaction (10.3)) is only very weakly quenched by N_2 or O_2. In the upper stratosphere, the collisional lifetime is of the order of tens to hundreds of seconds. Although the radiative lifetime is also very long (44 min) for the highly forbidden $O_2(^1\Delta_g \rightarrow {}^3\Sigma_g^-)$ Infrared Atmospheric Band at $\lambda = 1270\,\text{nm}$ (see Figure 10.2), a sensible fraction of $O_2(^1\Delta_g)$ can emit because of the inefficiency of physical deactivation. In fact, the Infrared Atmospheric Band is the most intense feature of the dayglow, the (0, 0) transition (observed at altitudes high enough to avoid self-absorption: see the final paragraph of Section 10.2.1) having an intensity of 20 MR. Consideration of the energy balance alone points to ozone photolysis as the source of $O_2(^1\Delta_g)$, since no species other than ozone absorbs enough sunlight at the altitudes where the dayglow emission originates. Peak concentrations of $O_2(^1\Delta_g)$, determined by rocket photometry, are typically 2×10^{10} molecule cm^{-3} at 50–60 km altitude. Concentrations of this magnitude mean that the metastable excited molecules are present at the parts per million level. Laboratory measurements of the quantum yield for singlet oxygen production in reaction (10.3), and of rate coefficients for quenching by atmospheric gases, may be put together with experimentally determined atmospheric ozone concentrations and solar irradiances to predict an $[O_2(^1\Delta_g)]$–altitude profile. The results of such a calculation are displayed as the solid line in Figure 10.4, while some rocket measurements of $[O_2(^1\Delta_g)]$ are shown as circles on the figure. Agreement of this kind may be taken as strong confirmation of the mechanism suggested for daytime excitation of $O_2(^1\Delta_g)$. After the Sun falls below the horizon, or is eclipsed, concentrations of $O_2(^1\Delta_g)$ fall rapidly. An analysis of

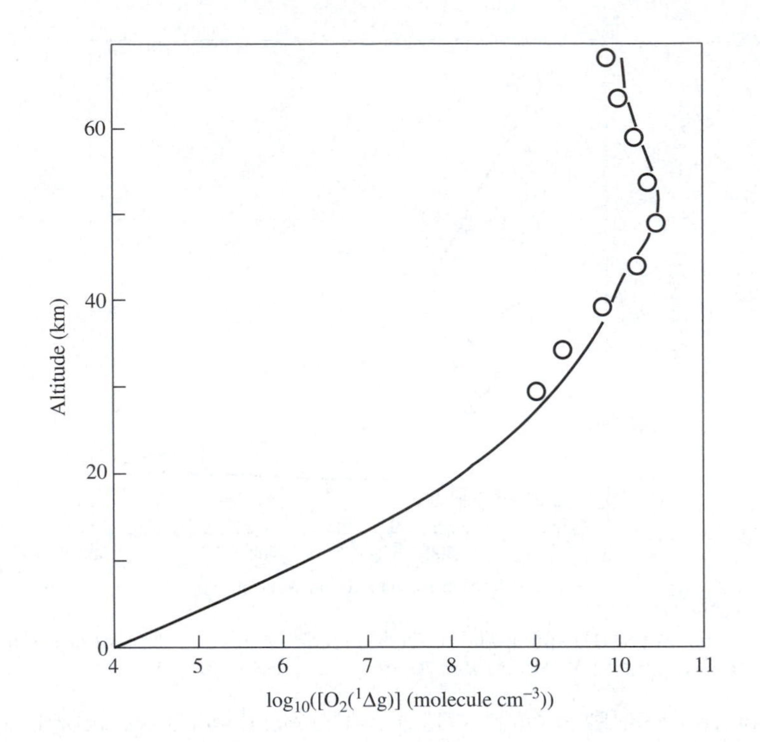

Figure 10.4 Atmospheric concentrations of $O_2(^1\Delta_g)$ during the day. Experimental values (circles) are derived from rocket measurements of the Infrared Atmospheric Band intensity in the dayglow, while the calculations (solid line) are based on laboratory kinetic and spectroscopic data. From P.J. Crutzen, I.T.N. Jones and R.P. Wayne, *J. Geophys. Res.*, 1971, **76**, 1490.

the intensity–time dependence in eclipses has been used to derive a composite quenching rate coefficient by atmospheric gases that is almost exactly identical to that calculated for air from the laboratory quenching data on the individual atmospheric gases. Confidence in the model is thus so great that dayglow measurements of the Infrared Atmospheric Band can be used to derive atmospheric ozone concentrations and profiles. One of the instruments (OSIRIS) on the ODIN satellite (launched 2001) routinely inverts the airglow intensities at $\lambda = 1270\,\text{nm}$, taking into account quenching rates, to derive ozone concentrations. In particular, OSIRIS is providing information on ozone depletion at Earth's poles and middle latitudes. The data are processed on the ground using tomographic techniques to provide three-dimensional worldwide maps of ozone concentration.

The second excited singlet of O_2, the $b^1\Sigma_g^+$ state (Figure 10.2), also produces a strong emission system in the dayglow. Just beyond the visible region, at $\lambda = 762\,\text{nm}$, the forbidden transition $O_2(^1\Sigma_g^+ \rightarrow {}^3\Sigma_3^-)$ is readily observed by rocket photometry. Radiation of this system is called the *atmospheric band.* Between 65 and 100 km, the dominant excitation mechanism is resonance absorption of solar radiation

$$O_2 + h\nu(\lambda = 762\,\text{nm}) \rightarrow O_2(^1\Sigma_g^+). \tag{10.13}$$

Above 100 km, $O_2(^1\Sigma_g^+)$ is mainly populated by a two-step process

$$O_2 + h\nu(\lambda = 175\text{ nm}) \rightarrow O(^1D) + O(^3P) \tag{10.14}$$

$$O(^1D) + O_2 \rightarrow O_2(^1\Sigma_g^+) + O(^3P). \tag{10.15}$$

In the first step, excited oxygen atoms are produced by O_2 photolysis, while in the second, energy is transferred from $O(^1D)$ to O_2. *Energy transfer* of this kind is yet another way in which atmospheric airglow may be excited. Below 65 km, reaction (10.15) is also responsible for exciting $O_2(^1\Sigma_g^+)$, but here the excited oxygen atoms are produced by photolysis of O_3, reaction (10.3).

The night-time persistence of excited atomic and molecular oxygen, as demonstrated by the nightglow, offers some difficulties in interpretation. Direct photochemical excitation is obviously excluded. So is indirect excitation by photochemically generated species, such as $O(^1D)$, whose lifetime against reaction is short. An obvious energetic reservoir for the oxygen species is ground-state atomic oxygen. Let us now examine how O atoms could populate the states observed and whether the intensities and altitude distributions are compatible with the proposed mechanisms.

Our starting point is the $O(^1S)$ state responsible for the green line of the nightglow. The accepted mechanism, suggested first by Barth, is another two-step process that starts with the formation of an energy-rich state of O_2, which we label as O_2*

$$O + O + M \rightarrow O_2^* + M. \tag{10.16}$$

A second step can form $O(^1S)$ if O_2* has enough energy as a combination of electronic and vibrational excitation, and $O_2(^1\Delta_g)$ if it has not. Thus, the reactions envisaged are

$$O_2^*(\upsilon'_{10.17}) + O \rightarrow O(^1S) + O_2(^3\Sigma_g^-) \tag{10.17}$$

$$O_2^*(\upsilon'_{10.18}) + O \rightarrow O(^1P) + O_2(^1\Delta_g) \tag{10.18}$$

where $\upsilon'_{10.17}$ is greater than $\upsilon'_{10.18}$. For the higher vibrational energies, the atomic O is excited to the ^1S state, and at lower vibrational energies the excess energy of reaction excites the $^1\Delta_g$ state of O_2. Extension of these ideas allows a coherent scheme to be formulated that explains the intensities of all the atomic and molecular oxygen emissions observed in the nightglow.

10.2.5 Atomic Sodium

Airglow emissions from atomic sodium and other metals are of interest because the *allowed*, resonance, transitions fall within the visible region and are therefore observable from the Earth's surface. Metallic ions have been observed consistently in the mesosphere and above by rocket-borne mass spectrometers. *Meteor ablation* (wearing away of the meteor by friction as it passes through the atmosphere) is the probable source of the metals. There are two principal origins of meteoroids in the atmosphere: dust trails produced by sublimating comets as they orbit the Sun that produce meteor showers such as the Perseids and Leonids; and fragments from the asteroid belt beyond Mars. Figure 10.5 shows concentration profiles for the most important metals; although Fe is rather more abundant than Na, it is the latter that has been studied in most detail.

The mesospheric layer of neutral sodium atoms lies roughly between 82 and 105 km, with peak concentrations being found typically at ~91 km. Normal peak concentrations are $\sim 5 \times 10^3$ atom cm^{-3}. Increases in concentration observed during meteor showers support the idea of a meteor ablation source for the atoms (see also Section 10.4.3). Dayglow emission of the sodium yellow $Na(^2P_{1/2,\,3/2} \rightarrow {}^2S)$ resonance doublet ($\lambda = 589.6, 589.0$ nm) is excited by direct absorption of solar radiation. Intensities are around 30 kR, corresponding to an excited state concentration of only $\sim 5 \times 10^{-4}$ atom cm^{-3}, which means that only about one atom in 10^7 is excited. Artificial excitation of the resonance emission provides a method for the measurement of concentration–altitude profiles of sodium from the ground. A dye laser is tuned to one of the resonance lines, and a short pulse of light directed into the atmosphere. Sodium atoms present in the atmosphere can resonantly absorb and re-emit the radiation. Examination of the intensity of the scattered light, and the time delay between outgoing and returning pulses, allows the profile to be derived. By analogy with radio-frequency radar sounding, the laser technique is called *lidar*. High-resolution lidar

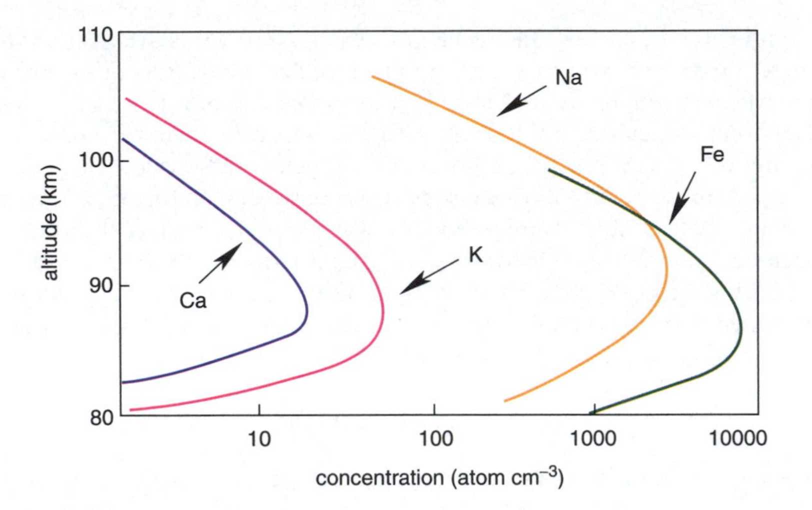

Figure 10.5 Vertical profiles of the annual mean concentrations of metal atoms at mid-latitudes. Source: *Atmospheric Chemistry of Meteoric Metals*, J.M.C. Plane, *Chem. Rev.* 2003, **103**, 4963.

measurements also allow wind and temperatures to be calculated for the mesosphere, in the same way that airglow emission spectra are used.

Nightglow intensities of the yellow sodium lines are much weaker ($\sim$100 R) than dayglow ones, but that they are present at all means that some reaction must be capable of exciting Na(^{2}P). Figure 10.6 illustrates an emission-rate profile measured by rocket: in addition, the figure shows a profile of ground-state Na measured at the launch site by lidar.

The appearance in the atmosphere of a highly reactive species such as atomic Na is rather surprising, and since the first observations of the layer in the 1950s has posed a considerable puzzle. Cyclic regeneration of free elemental sodium is demanded if sodium undergoes any reaction, regardless of the emission of light. An oxidation–reduction cycle, involving the species NaO, is the only plausible scheme. The cycle starts with the oxidation of sodium by O_3

$$\mathrm{Na} + \mathrm{O_3} \rightarrow \mathrm{NaO}(^2\Sigma^+) + \mathrm{O_2}, \tag{10.19}$$

and this reaction is followed by reduction of NaO by atomic oxygen to yield either excited, ^{2}P, or ground-state, ^{2}S, sodium atoms

$$\mathrm{NaO}(^2\Sigma^+) + \mathrm{O} \xrightarrow{\alpha} \mathrm{Na}(^2\mathrm{P}) + \mathrm{O_2} \tag{10.20}$$

$$\mathrm{NaO}(^2\Sigma^+) + \mathrm{O} \xrightarrow{1-\alpha} \mathrm{Na}(^2\mathrm{S}) + \mathrm{O_2}. \tag{10.21}$$

Involvement of ozone is inevitable in view of the emission intensity. Ozone is almost certainly the only reactive species in sufficient abundance at 90 km that could give an energy-rich product (NaO). Regeneration of atomic sodium requires a reduction process that is at least as fast as the oxidation reaction in the mesosphere, and that is sufficiently exothermic to excite Na(^{2}P). These conditions can be met if the reactions are of the electron-jump type; the participation of the electron-jump mechanism in reaction (10.19) as well can explain the very large rate coefficient for the process, which is about twice the limit allowed by hard-sphere collision theory.

The NaO is formed in reaction (10.19) almost exclusively in its first excited state ($^2\Sigma^+$), and not in the ground ($^2\Pi$) state. This is an important factor in the excitation mechanism, because for the

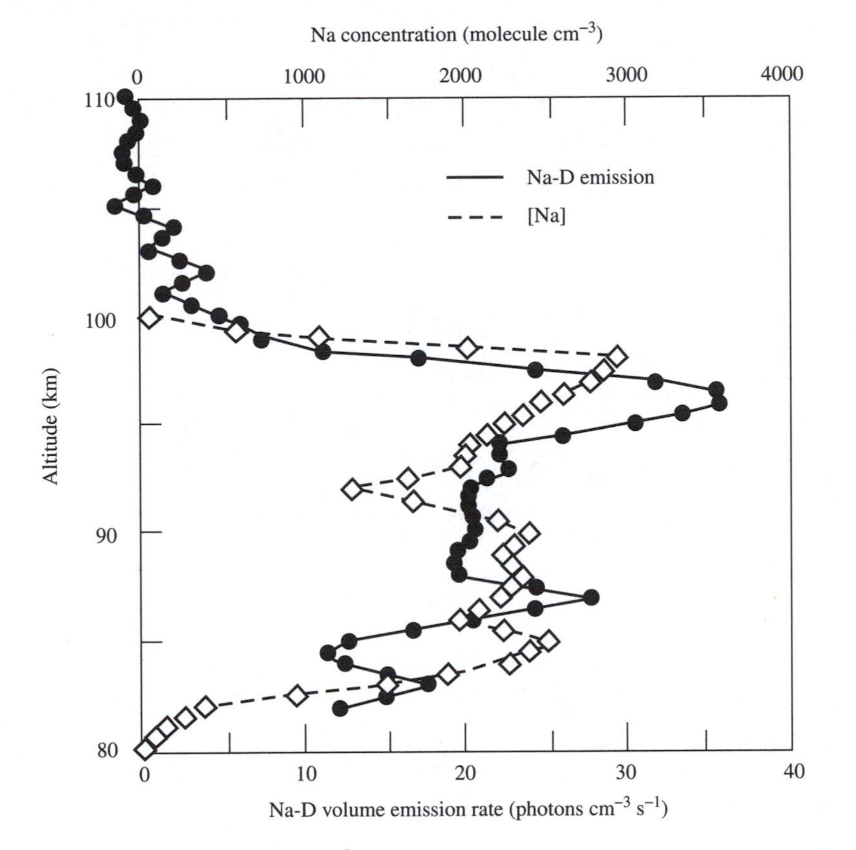

Figure 10.6 Volume emission rate from Na(^{2}P) measured by rocket, and the ground-state Na concentration measured from the ground by lidar. From H. Takahasi, S.M.L. Melo, B.R. Clemesha, D.M. Simonich, J. Stegman and G. Witt, *J. Geophys. Res.*, 1996, **101**, 4033.

excited-state NaO, the measured branching ratio α into the emitting ^{2}P state of Na is about 0.14, consistent with atmospheric measurements. For ground-state NaO, α is only $\sim$0.01, much too small for the mechanism to be capable of producing the observed atmospheric emission intensities.

A combination of laboratory kinetic studies and atmospheric modelling has now shown that sodium undergoes a quite complex series of regenerative cycles involving both neutral and ionic species. Figure 10.7 shows the major reactions in flow-diagram form. Species such as NaO_2, for example, participate in the cycling process. The major neutral sink is sodium bicarbonate, $NaHCO_3$, which is formed by a sequence of reactions involving O_3, H_2O, and CO_2. However, $NaHCO_3$, as well as NaO_2 and NaOH, are cycled back to atomic Na by reactions involving O and H. Note that these reactions represent *reductions*: although most of the atmosphere is an oxidizing medium, at altitudes between 80 and 100 km the atmosphere becomes strongly reducing because of a significant increase in atomic O and H relative to O_3 and O_2. At 90 km, almost all of the sodium is in the form of free atoms, while above 100 km, photoionization and charge-transfer from ambient ions such as NO^+ (see Section 10.4.2) produce Na^+, which is cycled back to Na by forming clusters that then undergo dissociative recombination in reaction with electrons.

Figure 10.8 shows how a state-of-the-art rocket airglow measurement may be used to establish concentrations of minor constituents in the mesosphere. In the experiments summarized in this figure and Figure 10.6, a total of ten airglow photometers (six forward-looking, and four

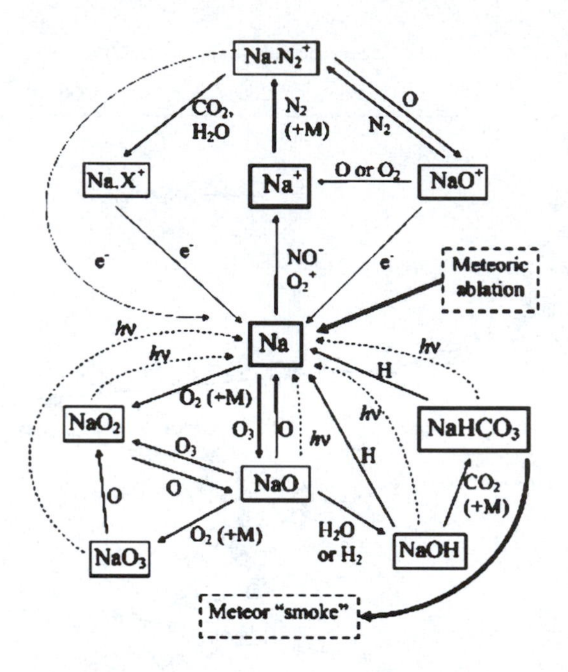

Figure 10.7 Flow diagram of major processes in the mesospheric chemistry of neutral and ionized sodium. Arrows with solid lines indicate gas-phase reactions, while photochemical processes are linked by broken thin lines. The lines are black if kinetic data have been obtained experimentally, grey if not. The major sodium species are surrounded by thick-edged boxes. 'Meteor smoke' is formed by the condensation of $NaHCO_3$ onto metal silicate smoke particles, which is thus a sink process. Source as for Figure 10.5.

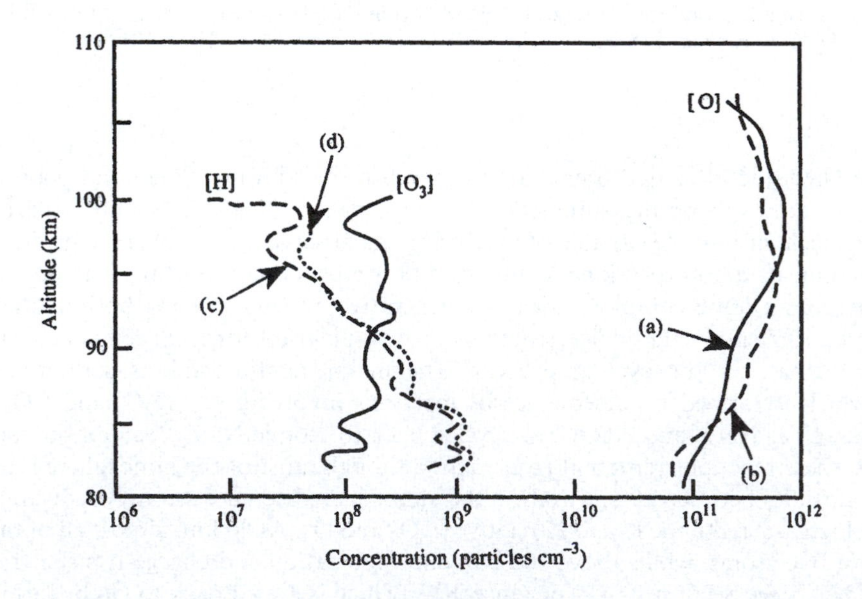

Figure 10.8 Concentrations of O_3, O, and H inferred from simultaneous measurements of $Na(^2P)$, $O_2(^1\Sigma_g^+)$ Atmospheric Band, and OH (8,3) Meinel Band intensities. Curve (a) is the O-atom profile derived from the intensity of the Atmospheric Band, while curve (b) is that predicted by a model. Curve (c) is the H-atom profile obtained from the intensity of the Atmospheric Band, and curve (d) is that derived from the OH band. Source as for Figure 10.6.

side-looking) were carried on the same rocket. The emissions observed included the OI 557.5 nm line, O_2 Atmospheric Band, Na D lines, and OH Meinel (8,3) Band (see Section 10.2.6). A lidar system operating at the ground at the launch site was used to measure Na atom concentrations at the time of the rocket experiment. Figure 10.8 shows the concentrations of O, H, and O_3 inferred from the measurements. Note that these determinations clearly support the view that $[O] \gg [O_3]$, which has been a feature of several points of our discussion of the mesosphere.

An interesting phenomenon associated with meteor-train luminosity may have one explanation in the cyclic chain processes (10.19) and (10.20). Most meteors that produce a visible train on entering the Earth's atmosphere show a very short-lived luminosity, following the impact of the body with the atmosphere. In about one in 125 000 visual meteor events, however, the train lasts for more than ten minutes; enduring luminosity for over an hour has been recorded. Emissions of sodium, iron and magnesium (as well as, perhaps, the O_2 atmospheric band) have been identified in these persistent trains. The afterglows are most strongly emitted from altitudes of $\sim$90 km. Oxidation and reduction in a cycle, as proposed for sodium in the airglow, seems an attractive way of explaining long-lasting excitation of additional sodium ablated off a meteor.

10.2.6 Hydroxyl Radicals

Vibrational emission phenomena are ordinarily thought of in connection with thermally equilibrated radiative transfer in the atmosphere (Sections 2.3.1 and 2.3.2). However, vibrationally *dis*-equilibrated OH radicals make a strong contribution to the longwave visible and near-infrared airglow, and offer a rare example of vibrational, rather than electronic, airglow emission. The OH airglow has additional interest because, as we shall see later, it provided a hint of the catalytic chains that are now known to be so important in atmospheric ozone chemistry (Section 9.3). Vibrational–rotational transitions among the nine lowest vibrational levels in the ground electronic state of OH give rise to the observed emission system, known as the *Meinel Bands* of OH. Transitions are observed corresponding not only to the allowed $\Delta\upsilon = 1$ fundamental, but also to the forbidden $\Delta\upsilon > 1$ overtones. Fundamentals lie in the wavelength range $\sim$4500 nm ($\upsilon = 9 \rightarrow 8$) to $\sim$2800 nm ($\upsilon = 1 \rightarrow 0$), and are intrinsically the strongest features, but are relatively difficult to detect because they lie well into the infrared. Overtone transitions of the $\Delta\upsilon = 4, 5...$ series lie at $\lambda < 1000$ nm, and, although much weaker because of their forbidden nature, they are more readily observed than the fundamentals without interference from the thermal background of atmosphere or detectors. Even $\Delta\upsilon = 9$, for $\upsilon = 9 \rightarrow 0$, is detectable just in the ultraviolet ($\lambda \sim 382$ nm), although the intensity of the band is nearly eight orders of magnitude weaker than the bands of the $\Delta\upsilon = 1$ series.

Evaluation of possible excitation mechanisms starts from the apparent sharp cut-off in excitation at $\upsilon = 9$, corresponding to $\sim$312 kJ mol^{-1}, with no emission from $\upsilon = 10$ ($\sim$337 kJ mol^{-1}) being seen. A reaction is needed for excitation whose exothermicity lies between 312 and 337 kJ mol^{-1}. Atomic hydrogen reacts with ozone to form OH in exactly this way

$$H + O_3 \rightarrow OH + O_2 + 322 \text{ kJ mol}^{-1}. \tag{10.22}$$

Laboratory investigations show that this reaction is indeed strongly chemiluminescent, with OH ($\upsilon < 9$) being the emitter. The immediate question for atmospheric excitation is why reaction (10.22) does not consume all available H atoms if it is fast enough to give the observed intensities. The answer is that the reaction

$$O + OH \rightarrow O_2 + H, \tag{10.23}$$

is fast enough to regenerate H atoms in the atmosphere. Indeed, the pair of reactions (10.22) and (10.23) is already familiar to us as the catalytic Cycle 1 shown in Section 9.3.

Determination of individual rotational-line intensities allows mesospheric temperatures to be calculated. One ground-based instrument employs spectral lines P1(2) at $\lambda = 840.0$ nm and P1(4) at $\lambda = 846.5$ nm in the $\upsilon = 6 \rightarrow 2$ Meinel Band. Results from such experiments often show large changes in temperature (~25 K) during a single night that are due to wave and tidal activity. A nice illustration of how simultaneous measurements of several nightglow emissions can provide

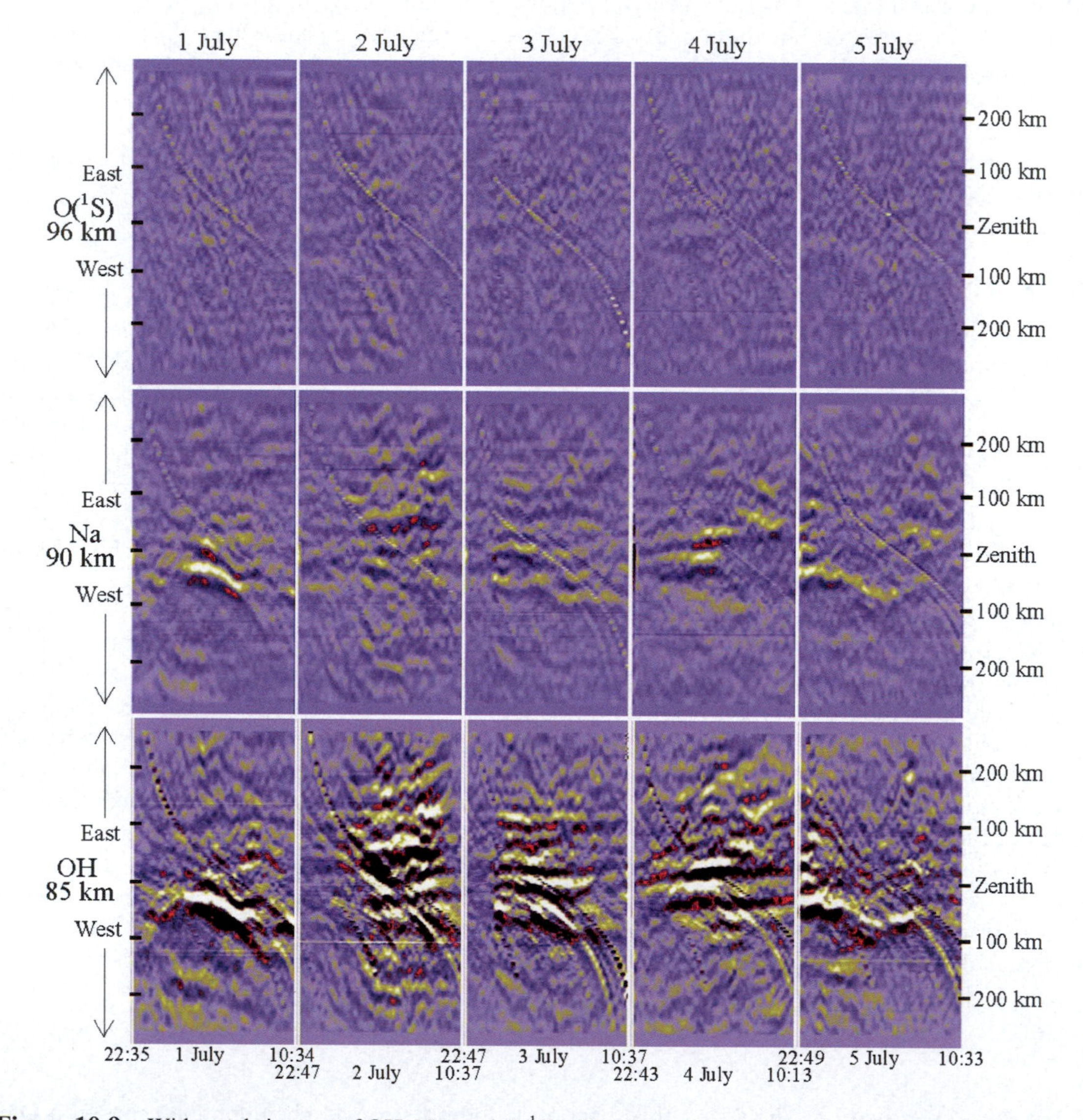

Figure 10.9 Wide-angle images of OH, Na, and O(^{1}S) nightglow emissions observed from Argentina during five successive nights. Time runs horizontally and the spatial (east–west) dimension is in the vertical with eastwards directed upwards and westwards downwards. The measuring observatory is situated along the middle of the plots at zenith, and the central position of the Andes range would be aligned horizontally and positioned near the 100-km line to the west. The stationary nature of the waves during the nights is clearly evident. Source: S. Smith, J. Baumgardner and M. Mendillo, *Geophys. Res. Letts.*, 2009, **36**, L08807.

information about mesospheric behaviour is contained in Figure 10.9. An instrument with a 'fish-eye' lens was used to gather nightglow radiation from the sky over El Leoncito Observatory, Argentina. In the figure, time increases along the horizontal axis and the zonal direction is the vertical dimension, with the observatory situated along the middle of the plots at zenith. Three of the emission systems we have been discussing were studied: OH (695–1050 nm), Na (589.3 nm) and $O(^1S)$ (557.7 nm). The emissions originate at different altitudes (85, 90 and 96 km, respectively, for the maxima). An obvious wave-like pattern is seen in the bottom row (OH emission) of the figure, and it does not change dramatically from day to day. The changes in emission intensity that cause this pattern in the image result from a large and extensive stationary-wave structure in the atmosphere. The structure is oriented north–south and is parallel to the Andes range. The wave structure is clearly visible in the OH emission from 85 km altitude, but can also be seen in the $Na(^2P)$ and $O(^1S)$ emissions, increasingly weakly, and with small time shifts with respect to the OH emission. Taken together, the results suggest that the waves are generated in the mesosphere by the surface features of the Andes, and the formation and propagation of the waves can be probed by examining the diurnal evolution of the structure and its altitude dependence.

These last examples illustrate how airglow studies give an understanding of atmospheric processes not easily available from other types of investigation. Complemented by reliable laboratory experiments on excited species, airglow is thus a prime source of information about the atmosphere of Earth and the other planets.

10.3 IONS IN THE ATMOSPHERE

Charged particles represent only a minute fraction of the total mass of the Earth's atmosphere, but they play a crucial part in many geophysical phenomena, such as variations in the geomagnetic field, lightning, and auroras. Neutral species may be produced or destroyed in reactions involving ions or electrons, and ion chemistry would be important even in relation to neutral composition and behaviour alone. Communication by radio waves is now regarded as an essential feature of 'civilized' life, so that the reflection, absorption, and propagation through the charged atmosphere of radio-frequency electromagnetic waves is of vital importance.

Electrons are the particles immediately involved in interactions with radio propagation. Atmospheric ions and their chemistry, which are properly the subjects of our study, play a central role in determining the concentration and distribution of electrons. Our survey of atmospheric chemistry would evidently be incomplete without some discussion of radio propagation, especially as radio experiments have given so much information about atmospheric ions. Section 10.3.3 is devoted to an explanation of the radio studies, and it follows a very brief introduction to two geophysical manifestations of charged species that are of particular historical interest.

Ions and electrons are most abundant at altitudes greater than $\sim$60 km (*i.e.* within the mesosphere or above). Several factors contribute to this high altitude distribution, as discussed at the beginning of Section 10.1 Free electrons at concentrations sufficient to affect radio-wave propagation are certainly only located at levels above 60 km, and this region has conventionally been called the ionosphere. Peak instantaneous electron densities can exceed 10^6 electrons cm^{-3} at $\sim$300 km, and drop to near zero at 60 km. Total charged-particle concentrations (positive ions + negative ions + electrons) also fall to a minimum of a few hundred per cm^3 at $\sim$60 km, but increase a little in the stratosphere.

10.3.1 Aurora

Nature provides in the high atmosphere a phenomenon so obvious and striking that it has been the subject of scientific study for centuries. In regions not far from the geographic poles, a spectacular display of coloured lights can sometimes be seen at night. Complicated shapes are produced, which often move and change rapidly. The luminosity is given the name *aurora*, or 'dawn', with the qualifiers *borealis* ('northern') or *australis* ('southern') being added according to the polar region in which the display is seen.

Measurements of the angle of elevation of one auroral feature from two places simultaneously showed early on that most of the light comes from altitudes of around 100 km. Such observations thus gave Man some knowledge of the atmosphere at heights then inaccessible to *in-situ* experiment. Spectroscopically, the auroral light is not dissimilar to that seen in the laboratory from electric discharges through low-pressure air. Atomic and molecular emission systems of oxygen and nitrogen predominate, the *auroral green line* of atomic oxygen, $O(^1S \rightarrow {}^1D)$, at $\lambda = 557.7$ nm being especially pronounced (see Section 10.2.4). The excitation seems to involve bombardment of atmospheric gases by particles from above, followed by ionization. Neutralization by electron capture leads to excitation, and directly or indirectly to emission of visible radiation in reaction (10.11). Most auroral activity is observed in an oval belt around the geomagnetic pole, between 15° and 30° from it. This observation shows that the bombarding particles are charged species (*e.g.* electrons or protons) projected from the Sun that are deflected by the Earth's magnetic field as they approach us. The charged beam will impinge equally on day and night sides, and will be concentrated in defined regions, one round each pole. Violent changes in the observed intensity of auroral light, according to this picture, correspond to changes in the primary ionizing stream originating initially from the Sun. There is also a connection between *solar activity* and auroral phenomena. Solar activity is often assessed in terms of the number of sudden 'storms' that disturb the Sun's surface. These storms reveal themselves as intense localized bursts of H-alpha light ($\lambda = 656.3$ nm; the $n = 3 \rightarrow n = 2$ transition). *Solar flares* of this kind are accompanied by the liberation of huge amounts of energy as electromagnetic radiation (especially in the X-ray region) and as kinetic energy of particles (protons mainly, but also α-particles and heavier nuclei). It is generally believed that the energy is produced by the annihilation of magnetic fields associated with sunspots. Sunspots are areas, dark only relative to the solar disc, that appear from time to time on the Sun's surface, and move across the disc in $\sim$13.5 days (solar rotational period $\sim$27 days). A definite *sunspot cycle* occurs in which the number of spots changes with a period of eleven years; it is thought to represent some important oscillation in the structure of the Sun itself. One of several manifestations in our atmosphere of solar storms is the enhanced auroral activity mentioned earlier. A complication arises in that the ionizing particles are not simply the storm particles emitted by the Sun, partly demonstrated in that there is too great a time delay between the storm and the auroral response. One suggestion is that particles already trapped within the Earth's magnetic field are released and accelerated by the arrival of the storm particles. Whatever the detailed explanation, there is a statistical correlation between solar activity and frequency of auroras.

10.3.2 Geomagnetic Fluctuations

Studies of the Earth's magnetism provided further early evidence that incident radiation might ionize the atmosphere at great altitudes. Detailed investigations had shown that the apparent direction of the Earth's magnetic field fluctuated regularly throughout a day, and that the fluctuations were smaller in winter than in summer. In 1882, Balfour Stewart suggested that electric currents circulating in the upper atmosphere might be responsible for diurnal changes in field. The atmospheric gases, *if conducting* and made to move across the Earth's magnetic field, would generate and carry such currents, as in a giant dynamo. Currents of the order of tens of thousands of amperes are involved! If solar radiation were responsible for the ionization, then the day-to-night variations would be explicable, as would the seasonal dependence of the magnitude. Solar particles seem a less likely agent of ionization, since they would arrive at the atmosphere near the auroral zones, while the geomagnetic effect is spread over the entire Earth. Nevertheless, the diurnal excursions of magnetic field do show a sympathy with the sunspot cycle, the swing being nearly twice as great at sunspot maximum as at minimum. This cyclic behaviour is now interpreted in terms of the increased ionization from solar X-rays at times of high sunspot activity.

As early as 1740, large, rapid, and irregular changes had been noticed as an occasional feature of the geomagnetic field. *Magnetic storms* of this kind are often seen when auroral displays are in progress, and they are most intense within the auroral zones. They are not, however, exclusively confined to the auroral regions, so that the excitation of auroras and the increase of atmospheric conductivity do not have the same *immediate* cause, even though both are somehow connected with the solar cycle.

10.3.3 Radio Propagation

Marconi's success in 1901 in sending signals from England to America is often regarded as heralding the birth of radio as a means of communication. Radio waves are electromagnetic radiation and travel in nearly straight lines, as does light. Diffraction cannot possibly account for the travel of radio waves across the Atlantic, so that Marconi's feat demanded some explanation. An overhead reflector was suggested by Heaviside in England and by Kennelly in America: the hypothetical reflecting layer was called the *Heaviside layer*. Commercial communication links were set up after Marconi's experiment, and it was soon noticed that signal strengths varied in a regular way throughout the day, the season, and the solar cycle. Magnetic storms were found to be associated with disturbances of the diurnal variability and sometimes with disruption of communication. When radio broadcasting started in the early 1920s, it was found that, *at night*, the signal strengths at distances of about 100 km varied widely over a few minutes, sometimes disappearing completely. This *fading* of the signals was ascribed to interference effects between a reflected *sky wave* and the direct *ground wave*. Reflections of the frequencies used for broadcasting thus seemed possible even for nearly vertical incidence, but only at night. The phenomenon is now very apparent with the large number of transmitting stations. At night, medium-distance (up to a few thousand km) AM stations can be received only too readily, with resulting interference between stations. During the day, only relatively local signals can be received at the frequencies involved ($\sim$1 MHz). Such discoveries suggested that radio reflection is caused by ionized layers in the atmosphere, and that the ionization is at least influenced, and probably caused, by solar radiation. As we shall see shortly, ionization at low altitudes also leads to absorption of broadcast frequencies, and prevents reflection from higher altitudes during the day. At night, the absorption is much diminished, and the various fading and interference effects become pronounced.

The early observations of the behaviour of radio waves were of great importance, because they indicated that radio waves could be used as a tool to explore the nature of the upper atmosphere and the ionization in it. Until rockets and satellites were able to enter the ionosphere or to examine it from above, radio studies provided much the most detailed information about our ionosphere. A crucial experiment concerned the polarization of reflected waves produced by the Earth's magnetic field. The results showed that the charged particles causing reflection were electrons and not ions. Sir Edward Appleton discovered that radio waves of frequency greater than 4–5 MHz penetrate the Heaviside layer, but, instead of escaping to space, they are reflected by a higher layer. This layer was named the *Appleton layer* after its discoverer. Appleton himself used the symbols E and F for the electric fields in the lower and upper layers, and suggested that the layers be called by the same letters in order to accommodate other possible layers by neighbouring letters of the alphabet. When discussing the ionosphere, it is universal practice to use this nomenclature, with D, E, and F layers or regions being particularly important. Electron densities in the ionosphere have been most commonly measured from the ground by a device called an *ionosonde*. A pulse of radiation of

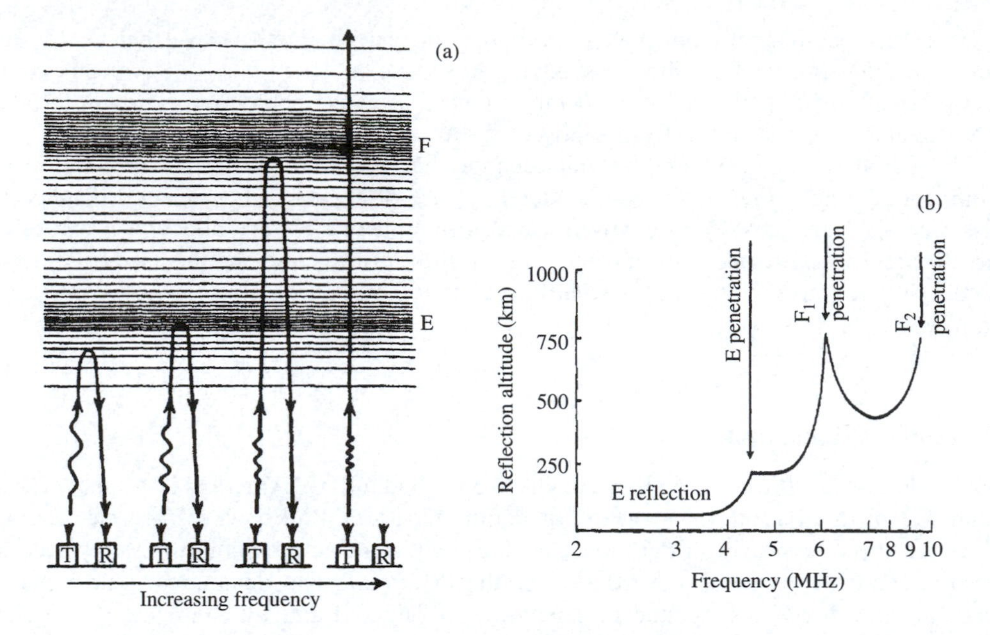

Figure 10.10 (a) Schematic representation of the operation of an ionosonde. T is the transmitter, and R the receiver. Horizontal shading indicates rate of change of ionization in the ionosphere. The four T–R pairs show, from left to right, what happens to radiowave pulses as the frequency increases. (b) Simplified and idealized ionogram obtained using the ionosonde technique of panel (a). Frequency of the emitted pulse of radiowaves is displayed on the horizontal scale, and the time delay for the return pulse is converted to effective altitude for the vertical scale. For frequencies less than about 4 MHz, reflection is from the E 'layer' at a height of about 120 km. At higher frequencies, the E-region is penetrated, and reflection is from the F_1 ledge, and (above ~6–7 MHz) the F_2 peak. Adapted from R.P. Wayne, *Chemistry of Atmospheres*, 3rd edn, OUP, Oxford, 2000.

known frequency is directed vertically upwards, and the time delay of any reflection is recorded automatically. The frequency of the pulse train is increased to permit penetration to greater altitudes (see below). Figure 10.10(a) illustrates the principle of the ionosonde, and Figure 10.10(b) is an ionogram obtained by day. At frequencies up to 4.4 MHz, reflection is by the E layer at an altitude of ~100 km. Wave penetration to the undisturbed F layer has a quite sharp onset at 4.4 MHz. Note that the F layer is partially split into two layers labelled F_1 and F_2: this is a day-time phenomenon, the F_1 'ledge' disappearing at night. The D-region is more involved with absorption, rather than reflection, of radio waves, and so must be investigated by different techniques. *Topside sounding* by satellite instruments now complements the data provided by ground-based ionosondes.

We have so far discussed the experimental results obtained on penetration, reflection, and absorption without explanation of why the phenomena occur, or why they are dependent on frequency, electron concentration, or altitude. Reflection itself is brought about by a process akin to the total internal reflection familiar in optics, although the detail is rather more subtle. An important conclusion is that *layers* of electrons are not necessary for reflection to occur, but rather only sufficient changes in densities. Nevertheless, the fairly sharp transitions between reflection altitudes at specific frequencies (Figure 10.10(b)) suggest that electron densities must themselves change rather sharply with altitude at those levels. Figure 10.11 gives some representative electron density profiles based on a variety of measurement techniques. The labels D, E, F (F_1, F_2) are now indeed seen to be associated with regions rather than layers, the boundaries between regions being altitudes where electron densities increase particularly rapidly with height. Day–night variations are shown for the

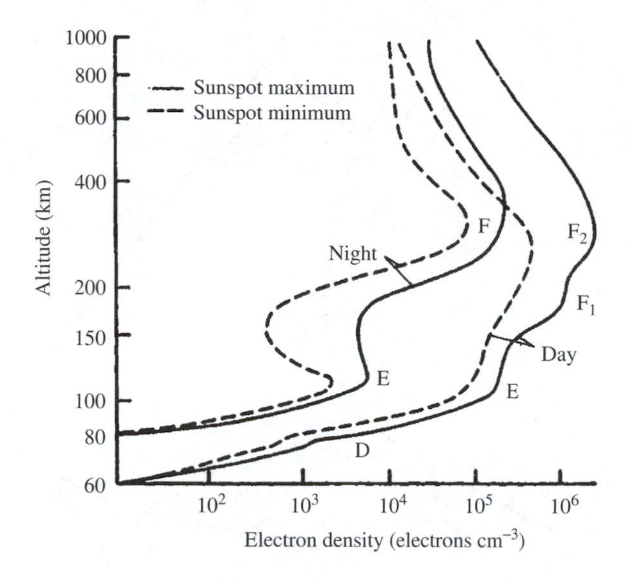

Figure 10.11 Typical electron density–altitude profiles for the mid-latitude ionosphere. From a wallchart prepared by W. Swider entitled *Aerospace Environment*, Air Force Geophysics Laboratory, Hanscom Air Force Base, Massachusetts.

two extremes of the sunspot cycle. Night-time electron densities are orders of magnitude smaller than daytime ones, and the distinction between F_1 and F_2 regions is lost. Electron densities in the D-region become near zero at night (although negative ions continue to be present in this region).

10.4 ION CHEMISTRY

The way in which radio waves are reflected by the ionosphere is just one means of identifying its regions. An equally valid division can be based on the chemical nature of the dominant ions and the chemical processes occurring at different altitudes.

Figure 10.12 shows ion compositions measured by mass spectrometry above about 100 km, together with total electron densities. The F-region (say above 150 km) is characterized by atomic ions: O^+ and N^+ are dominant throughout most of the region, although H^+ and He^+ are more abundant at the very highest altitudes. Molecular ions (NO^+, O_2^+) are the most important ions in the E-region (100–150 km), although total concentrations are lower than in the F-region. At lower altitudes, in the D-region, more complex molecular positive ions dominate, and negative ions are seen for the first time. We shall examine more carefully the composition and chemistry of each region in later sections. Here it is of interest to identify in outline some influences that lead to the compositions observed.

Gravitational separation (see final paragraph of Section 2.2) of ionic species and their neutral precursors obviously favours atomic over molecular ions at high levels in the heterosphere. The abundances of the various atomic ions above the F peak of electron density at $\sim$250 km fits this expectation well, with O^+ and N^+ being overtaken by the lightest ions H^+ and He^+ beyond about 1000 km. One point of interest concerns the scale heights (Section 2.1) for these ions. Electrons, being so light, could perhaps be expected to have near-infinite scale height, and to diffuse rapidly to the exosphere and thus escape. Electrostatic attraction to the positive ions in reality prevents this escape, but the effect is to make the scale height for the ions twice what it would be for the

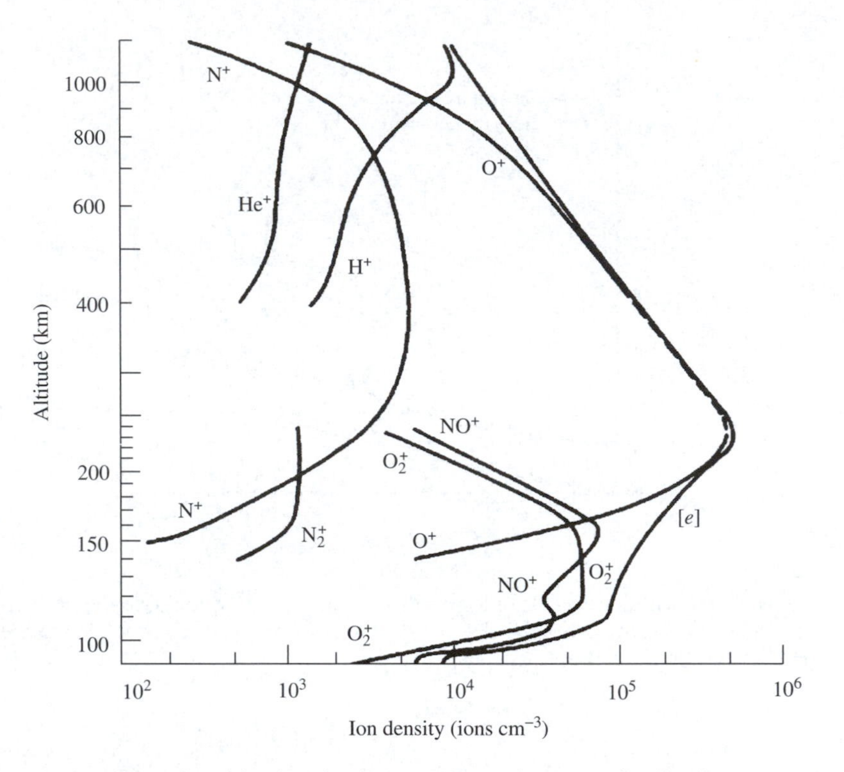

Figure 10.12 Composition of positive ions in the E- and F-regions. Ion distributions were obtained by mass spectrometer experiments during daytime and at solar minimum; the data are normalized to the measured electron-density distribution. From C.Y. Johnson, *J. Geophys. Res.*, 1966, **71**, 330.

corresponding neutral species. The scale height for the electrons themselves is determined by the concentration profiles of the most abundant ions: Figure 10.12 shows very well that [*e*] follows [O^+] closely where O^+ is dominant. Diffusion of the oppositely charged species is referred to as *ambipolar*, and it is of great importance in the F layer. The main distinction between the daytime F_1 and F_2 regions lies in the dominance of ambipolar diffusion over other loss processes in the F_2 region rather than in any fundamental chemical differences.

Ion-production processes also favour atomic-ion formation at high altitudes. Photoionization by extreme UV (EUV) is a major source of ions (see Section 10.4.1). The atoms N, O, and H have appreciably higher ionization potentials than the molecules NO or O_2. The shortest wavelength EUV is absorbed in the highest regions of the atmosphere, so that the remaining EUV can ionize only molecular species at lower altitudes (E, D-regions). Energies of the photons available at high altitudes (F-region) may be sufficient to dissociate a molecule and ionize one of the product atoms in a single step.

Altitude has a direct influence on chemical reactions through the pressure and hence the frequency of collisions. Ion–molecule reactions can convert the ions initially produced to different species that appear as *intermediate* and *terminal* ions. Although ion–molecule reactions are generally quite efficient because of long-range electrostatic attractions, they are obviously faster for a given ion concentration at lower altitudes where the neutral reactant is more abundant. A number of ionic processes of atmospheric importance are brought together in Table 10.1. Some, such as the exchange processes (10.26) and (10.27), have clear counterparts in neutral chemistry, while others, such as charge transfer, (10.24), do not.

Table 10.1 Some types of ionic process of atmospheric importance.

Process	*Examples*		*Type*
Charge transfer	$N_2^+ + O_2 \rightarrow N_2 + O_2^+$	(10.24)	
	$H^+ + O \rightarrow O^+ + H$	(10.25)	
Exchanges	$N_2^+ + O \rightarrow N(^*) + NO^+$	(10.26)	
	$O_2^+ + N_2 \rightarrow NO + NO^+$	(10.27)	
Recombinations	$O^+ + e \rightarrow O + h\nu$	(10.28)	Radiative
	$NO^+ + e \rightarrow N(^*) + O.$	(10.29)	Dissociative
	$NO^+ + NO_2^- \rightarrow NO + NO_2$	(10.30)	Ion–ion
Attachments	$e + O \rightarrow O^- + h\nu$	(10.31)	Radiative
	$e + O_2 + M \rightarrow O_2^- + M$	(10.32)	Three-body
	$e + O_3 \rightarrow O^- + O_2$	(10.33)	Dissociative
Detachments	$O_2^- + M \rightarrow O_2 + e$	(10.34)	Collisional
	$O_2^- + O \rightarrow O_3 + e$	(10.35)	Associative
	$O^- + h\nu \rightarrow O + e$	(10.36)	Radiative
Clustering	$O_2^+ + O_2 + M \rightarrow O_2^+.O_2 + M$	(10.37)	
	$H + (H_2O)_n + H_2O\ (+M) \rightarrow H^+(H_2O)_{n+1}\ (+M)$	(10.38)	
	$NO_3^- + H_2O + M \rightarrow NO_3^-.H_2O + M$	(10.39)	

Note: Products of exothermic processes may possess electronic as well as translational (and, where appropriate, vibrational and rotational) excitation. For example, the atomic nitrogen product of reactions (10.26) and (10.29) can be formed in the excited ^2D state. The radiative detachment process (10.36) is referred to as a *photodetachment*.

Conversion in the ionosphere of atomic ions to molecular ions is of critical importance in permitting charge neutralization, and thus in determining electron and ion concentrations throughout the atmosphere. For positive ions, charge is neutralized in the reactions shown only in the recombination processes (10.28) to (10.30). The energy of ionization is liberated when an electron and positive ion attempt to form a neutral species, and that energy must be dissipated if the product is not just to reionize immediately. Radiative recombination, process (10.28), is the only way in which atomic ions can be neutralized; it is extremely inefficient. Molecules can undergo dissociative recombination, as in reaction (10.29), where the fragments carry off the ionization energy as translational motion. Similarly, positive and negative molecular ions can interact (reaction (10.30)) to yield neutral particles that disperse the energy released as internal motions as well as kinetic energy of the products. A further mechanism for ion–ion recombination is recognized: a three-body association involving a neutral third body, M, that can dissipate excess energy in the usual way. Such ion–ion charge-annihilation processes are of particular importance at atmospheric altitudes less than about 60 km, since almost all the negative charge is carried by ions rather than electrons, although we shall see in later sections that the ions involved are considerably more complex than the NO^+ and NO_2^- represented in equation (10.30).

Attachment of electrons to neutral molecules to form negative ions requires the removal of excess energy in the same way that neutralization does. Radiative attachment, reaction (10.31), is once again much less efficient than the three-body or reactive processes, (10.32) and (10.33), and is of little atmospheric significance. Much more important is the three-body reaction, but since its rate depends on the square of the pressure of neutral constituents, it is confined to the D-region and below (as is the ozone reaction, (10.33)). Negative ions are consequently only found at low altitudes. Detachment processes, such as reactions (10.34)–(10.36), convert negative ions back to neutrals and electrons, so that the detailed balance between the number densities of negative ions and electrons depends on the competing rates of attachment, detachment, and ion–ion recombination.

The attractive forces between neutral molecules are not usually strong enough to favour formation of long-lived 'Van der Waals' species except at very low temperatures. For ions, however, the interactions with polarizable, or especially dipolar, molecules are much stronger, and 'cluster ions' are a feature of ion chemistry at low altitudes, as we shall see later. Some typical clustering processes are shown in reactions (10.37) to (10.39).

10.4.1 Ionization Mechanisms

Positive ions are formed in the atmosphere by three principal agencies: (i) solar radiation in the EUV and X-ray wavelength regions; (ii) galactic cosmic rays (GCR) and other galactic radiations; and (iii) precipitating energetic particles of solar origin and from the Earth's radiation belts. In the lower troposphere, radioactive emanation from rocks provides an ionization source at all times.

Photoionization, by EUV and X-rays, is the most important ion source above 60 km during the daytime with a 'quiet' Sun (*i.e.* in the absence of solar storms). Radiation of all wavelengths in the range reaches the F_2 region (>200 km), while the range 20–91 nm probably contributes to F_1 ionization ($\sim$140–200 km). E-region ($\sim$90–140 km) ionization comes from the more deeply penetrating part of the spectrum, EUV radiation between roughly 80 and 103 nm and X-rays from 1 to 10 nm wavelength. Several strong atomic lines are superposed on the solar continuum in the EUV region. Lower down in the atmosphere, in the D-region ($\sim$70–90 km), the only surviving ionizing radiations are EUV at $\lambda > 103$ nm and 'hard' X-rays ($\lambda \sim 0.2$–0.8 nm). The only strong atomic feature is H Lyman-α (Ly–α) at $\lambda = 121.6$ nm ($\equiv$10.19 eV); rather curiously, O_2 has a gap in its spectrum, where absorption is relatively weak, at exactly this wavelength, providing a 'window' that allows the radiation to penetrate more deeply than it would otherwise. The primary ions formed depend on the available wavelengths and the neutral precursors at any altitude. In the F-region, therefore, O^+, together with some N^+, are the dominant primary ions. Molecular ions become the major primary products in the E-region, O_2^+ being generated by solar atomic lines, and soft X-rays ionizing N_2 to make N_2^+. Only minor constituents such as NO or metal atoms can be ionized by the Ly-α that penetrates into the D-region. In addition, metastable *excited* O_2 in the $^1\Delta_g$ state (see Section 10.2.4) can be ionized in the wavelength range 102.7–111.8 nm. Hard X-rays can ionize O_2 and N_2; cosmic rays can ionize all constituents. Solar intensities in the EUV, and especially X-ray, regions are markedly dependent on the 11-year solar cycle and they also show an oscillatory variation over the 27-day rotational period of the Sun, so that the abundance and types of ions formed in the D-region also follow the solar cycle.

Ionization does not cease entirely at night, even though removal of the main source, solar radiation, does drastically reduce the ionization rate. Radiation from the illuminated portion of the atmosphere into the night sector may be achieved by resonance scattering (resonance absorption followed by fluorescence). Additional sources include starlight (stellar continuum radiation in the spectral interval 91.1–102.6 nm) in the E-region, and galactic cosmic rays (which continue to arrive at night, of course) in the D-region and below.

Corpuscular ionization—that is, ionization by impact with energetic particles from, or released by, the Sun—is a potential source of night- or day-time ionization. During 'disturbed' solar conditions, especially just after the maximum of the sunspot cycle, solar storms and flares (Section 10.3.1) are fairly frequent. Increases in ionospheric electron density are associated with the storms, and lead to a variety of ionospheric disturbances. *Sudden ionospheric disturbances* (SID) cause shortwave fadeout of radio signals due to absorption in the enhanced D-region (Section 10.3.3). The SID is produced by an intensification of hard X-ray emissions by several orders of magnitude, and is thus a photon effect. Corpuscular effects are generally delayed by at least several hours after the solar flare, and can persist for several days, in distinction to the flare itself that lasts for tens of minutes. One effect involves energetic electrons trapped in the Earth's radiation belts, which are slowly precipitated in the days following the storm to produce increased ionization in latitudes from approximately 45° to 72°. *Solar particle* (or *proton*) *events* (SPE: see Section 9.5.4), on the other hand, lead to enhanced ionization at high magnetic latitudes ($>60°$), and are due to energetic particles, mainly protons, ejected from the Sun. The delay in arrival of the particles is a result of their having been guided into a long spiral path by the interplanetary magnetic field. Radio blackouts lasting several days can follow SPEs; because the D-region enhancement covers the polar regions only, such radio disturbances are called *polar cap absorption* (PCA) events. In almost all disturbances (SID, electron, and SPE) of the D-region, the primary ions are virtually 100 per cent O_2^+, since the major constituent O_2 is capable of being ionized (NO^+, it will be remembered, is dominant in the quiet D-region).

10.4.2 The F-Region

F-region ion chemistry is relatively simple! It centres on the conversion of the primary ion O^+ to secondary molecular ions that can recombine with electrons: these are both points explained in the preceding parts of Section 10.4. Two pairs of processes illustrate the conversion and neutralization

$$O^+ + O_2 \rightarrow O + O_2^+ \tag{10.40}$$

$$O_2^+ + e \rightarrow O(^*) + O, \tag{10.41}$$

and

$$O^+ + N_2 \rightarrow N + NO^+ \tag{10.42}$$

$$NO^+ + e \rightarrow N(^*) + O. \tag{10.29}$$

Reaction (10.40) has the *effect* of charge transfer, although it may proceed partially through an exchange mechanism. Electronic excitation in one of the fragment atoms in reactions (10.41) or (10.29) can help to carry off excitation energy, and it is represented in the equations by the bracketed asterisk. Optical emission from the excited atoms is then sometimes observed as airglow, the subject of Section 10.2. At F-region altitudes, $[N_2] \gg [O_2]$ because photodissociation of O_2 is almost complete. Thus, although the rate constant for reaction (10.42) is about an order of magnitude smaller than that for reaction (10.40), the NO^+ route to neutralization is the more important.

The daytime peak in the F_2-region ion and electron concentrations is in part a result of the detailed charge-neutralization (*i.e.* loss) mechanism. Since the rate-determining step in the loss processes is conversion of O^+ to molecular ions, mainly in reaction (10.42), the loss rate is proportional to $[N_2]$. Ionization rates are proportional to [O], so that the steady-state electron concentration is proportional to $[O]/[N_2]$. But the concentration of N_2 falls off more rapidly than [O] with increasing altitude (gravitational separation, *etc.*), so that electron density in this steady-state picture *increases* with height. In reality, the chemical destruction of electrons and ions from any altitude is supplemented by transport loss. Diffusion rates increase with decreasing pressure (increasing altitude) so that electron concentrations ultimately fall. The position of the F_2 peak will thus occur where the chemical and diffusive loss rates are identical.

10.4.3 E-Region Processes

Molecular ions are the primary ionization products in the E-region. As discussed in Section 10.4.1, both O_2^+ and N_2^+ are produced from the major neutral atmospheric constituents. Observed ion distributions (Figure 10.12) show only a small contribution from N_2^+; this ion must therefore be consumed rapidly in secondary reactions. In fact, the reactions that represent losses of N_2^+ are also sources of the terminal ions

$$N_2^+ + O \rightarrow N(^*) + NO^+ \quad (10.26)$$

$$N_2^+ + O_2 \rightarrow N_2 + O_2^+. \quad (10.24)$$

Exchange with atomic oxygen in the first reaction is favoured at high altitudes, and charge transfer with O_2 at lower levels, following the $[O]/[O_2]$ profile in the atmosphere.

Neutral nitric oxide begins to play an important part in atmospheric chemistry at E-region altitudes. Two potential sources are the reactions

$$N(^2D) + O_2 \rightarrow NO + O \quad (10.43)$$

$$O_2^+ + N_2 \rightarrow NO + NO^+ \quad (10.27)$$

And since the reactant excited atomic nitrogen, $N(^2D)$, is formed in the dissociative recombination reaction (10.29), both these sources of neutral NO are in essence ionic processes. Once present, NO tends to act as a charge-transfer acceptor

$$O_2^+ + NO \rightarrow O_2 + NO^+, \quad (10.44)$$

because of its low ionization potential. Reactions such as (10.27) and (10.44) thus tend to increase the NO^+ concentration at the expense of O_2^+. Dissociative recombination of NO^+ in reaction (10.29), rather than of O_2^+ in (10.41), is the most important charge-neutralization step in the E-region, since $[e] \ll [NO]$, and most O_2^+ ions will collide with NO to transfer charge before they meet an electron.

From time to time, unusual propagation of short-wave radio signals suggests the presence of local areas of increased ionization in the E-region. These effects are known as *Sporadic E* phenomena. One manifestation of Sporadic E is long-distance reception of television pictures that have arrived by a reflected path that is usually absent. Narrow (1–3 km) localized layers of metal ions are often associated with sporadic E. Peak metal-ion concentrations can be two orders of magnitude higher than the molecular (NO^+, O_2^+) ion concentration in the vicinity of the layer. As explained in Section 10.2.5, these metals are brought in by incoming meteoroids ablating when they encounter atmospheric friction. Ionospheric chemistry is profoundly affected by the presence of metals. Metals have low ionization potentials ($\lesssim 7\,\text{eV}$) so that that they can become ionized by exothermic charge transfer from all molecular ions; *e.g.* with magnesium

$$\mathrm{Mg} + \mathrm{O_2^+}(\text{or } \mathrm{NO^+}) \rightarrow \mathrm{Mg^+} + \mathrm{O_2}(\text{or NO}). \tag{10.45}$$

Then, because they are monatomic, the positive charge-carrying metal ions are prevented from being neutralized, since the only electron recombination route is the very inefficient radiative process (see introduction to Section 10.4). Within the layers, therefore, electron (and ion) densities are abnormally high because there is no available loss pathway.

10.4.4 D-Region Positive-Ion Chemistry

Chemical complexity characterizes the D-region! Low temperatures (the lowest in the entire atmosphere), relatively high pressures, and a wide range of minor trace reactants permit a multitude of reactions. At the same time, *in-situ* experimental study of the D-region is difficult because ion and electron concentrations are low and variable; high pressures hinder sampling into mass spectrometers; and the large ions encountered have a tendency to fragment while being sampled. Nevertheless, a quite considerable understanding of D-region chemistry has emerged over the last 10–20 years, assisted by laboratory studies of the kinetics of potentially important ion–molecule reactions. Investigations of positive-ion compositions (in the altitude range 64–112 km) date from 1965. These earliest mass-spectrometric studies straight away showed that some unexpected ions were present at the lower altitudes, with mass numbers separated by 18 units: that is, the mass of H_2O. At first, there was much suspicion that contamination or some other artefact was responsible for the mass-spectrometric peaks, but it is now certain that the ions are ionospheric hydrated protons, $H^+(H_2O)_n$, with n ranging from 2 to at least 8, and occasionally even 20: H_3O^+ itself seems, at most, to be of minor importance. Figure 10.13 shows some typical results.

The growth of the water-cluster ions results from the sequential addition of H_2O to $H^+(H_2O)_2$

$$\mathrm{H^+(H_2O)_2} + \mathrm{H_2O}\ (+\mathrm{M}) \rightarrow \mathrm{H^+(H_2O)_3}\ (+\mathrm{M}), \tag{10.46}$$

and to each more hydrated product ion. A clear boundary, at ~82–85 km, is observed in most measured ion profiles, with the $H^+(H_2O)_n$ water-cluster ions being dominant below the boundary, and NO^+ and O_2^+ above it. At the same altitudes, most daytime rocket flights find that the electron density decreases by almost an order of magnitude within a height decrease of a few kilometres. Reduced photoionization rates cannot account for the abrupt decrease in electron densities at this 'ledge', which means that the changes must have been brought about by greatly increased loss rates at lower altitudes. Negative-ion formation accounts for some of the electron loss,

$$e + \mathrm{O_2} + \mathrm{M} \rightarrow \mathrm{O_2^-} + \mathrm{M}, \tag{10.32}$$

and, since it is a three-body process, its rate increases as the square of the pressure. Even more important, the large water-cluster ions are exceptionally good at dissipating recombination energy, since several polyatomic fragments are formed

$$\mathrm{H^+(H_2O)}_n + e \rightarrow \mathrm{H} + n\mathrm{H_2O}. \tag{10.47}$$

Dissociative recombination of $H^+(H_2O)_n$, with $n \sim 6$, is at least an order of magnitude faster than recombination with NO^+, reaction (10.29), at D-region temperatures (~200 K). Reaction (10.46) is not only a loss process for electrons, but also for cluster ions. It is thus at least self-consistent that

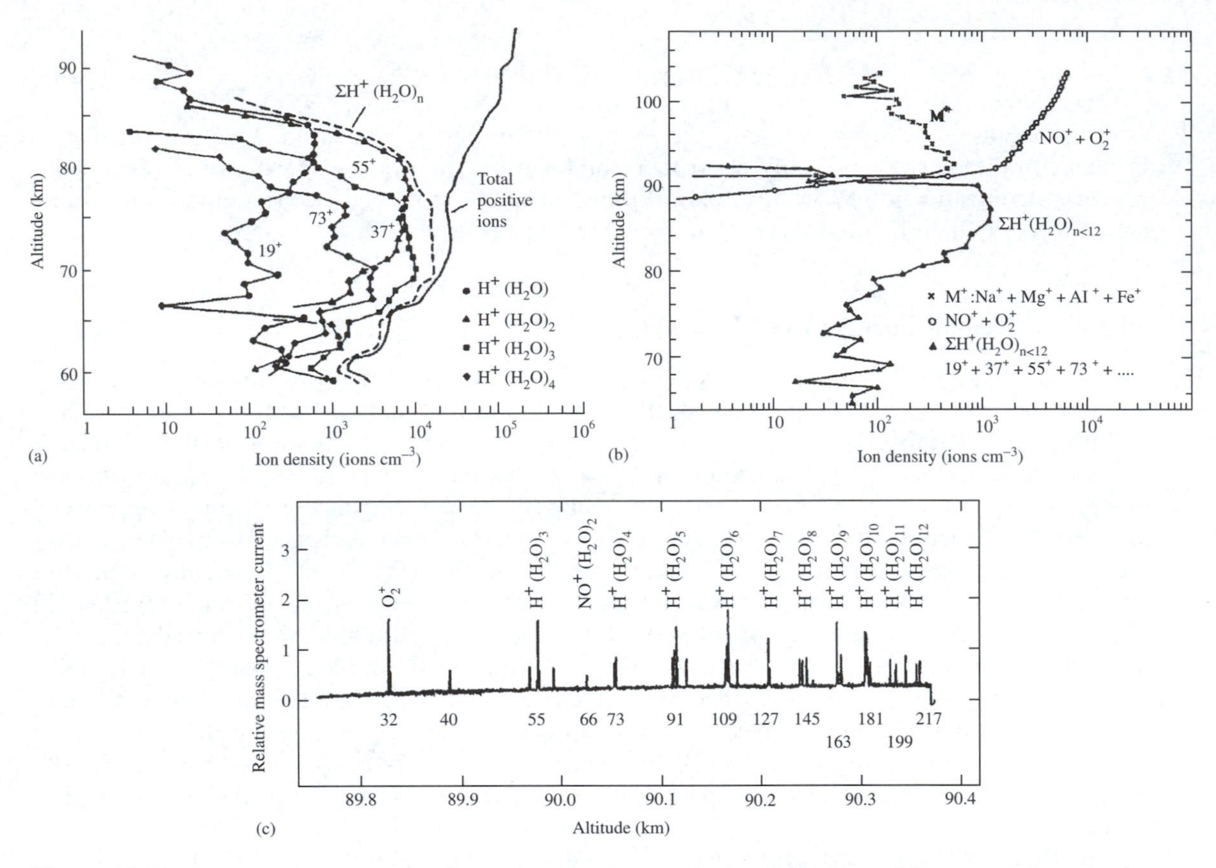

Figure 10.13 Results of some rocket-borne mass-spectrometric determinations of ion composition in the D-region: (a) above Red Lake, Ontario; the main proton hydrates are shown, and the decrease in concentration of these species above ~85 km is clearly marked; (b) above Kiruna, N. Sweden; the transition—here at ~90 km—between proton hydrates and NO^+, O_2^+ is even more evident; (c) a mass-spectral scan (from the Kiruna flight of (b)) just below the transition height showing that $H^+ (H_2O)_n$, $n=3$ to 12, are the dominant ions. From E. Kopp and U. Herrman, *Ann. Geophysicae*, 1984, **2**, 83, with kind permission of Springer Science and Business Media.

$H^+(H_2O)_n$ is absent when $[e]$ is relatively large ($\gtrsim$82 km) and that $[e]$ is small when $[H^+(H_2O)_n]$ is relatively large ($\lesssim$82 km).

There are, of course, many complexities and additional reactions that participate, and this is certainly a case where diagrams are worth many words that tend to be repetitive. In this spirit, we present Figure 10.14, which gives a 'simplified' overview of the major pathways in the ion chemistry of the D-region.

Before leaving the subject of the positive-ion chemistry of the D-region, there is one further phenomenon that should be described. *Noctilucent clouds* are shown in Figure 1.1 in the upper mesosphere, and proton-hydrate chemistry may be responsible for their nucleation. As the name implies, these clouds are only visible in the dark sky long after sunset: they are similar in appearance to thin cirrus clouds, but much higher and more tenuous. In order that they may still be lit up by the Sun, while the Earth's surface is in darkness, the noctilucent clouds must be situated at great height. The clouds lie at altitudes between 80 and 87 km, with the base most frequently near 82 km. Noctilucent clouds seem to be largely of summer–high-latitude occurrence. That is, they are formed in the regions of the atmosphere and the seasons of the year where atmospheric temperatures are

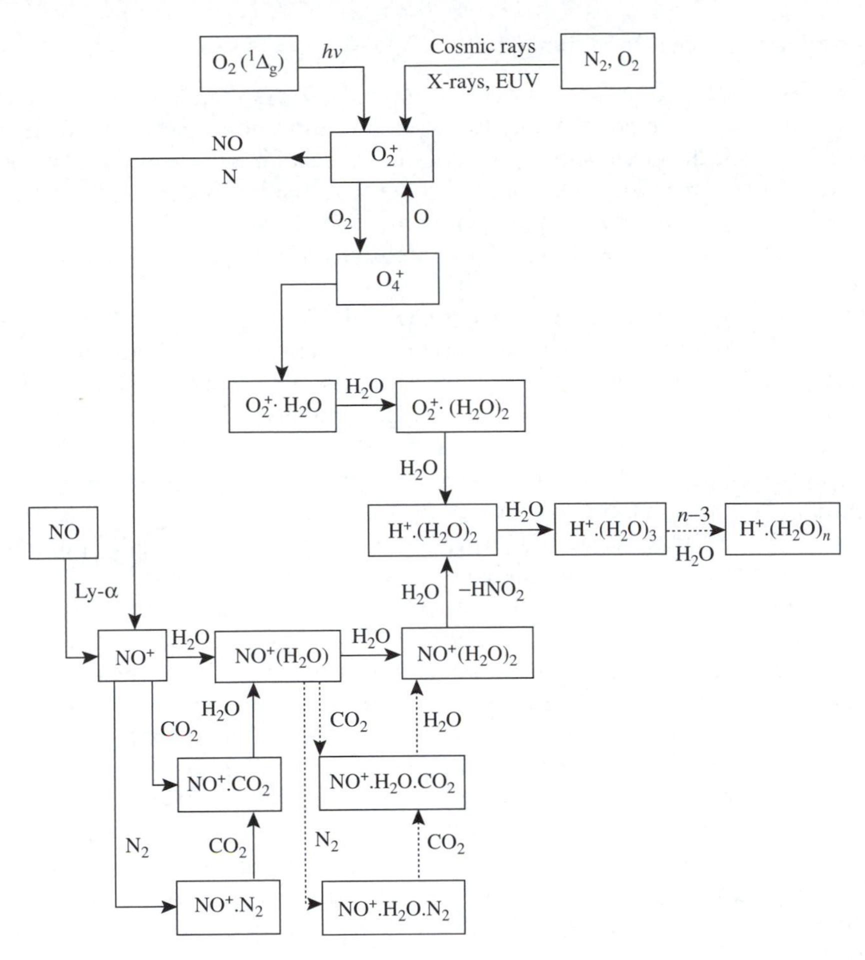

Figure 10.14 Reaction scheme for positive-ion chemistry in the D-region showing the major steps in the conversion of the primary ions, O_2^+ and NO^+, to the terminal ions, $H^+(H_2O)_n$. Adapted from E.E. Ferguson, F.C. Fehsenfeld and D.L. Albritton in *Gas Phase Ion Chemistry, Vol. 1*, ed. M.T. Bowers, Academic Press, New York, 1979.

the lowest—perhaps even as low as 100–120 K. They are also, it will be noticed, formed at altitudes close to the D-region transition in ionospheric composition and concentration. Various pieces of evidence, including direct rocket sampling, suggest that noctilucent clouds are composed of water-ice. However, even at the low temperatures involved, it is hard to see how ice crystals could be formed by conventional nucleation mechanisms (see Sections 1.3, 3.3.1 and 7.2.4). One hypothesis is that the ionic water-cluster species, known to be abundant at the cloud altitudes, are associated with the nucleation process. One possibility is that the dissociative recombination of $H^+(H_2O)_n$ in reaction (10.46) leaves a neutral cluster fragment $(H_2O)_m$ sufficiently large to start uninhibited growth. Whatever the detailed formulation, the salient feature of the proton hydrate idea is that the energy barrier that exists to nucleation of neutral molecules can be overcome by the electrostatic ion–dipole attractive forces in the protonic species. Mass-spectrometric rocket measurements of positive-ion composition have shown that there may be a depletion, or even complete disappearance, of the main positive proton hydrates as an intense noctilucent cloud is traversed. There is thus evidence for the attachment of the positive ions to the particles of the cloud.

10.4.5 D-Region Negative-Ion Chemistry

Rather few detailed mass-spectrometric measurements exist for D-region negative ions. One reason is that negative ions exist as important species only where total (ion + electron) concentrations are low ($z < 78$ km). Further, the negative-ion population consists of many different chemical species, so that any individual mass-spectrometric peak is weaker than the less distributed positive-ion peaks. Technical problems also interfere with the sampling and detection of negative ions. Notwithstanding the problems, an understanding, at least in outline, of D-region negative-ion chemistry has begun to emerge.

The terminal ions appear to be species such as CO_3^-, HCO_3^-, Cl^-, and the hydrates $NO_3^-(H_2O)_n$ and $CO_3^-(H_2O)_n$, while the primary negative ion is O_2^-, produced by three-body electron attachment (see Table 10.1, reaction (10.32)). Such a process will show a dependence of reaction rate on the square of the pressure, so that negative-ion production is confined to altitudes below 70–80 km. Dissociative attachment of electrons to ozone (Table 10.1, reaction (10.33)) is a minor source of negative ions; as with reaction (10.32), there is a strong altitude dependence because of the rapid increase in $[O_3]$ with decreasing altitude in this region. The transition from ions to electrons as dominant carriers of negative charge is sharpened by the steep increase in [O] above 75–80 km. Atomic oxygen inhibits the growth of stable negative ions, an effect that can be expressed (with some simplification) as a result of the associative detachment between O_2^- and O (Table 10.1, reaction (10.35)). The arguments for a sudden increase in negative-ion concentrations below a boundary altitude are thus very similar to those used to explain $H^+(H_2O)_n$ emergence at the same altitudes, but with the feature of reduced dissociative recombination being absent.

Electron affinities of the neutral precursors of simple atmospheric negative ions are generally rather small, and much less than the ionization potentials of similar neutral species. For O_2 itself, the electron affinity is ~ 0.44 eV (42.5 kJ mol^{-1}). The negative ions are thus relatively unstable and have a tendency to participate in reactions, such as (10.35), that lead back to the release of electrons. In competition with the loss processes, ion–molecule reactions with minor molecular species can lead to the terminal ions that are stable enough to resist electron detachment. For example, the electron detachment energy of NO_3^- is ~ 3.9 eV ($\equiv 380$ kJ mol^{-1}), one of the largest known for a simple negative ion.

Photodetachment (the analogue of photoionization) is possible with visible and even infrared radiation for ions with small electron affinity. With O_2 itself, near-infrared radiation can remove the electron

$$O_2^- + h\nu(\lambda \lesssim 2800\,\text{nm}) \rightarrow O_2 + e. \tag{10.48}$$

Daytime negative ion-to-electron concentration ratios are thus potentially affected by photodetachment processes, and there seems little doubt that the electron–ion transition boundary is shifted several kilometres to lower altitudes by day. A more detailed investigation of the rate at which the D-region negative ions build up at twilight, and electrons build up at dawn, reveals further subtleties. Effects of illumination are particularly well revealed by absorption of radio waves during PCA events (see Section 10.4.1). Absorption results from increased free-electron concentrations in the D-region. Formation of negative ions therefore reduces the influence of the PCA. Very large day–night modulations of the absorption are observed, with the anomalous effects virtually disappearing at night.

The progression from O_2^- to the terminal ions (NO_3^- and its hydrates) proceeds *via* a series of increasingly complex ions. Two threads can be identified in the conversion, one of which involves the ions O_4^-, CO_4^-, NO_3^{-*} and the other O_3^-, CO_3^-, NO_2^-. Figure 10.15 shows these pathways and the intermediate steps for D-region negative-ion chemistry in the same manner as the positive-ion chemistry was displayed in Figure 10.14. Note the two forms of the negative nitrate ion: 'ordinary'

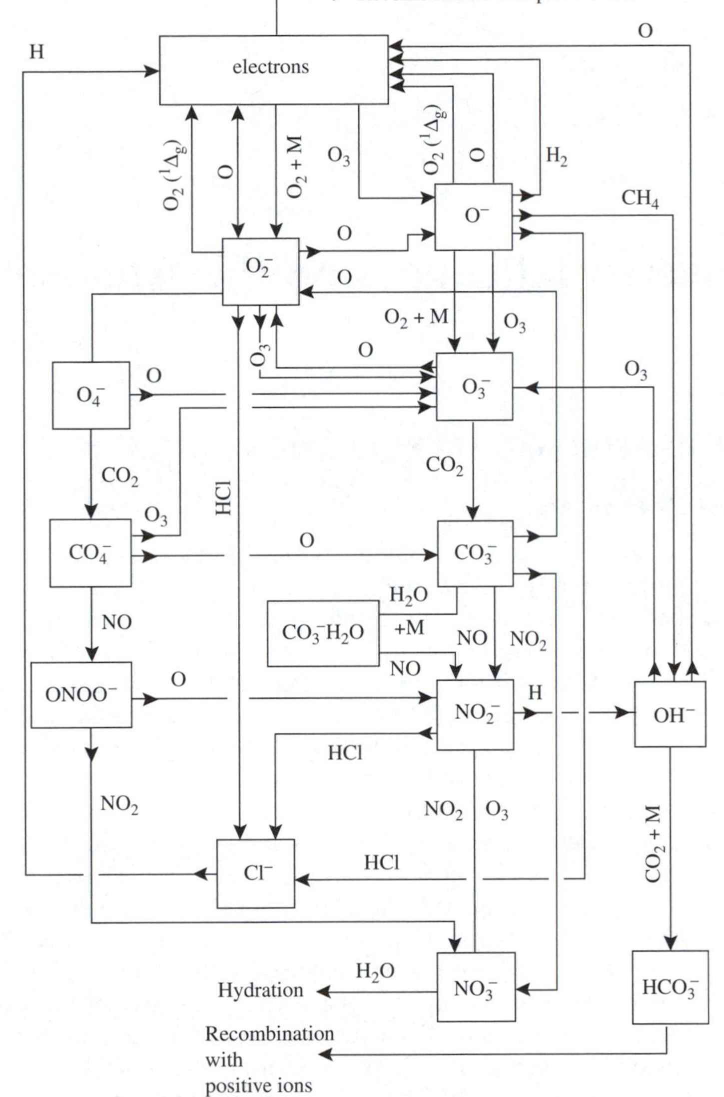

Figure 10.15 Reaction scheme for negative-ion chemistry in the D-region. O_2^- and O^- are the primary ions, while the terminal ions are species such as NO_3^-, CO_3^-, and their hydrates. Source as for Figure 10.14.

NO_3^- ($O^-.NO_2$) and NO_3^{-*} ($O_2^-.NO.$). At each stage of the conversion, the intermediate negative ions can be removed by reaction with atomic species or by photodetachment. Recent models of negative ion chemistry seem to mimic well what is known experimentally, and can therefore sensibly be used to predict what has not yet been observed.

With this brief introduction to negative-ion chemistry, we conclude our exploration of the higher reaches of the atmosphere, and in the next chapter examine the many ways in which Man can have an influence on the atmosphere, almost never, it has to be said, to the benefit of the environment.

CHAPTER 11

Man's Adverse Influences on the Atmosphere

11.1 LOCAL AND REGIONAL AIR POLLUTION

11.1.1 Clean and Polluted Air

Air is never perfectly 'clean' in the sense of containing only N_2, O_2, CO_2, H_2O, and the inert gases. Even before Man existed, there were many sources of the trace-constituent species, some of which would now be classified as pollutants. 'Natural air pollution' is caused by volcanic eruptions, breaking waves, windblown dust, forest fires, and pollens, dust, and terpenes from plants. Man has accentuated the frequency and intensity of release of some of these pollutants. Agricultural practices, in particular, have been a source of additional pollutant species ever since Man has been on Earth. 'Slash-and-burn' clearing of land adds greatly to the fire source; the removal of natural vegetation cover itself increases the rate of dust erosion.

Nitrogen fixed as fertilizer is in part returned to the atmosphere as N_2O, and alterations in the types and utilization of crops also affect the N_2O budget. Increased population of livestock and of rice-paddy cultivation imply an increase in CH_4 production. Destruction of forests, and their replacement by grasslands, may provide suitable habitats for termites that are also a potentially significant source of atmospheric CH_4 (Section 3.2.2). Human-accentuated forms of pollution, although significant, may seem minor compared with the releases of Man-made pollutants in industrial societies. Man seems capable of releasing—by accident or design—almost all known gaseous species to the atmosphere. Even substances such as lead, not commonly regarded as gaseous, can be produced as aerosols by Man's activities. It is not our purpose to catalogue here all the pollutants that have been, or might be, released. Rather, we wish to concentrate on pollution processes that are of rather widespread occurrence, with respect both to geographical location and to frequency. We have already seen (Section 9.3.1 and Table 7.1) that some trace gases whose release is due to Man (CF_2Cl_2), or accentuated by him (N_2O, CH_4), can have global consequences because of stratospheric interactions, and we shall extend the discussion in Section 11.2. In the troposphere, the most important forms of pollution seem to be related to the combustion of fossil fuels. It is these types of pollution that will provide the emphasis in the remainder of Section 11.1.

Atmospheric Chemistry
By Ann M. Holloway and Richard P. Wayne

Published by the Royal Society of Chemistry, www.rsc.org

Combustion of fossil fuels affects the atmosphere in three main ways: by the formation of carbon dioxide, by the release of substances such as sulfur dioxide and partially oxidized or unburnt fuel, and by the high-temperature fixation of atmospheric N_2 and O_2 to yield oxides of nitrogen. The carbon dioxide problem arises because, in the course of a few hundred years, appreciable fractions of the Earth's fuel deposits have been burnt to release carbon dioxide, thus reversing the chemistry that photosynthesis took millions of times as long to achieve. At present, atmospheric CO_2 concentrations are increasing by about one part in 300 a year, corresponding to about one-fifth of the annual release of carbon dioxide from the combustion of fossil fuels (see Figure 6.3). The potential climatic effects of increases in a 'greenhouse' gas are considered in relation to the evolving atmosphere (Section 11.4); changes in stratospheric temperature consequential on CO_2 increase can affect ozone chemistry (Chapter 9). Coal and petroleum products contain appreciable quantities of sulfur compounds. Combustion then leads to sulfur dioxide release. Particulate matter (soot), carbon monoxide, and unburnt hydrocarbons are also important minor by-products of the combustion process. We discussed sulfur dioxide chemistry in Section 8.4, and shall examine some consequences of SO_2 release in Sections 11.1.4 and 11.1.5. Sulfur dioxide and NO_x, formed when combustion is supported by air, contribute to the problems of acid rain (Section 11.1.5), and the oxides of nitrogen also play a special part in 'photochemical air pollution' (Section 11.1.6).

Pollutants released to the boundary layer may be quite rapidly removed by wet or dry deposition, so that they degrade air quality only near the source. Soot, SO_2, and malodorous compounds emitted from factories often fall into this category, and contribute to *local air pollution*. More widespread dispersal within the troposphere can lead to *regional pollution* of quite large geographical areas. If the entire atmosphere is affected, then the pollution is *global* in extent. Attempts to reduce the local impact of released pollutants can sometimes backfire by increasing the impact on the larger scale. For example, by building very tall chimney stacks for factories, local pollution can be much reduced. However, the removal of pollutants at higher altitudes may be a far slower process than near the surface, and on the regional scale the pollution problem may be exacerbated since the pollutants can now affect more distant regions downwind.

The importance of sensible legislation is paramount, and the complexity of the interactions requires scientific knowledge and insight. The economic, as well as the ecological, costs of many forms of pollution are potentially enormous. Control of emission is unfortunately also costly. Consider, for example, the emission of SO_2 and NO_2 by power stations using fossil fuels. Fuels can be desulfurized, or low-sulfur fuels used, and effluent stack gases can have their SO_2 and NO_x content reduced before release. But these processes are expensive: the cost of electric power would increase by at least 10 per cent if such measures were introduced. Alternative energy sources, such as nuclear fuels, are currently unpopular with the general public precisely because of the pollution dangers. Until the dream of fusion power becomes reality, some conflict seems bound to arise between economic and environmental pressures. Of course, combustion of fossil fuels will last for only a brief period in Man's history. No energy-conservation plan could reasonably envisage large fractions of our energy needs being met by fossil fuels alone for more than a few hundred years. By that time, an alternative energy source or a completely new life-style must have been found.

While on the subject of reduction of pollution, control measures, and policy making, it may be appropriate here to introduce the concept of *critical load*. The critical load may be defined as 'the highest load that will not bring about chemical changes that would produce long-term harmful effects on the most sensitive ecological systems'. It is expressed as a quantity of atmospheric pollutant deposited on unit area. Such a load thus represents a threshold quantity, based on a dose–response relationship. There are obviously attractions in using critical loads for formulating policy decisions, rather than adopting arbitrary concentrations or 'air-quality standards' as the basis of legislation.

11.1.2 Effects of Pollution

Air pollution is seen as a growing threat to our welfare, and especially to human health, as ever-increasing quantities of contaminants are delivered to the constant-sized atmosphere. An average adult male gets through about 13.5 kg of air a day, compared with about 1.2 kg of food and 2 kg of water. The quality of the air breathed is therefore at least as important as the cleanliness of our food and water. Air pollutants can exert an influence on health in two ways. First, they may be physiologically toxic, and secondly they may possess a nuisance value. Thus pollution can, by being smelly or dirty, have a health effect far beyond that of simple poisoning.

One of the major difficulties we face in assessing the impact of a pollutant on *human* health concerns the experiments that are possible. It is obviously out of the question to perform clinical tests of human response to any substance at concentrations that could conceivably be toxic. We have, therefore, to fall back either on the effects of accidental exposure of small groups to high doses, or to statistical correlations between mortality or morbidity and (chronic) exposure to low doses: such studies form the basis of *epidemiology*. By far the most hazardous pollution is that inflicted on self (and neighbours) by tobacco smokers, and the difficulties of air-pollution epidemiology are compounded because of this factor. Indeed, the differences between smokers and non-smokers in virtually all health indicators are so large that it is often impossible to detect whether other forms of air pollution are presenting a health hazard. Cigarette smoke contains, in addition to carcinogens and nicotine, sufficient carbon monoxide to clearly worsen coronary heart disease and respiratory problems. Toxicological trials on animals are of only limited relevance, because we cannot be sure that human responses will match those of the species tested, even after due allowance has been made for the size of the species.

We shall consider the health effects of individual pollutants in subsequent sections, but there are some comments of general applicability that can be made here. The main route of entry into the body of air pollutants is *via* the respiratory system, although some solid material deposited there may subsequently be swept to the gastrointestinal system. Because of the mode of access, respiratory symptoms are the most common response produced by pollutants. Low concentrations of oxidants such as O_3 or NO_2 produce nasal and throat irritation, while slightly higher concentrations impair mechanical lung function. In healthy subjects, the change is reversible within a few hours. So long as the change in function is unaccompanied by symptoms or by decreased work capacity, the exposure probably does not correspond to a real impairment of health. Of course, in persons with underlying respiratory illness, the situation is altered, since the *normal* function may already limit activity. Further increases in irritant concentration (say > 0.3 ppm of ozone) lead to sufficient discomfort to restrict normal activity. Even in the healthy, the effects experienced depend on whether the subjects are exercising or at rest. Presumably, when exercising the expiratory flow rates are increased, and breathing is through the mouth, so that the pollutant is delivered deeper within the respiratory tree.

Particulate matter in the atmosphere can obviously present a challenge to the respiratory system. Recent epidemiological studies suggest that particles of diameter less than 10 μm (identified as PM10) can increase premature mortality by one per cent if their concentration increases by as little as $10\,\mu g\,m^{-3}$. The relation seems to hold in all geographical locations where the studies were conducted, regardless of whether the local pollution was predominantly of SO_2 and primary particles, or of the *precursors* of photochemical oxidants and secondary particles (see Section 11.1.3). Two possible interpretations of these observations are either that there are common toxic or mutagenic species found in most particles, or that inhalation leads to a general inflammatory response independent of the specific composition. Smaller particles can reach the lungs even more easily. The smallest particles are generally those produced in combustion and by gas-to-particle conversion (see Figure 1.4). If particle size is of great significance in determining the health hazard, then pollution control strategies may have to be examined particularly carefully. For example, there has been a great effort to reduce the emissions of particles from diesel motors, visible as black smoke from older engines. Although improvements in design have led to a substantial reduction in emissions measured as total mass and of numbers of particles in the size range of 0.05 to 1 μm, these reductions may have been gained at the expense of greatly *increased* numbers of emitted particles of diameters less than about 0.05 μm. As a result, the reduction in atmospheric pollutant load (as well as perceived 'dirtiness' of the engine) may be somewhat illusory in terms of benefits to the health of the population.

Living species other than Man may be at risk from Man-made pollutants. As we shall see in Section 11.1.5, fish seem to be highly sensitive to the pH of their environment, and release of pollutants tending to increase acidity can have serious consequences for freshwater fish populations. Acute and chronic injury to the leaves of trees and plants can be caused by a number of pollutant gases. Sulfur dioxide, and the oxidants O_3 and NO_2, all cause damage: they are said to be *phytotoxic*. Combinations of the gases seem particularly harmful. Economic consequences in terms of lost foodstuffs or forests can be very great indeed. The indirect effects of reduced nitrogen fixation and CO_2 turnover, and of interference with entire ecosystems, could be of even greater long-term importance.

Economic loss can also be sustained by damage to inanimate materials. Many organic substances are susceptible to attack by oxidants such as ozone, especially if they contain double bonds. Elastomers such as natural and synthetic rubber, paints, and dyes are attacked. Chemical measures introduced to prevent damage may themselves constitute a major cost in production. Structural damage can be caused to buildings by species such as sulfur dioxide or sulfuric acid, which attack carbonates present in the stone. Compounds of larger volume are formed, which lead to flaking of the stonework. Disfigurement by blackening of stone surfaces may be a result of simple deposition of particles, but it may also involve a more complex organic process that first requires the absorption of SO_2.

Aerosols reduce visibility in their own right; in addition, particles may act as condensation nuclei to cause or aggravate water fogs. Apart from the 'nuisance' aspect of reduced visibility, it is also evident that limited visual range is a contributing cause of automobile and aircraft accidents.

Releases of species such as CO or CH_4, or alteration of their natural production rates by Man, seem to exert their main effects in an indirect way. Of course, carbon monoxide is toxic at high concentrations, and may be implicated in traffic accidents at rush-hour periods when urban atmospheric concentrations (up to 85 ppm) exceed the amounts known to degrade behavioural performance. However, average tropospheric concentrations ($\sim$0.2 ppm even in the industrialized Northern Hemisphere) are too low to elicit a direct response. Much more important is the reduction in tropospheric [OH] that is likely to follow increased CO emission rates. The reaction between CO and OH (reaction (8.6), Section 8.2.2) is a major loss process for OH ($\sim$70 per cent), so that [OH] is very sensitive to [CO]. Since the remaining 30 per cent of OH reacts with CH_4 in the 'clean' atmosphere, and CO is the product of oxidation, increases in atmospheric CH_4 also reduce hydroxyl radical levels in the troposphere. Smaller tropospheric [OH] in turn has consequences for the concentrations of a wide variety of natural and Man-made trace species, since reaction with OH is often the principal scavenging mechanism. As we have pointed out frequently, the possible responses then include not only changes in the troposphere but also modifications to stratospheric chemistry and global climate.

11.1.3 Primary and Secondary Pollutants

Pollutants may be grouped into two categories: primary and secondary. *Primary pollutants* are the chemical species emitted directly from identifiable sources. *Secondary pollutants*, on the

Table 11.1 Global Man-made and natural emissions of various species.

	Emission estimate (10^9 kg yr^{-1})				
Species	*Man-made: Combustion and industry*	*Biomass burning*	*Cultivated soils*	*Total Man-made*	*Natural*
CO_2				2.6×10^7	5.5×10^8
CH_4	100	40	60	375	160
CO	1315	500		1815	430
SO_2	140	6		146	~15
H_2S				4	1–2
N_2O	3.7	1.4	10	16	
NO_x (as NO)	51	17	13	85	26
NH_3				30	24
Particles	45	60		305	15

Based on tables presented in J. H Seinfeld and S. N. Pandis *Atmospheric Chemistry and Physics*, John Wiley, 1998. Natural CO_2 emission estimates are assumed to be the primary production from land (2.2×10^{17} kg yr^{-1}) and exchanged from the oceans (3.3×10^{17} kg yr^{-1}).

other hand, are species formed from the primary pollutants by chemical transformation. Adverse effects of pollution are often associated more with the secondary than with the primary pollutants. For example, although atmospheric sulfur dioxide has itself many harmful effects, the sulfuric acid formed as a secondary pollutant by oxidation of SO_2 is even more damaging to the environment.

Table 11.1 compares Man-made with natural emission rates of some of the most important species. Man-accentuated processes are not included, so that the total contribution of Man's activities to species such as CH_4, CO, or NO, may be much higher. Man evidently releases very large quantities of material to the atmosphere. The units are 10^9 kg yr^{-1}—that is, millions of tonnes annually. For most species, the 'anthropogenic' emissions are a minor, but significant, contribution to the global total budget. In the case of sulfur dioxide, Man produces more than the natural sources. Overall sulfur emissions include H_2S, COS, CS_2, and the organic sulfides, as well as SO_2. As a world average, Man contributes about 40 per cent, the rest coming from biogenic sources, sea-spray, and volcanoes, but in the industrial Northern Hemisphere, Man dominates over the sum of natural sulfur emissions.

Fuel combustion is by far the largest source of the oxidized species (SO_x, NO_x, CO). Coal contains up to 2.5 per cent of sulfur by weight. Transport (especially private cars and light-duty vehicles using petroleum fuel) is the single most polluting activity in the USA and other heavily urbanized parts of the world; the quantity of CO and unburnt hydrocarbons emitted reveals something about the efficiency of the internal combustion engine! That the sulfur emission is not higher is a result of desulfurization of the fuel. Motor spirit ('gasoline') contains between 0.026 per cent (USA Premium grade) and 0.040 per cent (UK) of sulfur, as compared with up to several per cent in the crude oil (as much as 1.8 per cent in residual fuel oil). Industrial operations such as chemical manufacturing, metal smelting and refining, and mineral extraction, are further significant sources of gaseous and particulate pollutants, and deliberate biomass burning adds further to the burden.

11.1.4 Smoke and Sulfur Pollution

Air pollution from the burning of coal has been a problem for centuries. At the beginning of the 14th century, Edward I forbade the use of coal because of the smell and smoke it produced. London, although not unique in suffering from the combination of smoke and SO_2 pollution produced by coal combustion, had severe problems well into the 20th century. London 'pea-souper' fogs were a recurrent nightmare in Victorian London (of course, the Hollywood version includes Sherlock Holmes and Jack the Ripper prowling through the nearly impenetrable gloom). The word *smog* was coined in 1905 to describe the combination of smoke and fog that was so disastrous.[i] Almost all heavily industrialized cities suffered to some extent. Probably the special place of London in smog history arose because of the size of the city and the scale of industrialization at a time when possible control measures were unknown. British bituminous coal is high in sulfur content, and the tars and hydrocarbons make for a high smoke yield. Such a combination is dramatically effective in fog nucleation in a climate already humid and possibly supersaturated. In 1952, a tragic air pollution episode occurred in London, as a result of which more than 4000 people died.

The deaths that followed acute smog episodes usually involved those with pre-existing heart and respiratory problems, primarily the elderly. It is recognized, however, that human health can suffer as a result of slower subtle effects, especially on the respiratory system, produced at lower air pollution levels. Sulfur dioxide is itself a respiratory irritant, the effects appearing at concentrations above about 1 ppm, substantially larger than the largest *now* recorded, even for highly industrialized urban areas. Below about 25 ppm, irritation is confined to the upper respiratory tract. The situation is greatly altered if particles (soot) are also present in the pollution. The lower part of the respiratory tract may then be involved at the lower concentration levels. A three- or four-fold potentiation of the irritant response to SO_2 results from the presence of particulate matter. Chronic and acute bronchitis, pleurisy, and emphysema are all produced by SO_2-containing smoke such as that generated in the combustion of bituminous coal. Part of the potentiation may be caused by the delivery and retention in the respiratory system of substances, including SO_2, absorbed on soot particles. In addition, as we saw in Section 8.4.2, SO_2 oxidation to H_2SO_4 (or SO_3) occurs efficiently on aerosol surfaces; in heavily polluted atmospheres, the presence of metals may enhance catalytic effects.

Shortly after the 1952 London smog disaster, decisive action was taken in Britain to alleviate pollution. Controls were placed on the type of fuel burnt and the kinds of smoke that might be emitted. A ban on all but 'smokeless' fuels in urban areas has been particularly effective in reducing the emissions of particulate matter, if not of sulfur dioxide. Although there were a few further serious episodes in the period immediately following the legislation, pollution disasters related to smog now appear to have become a thing of the past in London, and there seems to be a continuing downward trend in emissions. During the period 1970–2004, SO_2 concentrations measured in London decreased by a factor of roughly seven, and smoke by a factor as large as 34. Most other cities and regions report a similar success with smoke control. However, as the coal sources of SO_2 and smoke have diminished, emissions of pollutants, including particles, from road traffic have increased. Catalytic converters (see Section 11.1.6) have greatly decreased the emissions of gaseous species from motor vehicles, but the increased market share of diesel cars—which also has a beneficial effect on gas-phase emissions—may have led to less improvement in the emissions of particles. There is the further complication of the influence of particle *size* on health effects.

Despite the great improvements made in emissions in North America and Europe, in particular, sulfur dioxide remains a problem, as the emission inventories of Table 11.1 show. Developing nations may now be major contributors to the atmospheric load of SO_2 (and other pollutants), and the situation might well worsen. The most serious consequences may be due in part to the acidity of precipitation, and we turn now to this question.

[i] *Smog* is now used to describe any smoky or hazy pollution of the atmosphere, and includes the conditions encountered in Los Angeles (Section 11.1.6). A suitable qualifier, such as 'London', 'classical', or 'Los Angeles, 'photochemical', is often used to provide greater clarity of expression.

11.1.5 Acid Rain

Natural precipitation—rain and snow—is slightly acid, in part because carbon dioxide is dissolved in the falling droplets. Indeed, the acidity forms part of the geochemical weathering cycle, as discussed in Section 6.1. However, 'carbonic acid' (H_2CO_3) is a weak acid, so that the pH in water saturated with CO_2 is limited to minimum values of $\sim$5.6. With the presence of other naturally generated acidic gases, such as NO_x and SO_2, and in the absence of any neutralizing ions, the pH of natural rainwater is probably about 5.0, an estimate supported by examination of rain in remote and unpolluted regions. Over the last few decades, rainwater of much greater acidity (lower pH) has been of widespread occurrence in industrial areas, and the acids involved have mostly been 'strong' ones such as sulfuric, nitric, and hydrochloric. Acidic rain is potentially damaging to the environment, the two most serious influences appearing to be on freshwater fish and on forest ecology. Strong acids such as H_2SO_4 and HNO_3 have their origin in gaseous SO_2 and NO_2, while HCl may be produced by the reaction of H_2SO_4 with atmospheric NaCl of marine origin. We shall show shortly that most of the load of strong acid is a consequence of fossil-fuel combustion.

Two regions have been particularly badly affected by the acid-rain problem. They are the north-eastern United States and neighbouring parts of Canada, and the Scandinavian countries, particularly Sweden and southern Norway. 'Fossil' precipitation is sometimes preserved as glacial ice, and it indicates that the pH was generally greater than 5 outside urban areas prior to about 1930. Direct records in north-west Europe show that precipitation then became increasingly acid and that this acidity was more widespread geographically. In some parts of Scandinavia, the H^+ concentration in precipitation increased by a factor of more than 200 over the last two decades. Average pH levels in the north-eastern United States dropped to between 4.0 and 4.2, but values as low as 2.1 were recorded for individual storms. The greatest increase in acidity of precipitation in the USA appears to have taken place some time between 1930 and 1950. On the other hand, acid rain is no new phenomenon. In the early part of the 20th century, acidity was recognized in the rainwater of the industrial northern cities of England.

Emission control has been effective in some regions. For example, legislation in the USA in 1990 demanded reductions in SO_2 and NO_x emissions, and in the eastern US, total sulfur deposition decreased by 44 per cent between 1990 and 2007, while total nitrogen deposition decreased by 25 per cent over the same time frame. In 2004, the pH of water in this region was 4.5–4.7: a recovery from the lowest levels, but still quite acid.

Let us now consider the damage done by acidity. Large areas of southern Scandinavia and the north-eastern USA—those very regions most affected by acid rain—are underlain by granite-type rocks. Surface waters in such areas contain little dissolved matter, and are poorly buffered (in distinction to those waters underlain by chalk or limestone). The lakes and rivers are thus particularly sensitive to the prevalent acid precipitation, and have undergone extensive ecological damage. Fish are vulnerable to acidification, to which they exhibit a direct toxic response. In addition, a variety of other aquatic organisms in the food webare adversely altered. Freshwater fish have become extinct, or have declined in number, in Sweden, Norway, Canada, and the USA. About 10 000 Swedish lakes have been acidified to a pH below 6.0, and 5000 to below 5.0. Fish populations have been seriously affected, with losses of trout and salmon being particularly heavy. Those two types of fish spawn in rivers and streams where pulses of acidity may occur just at the same period as the vulnerable stage of egg hatching. Ion separation in the freezing and thawing process can lead to the acidity being concentrated in the first portion of melt-water liberated in spring.

The other potential effect of acid rain that we shall consider is that on land vegetation. Interpretation of response is difficult because the gaseous pollutants such as SO_2, NO_x, or O_3 have an adverse effect on plants. Soil may also be affected, *via* leaching of inorganic ions, reduced nitrogen availability, and decreased soil respiration. Conversely, acid rain may actually supply needed sulfur or nitrogen to soils deficient in these species, so that the damaging effects can be masked by nutritional benefits. Forest damage, in particular, has given rise to special concern, above all in Germany, where the possibility of acid rain leading to forest decline excited strong emotions. That this decline is caused mainly by acid precipitation, or the gaseous pollutants, acting alone or in combination does, however, look much less likely than was once supposed.

11.1.6 Photochemical Ozone and Smog

The London type of air pollution, characterized by particulate matter and SO_x (Section 11.1.4) has been recognized for centuries. In the mid-1940s, the effects of a new kind of oxidizing air pollution, which caused eye irritation, plant damage, and visibility degradation, became evident in Los Angeles. The oxidants include ozone, nitrogen dioxide, and peroxyacetyl nitrate (PAN: see Section 8.2.6), and they are formed photochemically by the action of solar radiation on mixtures of NO_x and hydrocarbons (HC) in air. Because of the region where it was first observed, and because of its photochemical origin, oxidizing pollution of the kind described is called *Los Angeles smog* or *photochemical smog*. Very many other cities, especially in the south-west USA, but also in Greece, Israel, Japan, Australia, and even in the UK, have subsequently been found to suffer from photochemical pollution. Photochemical smog has, however, been a particularly serious problem in Los Angeles and Southern California, and there are several contributing factors. The Los Angeles basin faces the Pacific Ocean to the south-west, and is otherwise almost enclosed by mountain ranges. This topography results in frequent temperature inversions in the boundary layer (Section 8.1.1) and lower troposphere, and pollutants are trapped within the basin. Intensely sunny days are frequent, thus promoting photochemical processes. Last, but not least, there is a very high density of automobiles that are thought to be the most important source of primary pollutants. Curiously, it turns out that photochemical 'pollution' is not really a new phenomenon at all! San Pedro Bay was named the 'Bay of Smokes' in 1542, and eye irritation was first recorded in Los Angeles by 1868. The blue haze of the Smoky Mountains—a tourist attraction—probably owes its existence to a biogenic variant of the automobile problem (see Section 8.2.7). The severity and incidence of pollution have, however, grown out of all recognition from the early precursors of photochemical smog, and the growth has paralleled the vast expansion of use of the internal combustion engine (especially for light motor vehicles) over the last sixty or seventy years.

What is observed on a smoggy day? Nitric oxide (NO) concentrations build up during the night and during the early-morning period of heavy commuter traffic. After dawn, NO becomes replaced by NO_2, and ozone is generated. By noon, there are high concentrations of ozone and nitrogen dioxide in the atmosphere, there is a brown haze because particles are present, and the eyes run because PAN, a powerful lachrymator, is formed. Figure 11.1 shows the interrelations between $NO–NO_2–O_3$ concentrations during an air-pollution episode.

We must now consider the chemistry that gives rise to the observed behaviour. The primary pollutants in automobile exhaust are NO_x (mainly NO) from the high-temperature combustion, carbon monoxide, partially oxidized and unburnt hydrocarbons (HCs), and sulfur dioxide from sulfur-containing fuels. It is these species that then undergo photochemical transformation to ozone, nitrogen dioxide, aldehydes, ketones and acids, PAN, and inorganic and organic aerosols. As we have already emphasized in Section 5.2.1, NO_2 photolysis followed by $O + O_2$ combination is the only known tropospheric source of O_3

$$NO_2 + h\nu \rightarrow O + NO \tag{11.1}$$

$$O + O_2 + M \rightarrow O_3 + M. \tag{11.2}$$

A route must therefore be found for conversion of the primary species NO to NO_2, before ozone can be formed. Inorganic chemistry on its own seems unable to bring about the oxidation. Three-body reaction

$$NO + NO + O_2 \rightarrow 2NO_2 \tag{11.3}$$

is far too slow at the concentrations of NO present (its rate is proportional to $[NO]^2$).

Reaction of NO with O_3,

$$NO + O_3 \rightarrow NO_2 + O_2, \tag{11.4}$$

would be fast enough to convert NO to NO_2, but, paradoxically, requires ozone to be available already (and the sequence (11.4), (11.1), (11.2) leaves ozone concentrations unaltered). Obviously, the organic species released in the exhaust gases must play a part in the oxidation process and in ozone formation. This result is confirmed in test-chamber experiments where the organic components are omitted from the gas mixture: no ozone is formed. The part played by reaction (11.4) seems to be to prevent O_3 build-up until almost all free NO has been consumed (and converted to NO_2), as suggested by the curves of Figure 11.1.

We do, of course, already know processes that convert NO to NO_2 in the unpolluted troposphere. They include the reactions with peroxy radicals

$$HO_2 + NO \rightarrow OH + NO_2 \tag{11.5}$$

$$CH_3O_2 + NO \rightarrow CH_3O + NO_2 \tag{11.6}$$

$$RO_2 + NO \rightarrow RO + NO_2 \tag{11.7}$$

These reactions do, indeed, seem to be the critical ones for oxidizing NO in photochemical smog. Smog chemistry is, then, a grotesquely exaggerated form of the oxidation and transformation

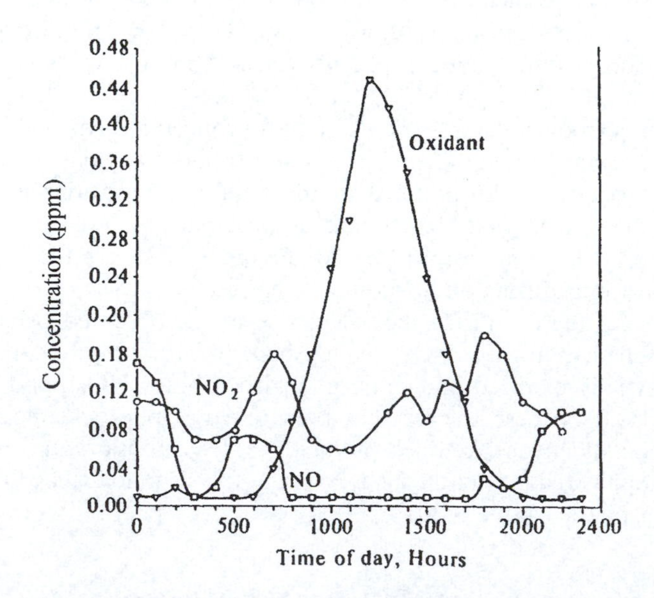

Figure 11.1 Variations in concentration of oxidant (mainly ozone) and oxides of nitrogen during the course of a smoggy day in Southern California. From B.J. Finlayson-Pitts and J.N. Pitts, Jr. *Adv. Environ. Sci. Technol.*, 1977, **7**, 75.

chemistry (Section 8.2.2) of the unperturbed troposphere. Higher concentrations of primary species (NO_x, HCs) are present in the polluted atmosphere, and perhaps a wider variety of saturated, unsaturated, and aromatic HCs is liberated. But the oxidation chain is still carried by OH, HO_2, and organic oxy- and peroxy-radicals, as in the natural troposphere. We may emphasize the conversion of NO to NO_2 in the chain process by writing the reaction sequence following attack of OH on an alkane, RCH_3

$$OH + RCH_3 \rightarrow H_2O + RCH_2 \tag{11.8}$$

$$RCH_2 + O_2 \rightarrow RCH_2O_2 \tag{11.9}$$

$$RCH_2O_2 + NO \rightarrow RCH_2O + NO_2 \tag{11.10}$$

$$RCH_2O + O_2 \rightarrow RCHO + HO_2 \tag{11.11}$$

$$HO_2 + NO \rightarrow OH + NO_2 \tag{11.5}$$

$$\text{Net} \qquad RCH_3 + 2NO + 2O_2 \rightarrow RCHO + 2NO_2 + H_2O.$$

Reactions (11.9), (11.10) and (11.11) are the analogues for RCH_2 radicals of equations (8.21), (8.20) and (8.23) in Chapter 8.

Attack of OH on RCHO continues the hydrocarbon oxidation by yielding carbonyl radicals

$$OH + RCHO \rightarrow RCO + H_2O, \tag{11.12}$$

that can lead, directly or indirectly, to carbon monoxide and the radical R possessing one less carbon atom than the starting hydrocarbon. Acids, RCOOH, are a minor product from RCO radicals, but they, together with the aldehydes, are found in photochemical smog. The degradation from RCH_2 to R radicals is accompanied by formation of aldehydes and acids with all possible numbers of carbon atoms, down to the first members, HCHO (formaldehyde) and HCOOH (formic acid).

Carbon monoxide is the final oxidation product of the organic chain, but is itself oxidized by OH. The sequence

$$OH + CO \rightarrow H + CO_2 \tag{11.13}$$

$$H + O_2 + M \rightarrow HO_2 + M \tag{11.14}$$

$$HO_2 + NO \rightarrow OH + NO_2 \tag{11.5}$$

$$\text{Net} \qquad NO + O_2 + CO \rightarrow NO_2 + CO_2$$

shows how this oxidation of CO is yet again accompanied by the oxidation of NO to NO_2.

Although we have illustrated the oxidation chain with an alkane as fuel, alkenes (olefins) react with OH radicals even faster, at rates approaching the collision- or diffusion-controlled limit. The initial attack appears to be the addition of OH to the double bond (Section 8.2.5), with the major products being the appropriate aldehydes and ketones (CH_3CHO from C_2H_4, and $C_2H_5CHO + CH_3COCH_3$ from C_3H_6). Aromatic compounds (Section 8.2.7) constitute a significant fraction of the reactive hydrocarbons in automobile exhaust gases. Hydroxyl radical attack is rapid, and the products are of particular interest because they may include long-chain oxygenated compounds that can be involved in aerosol formation.

The general mechanism of hydrocarbon oxidation, and of NO to NO_2 conversion, in a free-radical chain reaction seems well established. Hydroxyl radicals (and probably HO_2) attack the organic 'fuel' to propagate the chain. We must therefore now consider the origin of radicals in polluted atmospheres. The ozone photochemical source (Section 8.2.1) that is important in the natural troposphere may be supplemented by several other processes. Of these, two are of particular interest, since they involve species detected in photochemical smog. Nitrous acid (HONO) can be formed in the process

$$NO + NO_2 + H_2O \rightarrow 2HONO \tag{11.15}$$

by either a homogeneous or a heterogeneous route. The molecule is photolysed at relatively long wavelengths ($\lambda < 400$ nm) that reach ground level

$$HONO + h\nu \rightarrow OH + NO. \tag{11.16}$$

One oxidized molecule, NO_2, is lost in reaction (11.15), but two HONO molecules, and thus potentially two chain-initiating OH radicals, are created. Aldehydes may provide an important entry into the radical chain. One channel for the photodissociation of formaldehyde (at $\lambda < 340$ nm) yields H and HCO radicals. Both these radicals are converted to HO_2 in the presence of O_2, so that the photochemical initiation steps can be represented by the sequence

$$HCHO + h\nu \rightarrow H + HCO \tag{11.17}$$

$$HCO + O_2 \rightarrow CO + HO_2 \tag{11.18}$$

$$H + O_2 + M \rightarrow HO_2 + M. \tag{11.19}$$

Aldehydes are also implicated in the formation of peroxyacetyl nitrate, an important component of photochemical smog. Carbonyl compounds, either emitted as primary pollutants, or produced *via* processes such as (11.11) as oxidation intermediates, can be converted to acyl radicals in reaction with OH, and thence to peroxyacyl radicals

$$OH + RCHO \rightarrow RCO + H_2O \tag{11.20}$$

$$RCO + O_2 \rightarrow RCO.O_2. \tag{11.21}$$

Addition of NO_2 to $RCO.O_2$ then yields a peroxyacyl nitrate

$$R.CO.O_2 + NO_2 \rightarrow RCO.O_2.NO_2. \tag{11.22}$$

Formation of PAN itself from acetaldehyde is shown in Section 8.2.6.

Many of the undesirable effects of photochemical smog arise from the presence of suspended particulate matter. Nearly half the aerosol mass can be organic in severe photochemical smog, and of this organic fraction, 95 per cent is secondary in origin. A variety of long-chain aliphatic and aromatic compounds is found, together with oxygenated species such as acids, esters, aldehydes, ketones, and peroxides. Amongst the species are the *polycyclic aromatic hydrocarbons* (PAHs) to be discussed in Section 11.1.7; they include such potent carcinogens as benzo[a]pyrene. The mechanisms leading to the formation and growth of particles are still not entirely clear. As explained in Section 8.2.7, it has long been known that oxidation by ozone of hydrocarbons such as terpenes leads to polymerization and aerosol formation. Indeed, the 'natural' photochemical smog of California or the Smoky Mountains, alluded to at the beginning of this section, is ascribed to the reaction between oils from pine forests or citrus groves with ozone naturally present in the troposphere (Section 8.2.3).

Inorganic aerosol in photochemical smog includes sulfate, nitrate, and ammonium ions, as well as a variety of trace metals. Sulfuric acid is formed (Section 8.4) from SO_2 released by combustion of sulfur-containing fuels, while nitric acid involves the usual reaction with OH

$$OH + NO_2 + M \rightarrow HNO_3 + M. \quad (11.23)$$

Ammonia is assigned a prominent role in the neutralization of the acids, especially when there are high local concentrations produced by primary sources (*e.g.* cattle stations), as in California.

Photochemical air pollution degrades the 'quality' of the environment in all the ways outlined in Section 11.1.2. Human health is affected primarily by the oxidant species such as ozone, but PAN, NO_2, and the aerosols are also harmful. Impairment in physical performance has been demonstrated at oxidant levels (ozone + PAN) above 150 ppb. Attacks in asthmatics are exacerbated at 250 ppb of oxidant (but the same level may have no effect on healthy persons). In general, it seems that discomfort can be perceived—as chest pains, cough, and headache—for concentrations of oxidant beyond 250–300 ppb. The European Union has set limits of 50 ppb for an eight-hour exposure, and 89 ppb for a one-hour exposure. World Health Organization (WHO) guidelines for human health are essentially the same for the eight-hour exposure, while the current USA 'air quality standard' quotes a limiting ozone concentration of 75 ppm over an eight-hour period (70 ppb in California). Reference to Figure 11.1 shows that the ozone concentrations in the episodes represented greatly exceeded the standards throughout the period 10 a.m. to 6 p.m., and that the levels around midday were sufficient to have noticeable effects on health. In Great Britain, meteorological conditions rarely favour serious smog formation, although atmospheric ozone has exceeded the USA air quality standard on several occasions since regular monitoring began in 1970. During the exceptionally hot summer of 1976, photochemical pollution was enhanced, and between 22 June and 17 July rural hourly-mean ozone levels exceeded 250 ppb. At one rural site, ozone concentrations were in excess of 100 ppb over at least eight hours for 18 consecutive days of the 21 days of the episode. Similar levels were measured during the heatwave of mid-August 2003.

Peroxyacetylnitrate, PAN, is a powerful lachrymator, in addition to its oxidant effects on the respiratory system. That is, it causes intense irritation of the eye, with consequent tear formation. Irritation increases steadily for oxidant concentrations between 100 and 450 ppb, although ozone on its own is not an eye irritant. We cannot be clear whether eye irritation constitutes a real impairment of public health since it is reversible, and there is no proven association between pollution-induced irritation and chronic eye damage. Nevertheless, the effect is undoubtedly unpleasant, and is perhaps the most obviously perceivable nuisance aspect of being exposed to photochemical air pollution.

Vegetation is easily harmed by photochemical air pollution. Once again, the main agents of damage are ozone and PAN, which is one of the most phytotoxic substances known. Plants respond to the oxidants by first increasing their cell-membrane permeability. Higher doses lead to cellular and biochemical changes with visible leaf injury, leaf drop, and reduced vigour and growth, and finally death.

Can anything be done to reduce and control the formation of photochemical smog? Photochemical ozone concentrations can, in principle, be decreased by reductions in HC and other VOC emissions and by reductions in NO_x. Unfortunately, the most effective control strategy can be elusive, because of a non-linear (and sometimes inverse: see later) dependence of ozone production on precursor emissions. Various photochemical air-quality models have been developed that seek to predict ozone concentrations and their response to different control measures. Such efforts encounter difficulties associated with the variability in place and time of source and sink terms, with the transport and age of the air sample, and with the formulation and detailed kinetics of the chemistry. It is necessary both to assess the impact of a mix of anthropogenic VOCs with widely differing reactivities, and to evaluate that impact in relation to ozone produced by biogenic hydrocarbons.

A first step in understanding the effects of anthropogenic releases is to develop *reactivity scales* that compare different VOCs. Because of the inherent non-linearity of the system, attention is often directed to some measure of *incremental reactivity*, R_i, defined for a specific VOC as

$$R_i = \frac{-\Delta[O_3]}{\Delta[V_i]} \tag{11.24}$$

where $-\Delta[O_3]$ is the ozone response in ppb and $\Delta[V_i]$ is the incremental increase of reactive VOC i in mass units. It may be convenient to express the incremental reactivity relative to some chosen compound. The *photochemical ozone creation potential* (POCP), P, uses ethylene as the standard for comparison, and is defined as

$$P = \frac{R_i}{R_{ethene}} \times 100 \tag{11.25}$$

One of the most versatile methods for evaluating POCPs uses an explicit chemical mechanism in a model, with rate coefficients for each key reaction being derived from laboratory experiments where possible. However, it is much more straightforward to assemble a reactivity scale that is based on the rate coefficients for reaction of OH with the VOC. The problem is that the ability to produce ozone depends not only on the rate of initiation of oxidation chains, but also on the efficiency with which the subsequent steps proceed. Despite these drawbacks, the method provides a first estimate of the POCP. The actual ozone-production rates will depend on the concentrations of the individual hydrocarbons present in the atmosphere. Table 11.2 presents selected data for ozone-production rates calculated from the rate coefficients and a series of HCs, for which concentration measurements are available for rural sites in the UK. It is noteworthy that alkenes sit in the top four positions, and dominate the list (although aromatic hydrocarbons also make a strong showing). Isoprene is high in the list, not because its concentration is high, but because it is so reactive. The only alkane in this list is iso-pentane: it owes its position to a high ambient concentration rather

Table 11.2 Contributions to summertime ozone production of some selected hydrocarbons at rural sites in the UK.[a]

Hydrocarbon	*Mean concentration /ppb*	$10^{12}\,k_{OH}$ /cm^3 $molecule^{-1}$ s^{-1}	O_3 *production rate /ppb* h^{-1}
Iso-butene	0.21	51.4	0.387
Propene	0.27	26.3	0.256
Ethene	0.67	8.52	0.206
Isoprene	0.05	101	0.182
1,2,4-Trimethylbenzene	0.15	32.5	0.176
(*m*- + *p*-) Xylene	0.21	19	0.144
1,3,5-Trimethylbenzene	0.06	57.5	0.128
E-but-2-ene	0.05	64	0.115
Toluene	0.46	5.96	0.099
E-pent-2-ene	0.04	66.9	0.097
1,3-Butadiene	0.04	66.6	0.096
Z-but-2-ene	0.04	56.4	0.081
Iso-pentane	0.52	3.9	0.073
o-Xylene	0.14	13.7	0.069
But-1-ene	0.05	31.4	0.057

[a]From R. G. Derwent, in C. N. Hewitt, (ed.) *Reactive Hydrocarbons in the Atmosphere*, Academic Press, San Diego, 1999.

than to a high intrinsic reactivity. Armed with this kind of information, it becomes possible to discover which sources of VOCs make the most damaging contribution, and which might therefore be most susceptible to effective control.

Attacking the primary target of motor vehicles, the State of California, one of the worst afflicted areas, has taken the lead in recommending and enforcing legislation. One of the first moves, in 1961, was to require the installation of positive crankcase ventilation on new and used cars, followed by the approval of catalytic converters in 1964–6 to reduce hydrocarbon and carbon monoxide emissions. In 1966, an alternate 'lean-burn' method of reducing HC and CO emissions was implemented that did not require catalytic devices or afterburners.

Control of NO_x emissions is no less essential than that of HCs, but there is considerable debate as to the degree of control that is appropriate for any particular area. The nature of the problem can be seen from the consequences of the 1966 California legislation, which led to *increased* NO_x levels. The legislation did have the required effect of reducing average ozone levels in downtown Los Angeles, but, unfortunately, oxidant levels downwind of the central area actually increased. Ozone concentrations near the release area were decreased because NO—the main component of the increased NO_x emission—rapidly destroys O_3 (reaction (11.4)). While the air is being transported downwind, the NO_2 product is photolysed, and ultimately produces ozone again.

The present strategy is to reduce vehicle emissions by catalytic conversion of the exhaust gases. By 1975, catalytic converters were virtually universal in the USA, and stringent standards for HC and CO emissions could be met without greatly impairing engine performance or economy. An essential requirement of catalytic afterburners is that they have a long enough lifetime, and that they should be effective against partially oxidized species. Indeed, incomplete oxidation of hydrocarbons over an inefficient catalyst could aggravate the pollution problem by producing aldehydes, which are more reactive in initiating and promoting smog than the parent hydrocarbons. Since 1993, catalytic converters have been mandatory in the nations of the European Union. Typical converters consist of a layer of precious metal (such as platinum or rhodium) coated on an alumina substrate, which acts as a *three-way catalyst*. The 'three-way' aspect is that, by causing all the oxidants in the exhaust gases to react with all the reductants, the NO, CO, and HC pollutants are converted simultaneously (to N_2, CO_2, and $CO_2 + H_2O$, respectively). A conversion efficiency approaching 90 per cent for all three gases is possible, but only if the air:fuel ratio is kept within the strict limits of (14.7 ± 0.1):1. To maintain the correct ratio, an O_2 sensor is often provided in the exhaust flow, and it sends a feedback control signal to the fuel metering device. That the measures are effective seems in little doubt. HC and NO_x emissions from petrol-engine cars each decreased by a factor of ten for vehicles subject to the European Union directive, and CO emissions decreased by a factor of three.

11.1.7 Polycyclic Aromatic Hydrocarbons (PAHs)

Atmospheric constituents that could be carcinogenic or mutagenic, rather than simply toxic, must come under special scrutiny. One of the first classes of atmospheric species that were shown to be carcinogenic were the *polycyclic aromatic hydrocarbons* (PAHs). A small contribution to the atmospheric load of PAHs may come from natural sources such as forest fires or volcanoes, but the predominant sources are anthropogenic.

PAHs are hydrocarbons that consist of two or more benzene rings fused together in a variety of ways as linear, angled, or clustered structures. More than 100 different PAH species have been identified in the atmosphere, ranging from naphthalene (two benzene rings; relative molecular mass 128) to coronene (seven rings; RMM 300). Figure 11.2 illustrates some of the structures. The largest PAHs, with five, six, and seven aromatic rings, are found in the atmosphere predominantly as aerosols, while naphthalene exists exclusively in the gas phase. The compounds may also become

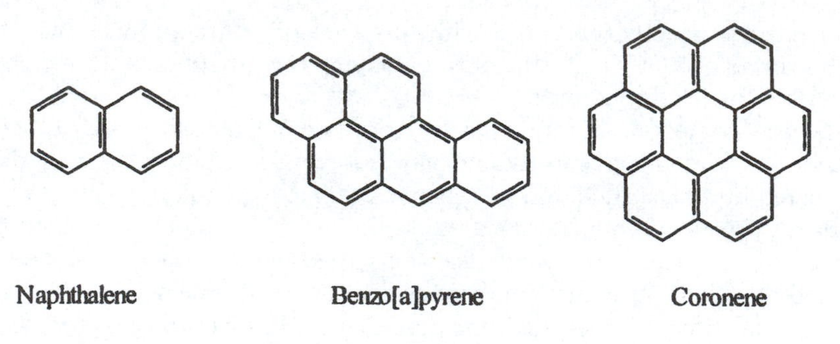

Figure 11.2 Structures of some atmospherically important polycyclic aromatic hydrocarbons.

adsorbed on the surfaces of other aerosol particles, such as those generated in combustion processes, and products more polar and reactive than the parent PAHs can be formed.

In general, the PAHs are liberated to the atmosphere as a result of incomplete combustion. Anthropogenic sources include motor vehicles (both diesel and petrol), stationary power plants (coal and oil fired), domestic (coal and wood burning, tobacco smoke), as well as deliberate biomass burning (see Section 11.1.8). The compounds are thus ubiquitous in our atmospheric environment. The importance of the compounds resides largely in their impact on human health. As long ago as 1942, it was discovered that the organic extract from particles in the ambient air could produce cancer in experimental animals, and benzo[a]pyrene was subsequently identified as one of the causative agents. Benzo[a]pyrene in tobacco smoke is also often cited as the reason that cigarette smoking produces cancer, although many other carcinogenic compounds, including other PAHs, are implicated in reality. Of course, PAHs that are present as aerosols or adsorbed on other particles—and cigarette smoke is a fine example of that kind of system—offer a special hazard for the reasons outlined in Section 11.1.2.

Despite the identification of carcinogenic PAHs in the atmosphere, epidemiological surveys suggested that other, unidentified, carcinogens must be present in some polluted atmospheres. PAHs such as benzo[a]pyrene are *promutagens* that require activation before a cancer can be produced. The activity of some air particles suggests that other species may be present that possess direct, and powerful, mutagenic properties. Considerable evidence now points to mono- and di-nitro derivatives of PAHs as these agents. Several such compounds (for example, 1-nitropyrene and 3-nitrofluoranthene) have been identified in the particles collected from several combustion sources. Even more important may be secondary species, such as 2-nitropyrene and 2-nitro-fluoranthene that are found in the atmosphere of widespread geographical locations, even though they are not emitted significantly by combustion. Atmospheric chemical transformation of primary PAHs is the probable source of these nitro-compounds. Radical reactions in the gas phase effect the conversions when NO_2 is present in the atmosphere. The initiator is OH during the day, as usual. However, NO_3 can effect conversion at night, and a high degree of mutagenicity of air downwind of Los Angeles has been attributed to nitronaphthalenes and methylnitronaphthalenes generated by the initial attack of NO_3 on naphthalene and methylnaphthalene.

11.1.8 Biomass Burning

Man burns plant matter for a variety of reasons, and *biomass burning* is an accepted aspect of land cultivation in many parts of the world. The motivations for biomass burning include forest clearing, pest control, energy production, nutrient mobilization, and the like. This intentional burning, most of which takes place in the tropics, is a major contributor of anthropogenic combustion products to the atmosphere, and Table 11.1 has an entry for release rates of several species resulting from biomass burning. It is immediately apparent that, for some compounds, the source is highly significant. Incidentally, although releases of NMHCs are not included in the table, biomass burning certainly contributes to these atmospheric VOCs: measurements of emissions from a variety of real fires show the presence of C_2H_6, C_2H_2, C_3H_8, C_3H_6, and n-C_4H_{10}.

Dry plant matter is mainly carbohydrate (empirical formula CH_2O), and burns to CO_2 and H_2O. Release of CO_2 itself may be of great significance, but here we are concerned with minor products of combustion. In addition to the carbohydrate, smaller quantities of other elements such as N, Cl, S, P, and K are present, and combustion allows volatilization of some of the compounds that are formed. Temperatures do not generally become high enough to oxidize atmospheric N_2 (in distinction to the conditions that often obtain in the combustion of fossil fuel). The nitrogen compounds emitted must therefore come from N in the fuel itself. Laboratory test fires show that H_2, CH_3Cl, N_2, COS, particulate matter, and several other minor compounds are emitted. In the early, flaming, stage of combustion, when temperatures are high, compounds of high oxidation state (CO_2, SO_2, NO_x) are emitted, while in the later, smouldering, stage, less oxidized compounds (CH_4, N_2O) are produced. According to such experiments, for every one kilogram of dry biomass burnt, emissions are about 0.5 g of N as NO_x, 42 g of C as CO, and 4 g of C as CH_4. Given the enormous amounts of biomass burned each year, these values show immediately why the process is so important. During the dry season, when it usually takes place, biomass burning can be so widespread and intense that it can easily be seen from space. Emissions from the fires can significantly enhance tropospheric ozone concentrations. Large-scale burning of the savanna in Africa, for example, is believed to produce as much regional-scale ozone as urban industrialized regions. The emissions may also be a major source of nitric acid in the tropics. The effects on atmospheric chemistry and climate are critically important.

11.2 MAN'S IMPACT ON THE GLOBAL STRATOSPHERE

In the early 1970s, it became evident that the stratosphere, where the air is both thin and stable against vertical mixing, might be vulnerable to human impact. Pollutants introduced into the stratosphere by Man would have a lifetime for physical removal by transport of several years, and might therefore build up to globally damaging levels. The situation is quite different from that in the troposphere or the boundary layer. Lifetimes of many pollutant species that have an impact there are small and effects are thus often localized. As we have already seen in Section 9.5, natural phenomena are able to perturb stratospheric ozone concentrations, so it seems entirely plausible that Man could contribute in a similar way. Indeed, the discussion of ozone variations and trends in Section 9.6 shows that there is clearly a depletion of ozone greater than can be attributed to any natural causes. Most of the chlorine implicated in the anomalous chemistry in polar regions described in Section 9.4 is in the stratosphere because of Man, but in 1970 the discovery and explanation of the Antarctic ozone 'hole' lay nearly fifteen years in the future. In the present chapter, we use the chemical knowledge gained in Chapter 9 to examine more closely the ways in which Man could (and has) upset the balance between ozone production and loss.

Initial concerns centred on the role of supersonic stratospheric transport (SST) aircraft that could emit H_2O, CO, and NO_x into the stratosphere. The Anglo-French Concorde was such an aircraft, and was in the late stages of prototype testing, and destined to enter service in 1976. The special feature of these aircraft is not that they are supersonic, but that for efficiency they fly in the stratosphere. Paul Crutzen, then working in Oxford, had drawn attention to the catalytic role of NO_x in the natural stratosphere, and Harold Johnston in California pointed out that artificial injection of NO_x could bring about a disproportionately large reduction in ozone that would not be calculable from a simple reaction stoicheiometry. To examine the problem, an enormous, and extraordinarily valuable, scientific research effort was put in motion, and it is hard to overestimate the advance in understanding of our atmosphere that has followed. One consequence of the SST studies was the identification of a series of potential modifiers of the ozone layer, some of which are anthropogenic. The agents considered include a BrO_x cycle, as well as changes in the NO_x, HO_x, and ClO_x catalytic cycles that we have already discussed for the natural stratosphere. Intensive agriculture, especially when it involves nitrogen-containing fertilizers, can lead to increased production of N_2O, one of the precursors of stratospheric NO_x. Methane, a precursor of stratospheric HO_x, has emission strengths that respond to several human activities. Other influences on ozone result from the release of infrared active gases that can modify stratospheric temperatures, and of species such as CO that can indirectly modify (*via* reaction with OH) stratospheric composition. Certain of the agents can be injected directly into the stratosphere, while others may be of tropospheric origin, but of sufficient tropospheric stability to be transported to the stratosphere. In the subsequent parts of Section 11.2, we look at some of the individual mechanisms by which stratospheric ozone can be influenced, and we also attempt to identify the effects of coupled perturbations by several agents. The most spectacular response of all, the development of ozone 'holes', warrants a section of its own (Section 11.3). First, however, we must see why possible changes to atmospheric ozone have aroused so much concern.

11.2.1 Consequences of Ozone Perturbation

A reduction in stratospheric ozone could have biological consequences because of increased intensities of UV-B (280–315 nm) that would reach the ground. Changes in ozone concentration could also have climatological effects because of altered stratospheric heating. Increased UV-B radiation could exert deleterious effects both on human beings and on all other plant or animal species. Fear of cancer has made human skin cancer the most publicized potential effect of ozone depletion. It should, however, be pointed out that some of the kinds of cancer that can definitely be attributed to sunlight (basal cell and squamous cell cancers) are not terribly dangerous, since, if caught in time, they may be successfully treated. There are much more dangerous cancers of the skin, the *melanomas*. The melanomas used to be relatively rare, and often found on parts of the body not exposed to sunlight. However, the incidence of melanomas seems to be increasing as humans deliberately spend more time in the sun, and a causal link with exposure to ultraviolet radiation is emerging. In particular, there appears to be an increase in the incidence of malignant melanoma in adults who suffered one or two episodes of sunburn when they were less than 11 years old.

Biological damage, other than the skin-cancer response, is quite clearly produced by radiation in the UV-B region. The so-called *action spectrum* for biological response generally falls off by four or five orders of magnitude from $\lambda = 280$ nm to $\lambda = 315$ nm. This range is, of course, exactly the one over which ozone absorption also falls off, and solar irradiance at the ground thus increases. Above all, the observations demonstrate how important the ozone layer is to survival of the biota on the surface of the planet. However, the details of the behaviour depend on the particular species or response being investigated.

Measurements of human skin cancer incidence and of UV-radiation dose have now been carried out together in several geographical areas, including Australia, the US, New Guinea, and Ireland. For the US at least, it seems certain that more than 90 per cent of skin cancer other than melanoma is associated with sunlight exposure, and that the damaging wavelengths are in the UV-B region most affected by changes in ozone concentration. Skin cancer incidence in mid-latitudes doubles for regions increasingly near the equator for each 8–11° of latitude, or about 1000 km. Incidence of the serious skin-cancer melanoma also increases with decreasing latitude, suggesting that UV radiation is a contributing factor, but this contribution is compounded by occupational exposure and other factors. Figure 11.3 shows data on melanoma mortality rates as a function of latitude: there

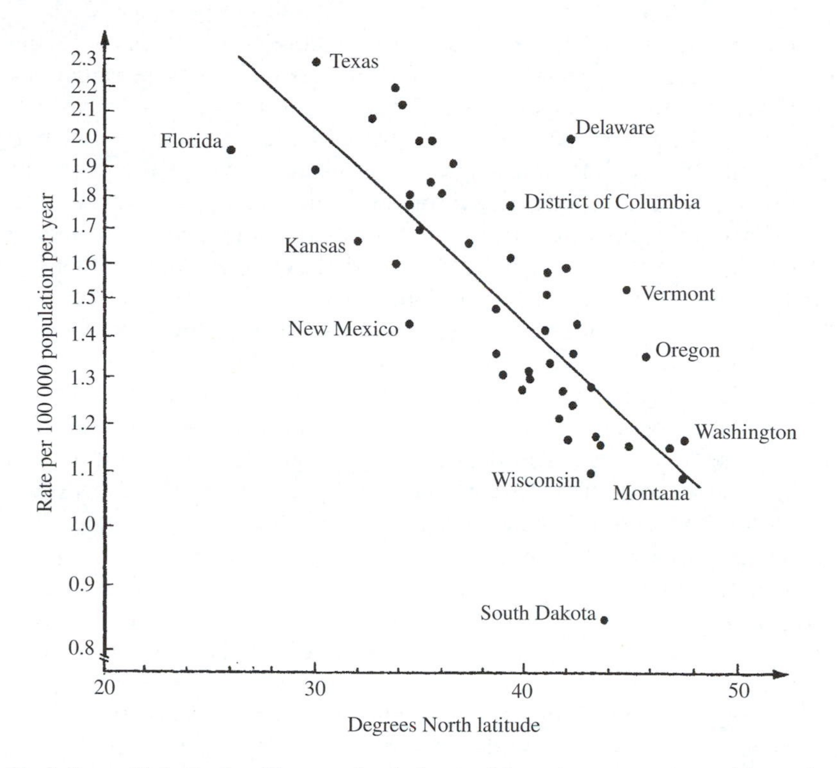

Figure 11.3 Variation with latitude of human death due to skin melanoma among white males in the United States excluding Alaska and Hawaii. From F.S. Rowland, Chapter 4.6 in *Light, Chemical Change and Life*, ed. J.D. Coyle, R.R. Hill and D.R. Roberts, Open University Press, 1982.

appears to be a statistical correlation, but not necessarily a causal one. In the US, a two per cent increase in UV-B is epidemiologically connected with a two to five per cent increase in basal skin cancer, and a four to ten per cent increase in (the more serious) squamous cell cancer.

A potentially hazardous response to increased ultraviolet exposure is that of the immune system of both animals and humans. Even a mild sunburn has been found to decrease the lymphocyte viability and function in humans. Exposure of animals to otherwise tolerable doses of ultraviolet radiation can produce changes in allergic reaction, skin-graft rejection rates, and so on; uninhibited growth of normally rejected skin tumours also occurs. It may be this suppression of the immune system that makes possible the subsequent development of cancers.

Possible effects of increased UV-B levels on species other than humans are rather difficult to evaluate. Certainly, many plants and animals are sensitive to the ultraviolet radiation dose that they receive, and many experiments have been performed to examine response as a function of dosage. Higher-order species depend on the results of competition among the lower orders, thus increasing the complexity of analysis. Small climatic temperature changes could also affect the competition. Ultraviolet irradiation of key cytoplasmic and nuclear constituents is definitely damaging, and animals and plants have developed both avoidance and repair mechanisms. Even the humble plankton may migrate to greater depths during the day and rise towards the surface at night. Since plants do not enjoy mobility, ultraviolet radiation can be avoided only through protective shielding (*e.g.* by waxes and flavenoid pigments) or by changes in the orientation of the leaves. Photosynthesis may ultimately prove to be the most critical element if the ozone shield is depleted, because of its general dependence on cellular integrity and specific dependence on UV-vulnerable enzymes and other proteins.

A great deal of effort has been directed towards understanding the effects that changes of ozone concentration might have on meteorology and on climate. Ozone affects the radiative balance of the atmosphere–Earth system through absorption of radiation in the ultraviolet, visible, and infrared spectral regions. It has long been recognized that the climatic effects of ozone change are highly dependent on the altitude at which the changes occur. Changes within the region near the tropopause are the most important. Climate behaviour is linked to ozone amounts in a highly complex manner. A loss of ozone in the lower stratosphere leads to an increase in visible and ultraviolet radiation reaching the troposphere, which tends to warm the climate. On the other hand, ozone also absorbs and emits infrared radiation, and is thus an active 'greenhouse' gas (see Section 2.3.2); loss of ozone will tend to *cool* the climate through the loss of greenhouse heating. Cooling of the lower stratosphere means that there is reduced emission of infrared radiation from stratosphere to troposphere. In simple models, at least, the net effect of the ozone loss is to cool the climate. As we have seen in Section 9.6, there is considerable evidence that ozone concentrations in the stratosphere have been decreasing over recent decades, almost certainly as a result of Man's activities. Several studies have now shown that there has been a negative trend in temperatures in the lower stratosphere of about 0.4 °C per decade over the same period.

One point of particular interest is that many of the species known to be implicated in stratospheric ozone depletion (CH_4, N_2O, the halocarbons: see the rest of Section 11.2) are also important contributors to the greenhouse effect. Global warming as a result of increased releases of such compounds is a very real concern, as explained in Section 11.4. But it seems that at least a part—perhaps 20 to 25 per cent—of the increased *direct* greenhouse heating by these gases is offset by the reduced heating by ozone that these gases also bring about *indirectly*.

One of the strongest arguments for avoiding what amounts to deliberate change in ozone levels is that anthropogenic perturbations, such as those to be discussed in the next sections, can lead over a few years to changes comparable with the variations occurring naturally over evolutionary periods (10^5 to 10^7 years). Avoidance responses in living organisms have been learnt over the course of evolution of the species. Most important, those responses are generally triggered by particular combinations of visible light intensity and temperature. So long as UV-B and visible light bear a constant intensity relationship, the longer-wavelength light is a measure of the ultraviolet. Depletion of the ozone layer changes the balance. Of all species, only humans could directly measure the UV-B intensities and adjust their life-style rapidly. Most flora and fauna have adapted to existence at particular latitudes, so that the 50-fold greater UV dose at the equator is of little interest to polar species: a factor of two increase might be intolerably large.

The conclusion that has to be drawn is that rather little is yet known about the physical effects or the ecological consequences of stratospheric ozone changes in either the downward or upward direction. It may well be that negative-feedback mechanisms make the atmosphere 'self-healing' and it may be that life can cope with changed ultraviolet intensities. However, the identification of areas where damage might be done, coupled with our lack of knowledge about them, demand that we do not make unwarranted perturbations of the ozone layer. In the sections that follow, we look at some of the ways in which Man is now presumed capable of modifying his ultraviolet screen. We start by considering a series of individual perturbations. These perturbations are hypothetical in the sense that they do not represent any change to the real atmosphere, in which several parameters vary simultaneously. They do, however, provide a basis for realistic 'scenarios' of simultaneous multiple perturbations.

11.2.2 Aircraft

Section 11.2 showed how concerns about the environmental impact of SST aircraft stimulated intense research on stratospheric processes that continues to this day. High-temperature combustion in air leads to formation of NO by 'fixing' of atmospheric N_2 and O_2, as well as the ordinary combustion products CO_2, CO, and H_2O. Injection of these species, especially the catalytically active NO, into the stratosphere could destroy ozone. Because of the vertical stability of the stratosphere, physical lifetimes against removal of injected species are considerable, ranging from 1–2 yr at 17 km to 2–4 yr at 25 km altitude. For the 100-passenger Concorde flying at 17 km, up to 440 kg h^{-1} of NO_x were emitted during flight, or about 1.1×10^5 molecule cm^{-2} s^{-1} averaged over the globe. Early estimates of growth of aircraft anticipated several hundred Concorde and US SSTs in use by 1990. Models envisaging 500 US SSTs emitting a total of 1.2×10^9 kg of NO per year at 20 km altitude predicted a reduction in the global ozone column of roughly 12 per cent, with a worst-case reduction of 25 per cent near the flight corridor. Subsequent models, based on improved chemistry, predicted much smaller depletions of ozone, as will appear later in this section. Equally important, the projected numbers of aircraft now seem like a fantasy. In reality, only 16 production Concordes were built, and they have all now been withdrawn from service. More recently there has been renewed and intensified examination of the impact that aircraft operations may have on the chemistry of the atmosphere and on climate. In part, this research has been stimulated by discussion of a proposed second-generation high-speed civil transport (HSCT) aircraft. However, it has also become apparent that the cruising altitude of *current* commercial (subsonic) aircraft is, for 30 to 40 per cent of the time, above the tropopause. Emissions of combustion products into the stratosphere should therefore already be considered. Military aircraft, often flying even higher, may make an additional contribution, although it is difficult to obtain quantitative information.

Alongside the revival of interest in the stratospheric effects of aircraft, there has been increased concern about what effect the release of aircraft effluents might have on the troposphere. The largest amounts of aircraft exhaust gases are released at altitudes between 10 and 12 km, a region that might encompass the upper troposphere and lower stratosphere (UTLS). As it happens, the UTLS is rather poorly understood in several ways, so that assessment of the effects of aircraft is made all the more difficult. Several research initiatives have recently sought to address the problems. One thing seems certain, regardless of the future of SSTs, and that is that the fleet of conventional aircraft will grow in the foreseeable future. In May 2009, during a period of economic downturn, the International Civil Aviation Organization (ICAO) forecast growth in passenger-km flown annually by a factor of 2–3 between 2006 and 2036, and an increase in fuel burned from the 2006 base of 174 Mtonne yr^{-1} by a factor not much smaller.

Figure 11.4 indicates some of the major impacts that aircraft emissions might have on the atmosphere. As we have come to expect in atmospheric chemistry, there are interactions and feedbacks between the various components of this diagram; aerosols and other particles are seen to be central to several of the phenomena. Certain chemical species emitted by aircraft engines are listed in Table 11.3. The *emission index* for each substance is listed for different flight conditions, and for typical modern engines. This index is the number of grams of a material emitted for each kilogram of fuel consumed. A short-haul flight such as London to Paris (346 km) in an Airbus A320 requires about 2.25×10^3 kg of fuel; a longer flight (say Los Angeles to Tokyo, 8753 km) in a 465-passenger Boeing 777 might need up to 8.72×10^4 kg, so that large quantities of the gaseous species are released. Table 11.3 has rather sparse data about the emissions of solid particles and liquid aerosols such as soot, directly formed sulfate particles, and oil. Although, as we shall see soon, such particles may be of great importance, there are considerable uncertainties about the emissions. Major improvements have been made in jet-engine technology since the 1960s. A gain in fuel-conversion efficiency of roughly a factor of two has been achieved as various types of turbofan have replaced turbojet types. These fan engines by-pass part of the gas flow around the main combustion chamber.

The importance of the NO_x that is released by aircraft has provided the focus so far of our discussion of impacts. The oxides of nitrogen have the potential of modifying the abundance and distribution of ozone in the UTLS and middle stratosphere, with the consequences, including

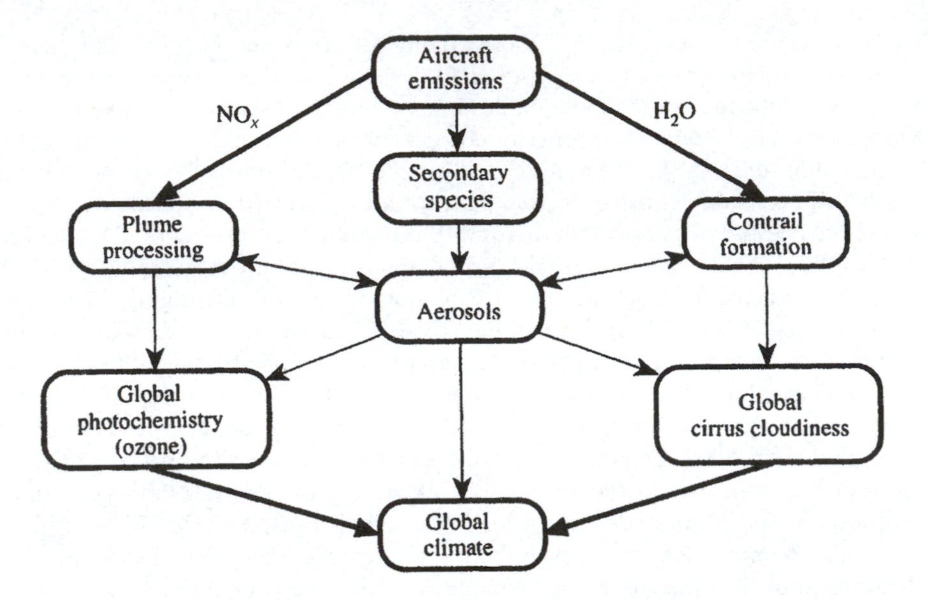

Figure 11.4 Potential atmospheric impacts of aircraft emissions. From *European Assessment of the Atmospheric Effects of Aircraft Emissions*, ed. G. Brasseur, G.T. Amanatidis and G. Angeletti, *Atmos. Environ.*, 1998, **32**, 2327.

Table 11.3 Typical emission indices for some products of combustion in aircraft engines.

	Emission index ($g\,kg^{-1}$ *fuel*)							
Species	*Idle*		*Take off*		*Climb out*		*Approach*	
Engine:	*GE90*[a]	*CFM-56*[b]	*GE90*	*CFM-56*	*GE90*	*CFM-56*	*GE90*	*CFM-56*
NO_x (as NO_2)	5.11	4.28	44.44	24.79	33.85	19.88	15.78	8.94
CO	40.59	18.5	0.07	1.1	0.07	1.1	2.29	2.8
Hydrocarbons	4.58	1.53	0.03	0.23	0.03	0.23	0.06	0.45
H_2O	1230	1230	1230	1230	1230	1230	1230	1230
CO_2	3157	3157	3157	3157	3157	3157	3157	3157
Particles	5.7[c]	8.2[c]	3.2[c]	5.6[c]				

Assembled from data of the engine manufacturers and available from the ICAO databank.
[a]GE90-110B1 as used in Boeing 777-200LR;
[b]CFM56-5 as used in Airbus A320-200;
[c]in units of 10^{15} particles per kg of fuel; for the particle size distributions encountered, 10^{15} particles per kg corresponds very roughly with a mass EI of 0.1 g per kg data from S.C. Herndon *et al.*, *Environ. Sci. Technol.* 2008, **42**, 1877–1883. For the CFM-56 engine the entries are measured values, but for the GE90 measurements are not available, and the values are the average for all engines investigated.

perturbation of climate, that were explained in Section 11.2.1. NO_x is one of the most important catalysts in ozone removal in the middle stratosphere (Section 9.3). In the UTLS, on the other hand, NO_x increases may lead to an *increase* in ozone concentrations. In the first instance, NO_x in the troposphere acts as a photochemical source of atomic oxygen, and thence of ozone, *via* the normal addition process (reaction (11.2)); this aspect is examined in much more detail in Section 8.2.3. and associated parts of Chapter 8. Even in the lower stratosphere, some O_3 generation may occur, depending on the concentrations of other radical species. A further reason for enhancement of ozone concentrations in the lower stratosphere is that ozone losses are dominated there by chlorine-radical catalysis (Section 9.3). NO_x is essential to the formation of one of the most important reservoir species, $ClONO_2$ (Section 9.3.2). Increased NO_x thus means reduced efficiency of chlorine losses. Models show that heterogeneous chemistry occurring on aerosol particles (Sections 9.4.4 and 9.4.5) may also considerably diminish the impact of NO_x injected by aircraft.

As indicated by Figure 11.4, increases in NO_x represent only one aspect of the possible effects of aircraft emissions (although the consequences are of particular relevance to our enquiry about stratospheric ozone because of the importance of this species in ozone chemistry). It is appropriate, therefore, to examine briefly the possible effects of the other engine effluents.

By far the largest outputs are of CO_2 and H_2O vapour. The potential impact of these aircraft emissions on climate are discussed in Section 11.4.5. In the context of stratospheric ozone chemistry, the release of water by high-flying aircraft might increase the frequency with which polar stratospheric clouds (Sections 9.4.2 and 9.4.4) occur. These clouds are an important component of the special and perturbed chemistry (Sections 9.4.4 and 9.4.5) that gives rise to the formation of the Antarctic (and probably Arctic) ozone holes. Sulfur dioxide in engine exhausts can be converted to sulfuric acid aerosol and, together with the sulfates, soot, and other solid and liquid particles, can have an effect in altering the radiative balance of the atmosphere as well as playing a role in cloud formation. In addition, the particles may enhance substantially the extent of heterogeneous chemical processing (Sections 8.6 and 9.3.4), whose place in the chemistry of the atmosphere has become well established.

11.2.3 Halocarbons: Basic Chemistry

Methyl chloride is the dominant naturally occurring chlorine compound to be found in the atmosphere. Global average mixing ratios are at present about 0.55 ppb (0.55×10^{-9}). At least some of that is produced by Man (by biomass burning and various industrial activities): the proportion depends on estimates of a potentially large natural source from tropical plants and their dead or senescent leaves. Meanwhile, total chlorine mixing ratios in the troposphere reached a peak of roughly 3.7 ppb in 1996, and were still well over 3.0 ppb in 2009. Almost all the difference between the contribution from CH_3Cl and this total is contributed by Man-made fluorinated chlorocarbons and some other chlorinated solvents (CH_3CCl_3 and CCl_4, for example).

The fluorinated hydrocarbons were developed in 1930 by the General Motors Research Laboratories in a search for a non-toxic, non-flammable refrigerant to replace the sulfur dioxide and ammonia that were then in use. Dichlorodifluoromethane, CF_2Cl_2, is a typical member of the class of compounds. 'Freon' and 'Arcton' are trade names for the CFCs. Chemical inertness has made the CFCs valuable as aerosol propellants, as blowing agents for plastic-foam production, and as solvents, in addition to their use as refrigerants. CFC production was until quite recently a world-wide industry with an estimated 3.6×10^8 kg of CFC-11 and 4.5×10^8 kg of CFC-12 being manufactured in 1988.[ii] From that time on, production decreased, and regulatory action (see Section 11.2.5) required that production should cease in the developed countries. The uses of the CFCs all lead ultimately to atmospheric release, since even 'hermetically sealed' refrigerators and closed-cell foams finally leak to the air. Ninety per cent or more of all the CFC-11 and CFC-12 produced up to 1996 is believed to have been released. Figure 11.5 is a pictorial representation of the contribution of different compounds of halogens to the Cl and Br found in the stratosphere for the year 2004. These compounds are often referred to as *ozone-depleting substances* (ODSs).

Meanwhile, in 1973 James Lovelock and his collaborators reported the presence of halogenated hydrocarbons in the troposphere. It soon became apparent that the quantities of the CFCs were

[ii] In the usual nomenclature, CFC (chlorofluorocarbon) is followed by a coded two- or three-digit number. The hundreds digit is the number of carbon atoms in the molecules less one, the tens is the number of hydrogen atoms plus one, the units is the number of fluorine atoms, and the residue is assumed to be chlorine. If the first digit is zero, it is dropped. Thus CFC-11 is $CFCl_3$, CFC-12 is CF_2Cl_2, and CFC–115 is CF_3CF_2Cl.

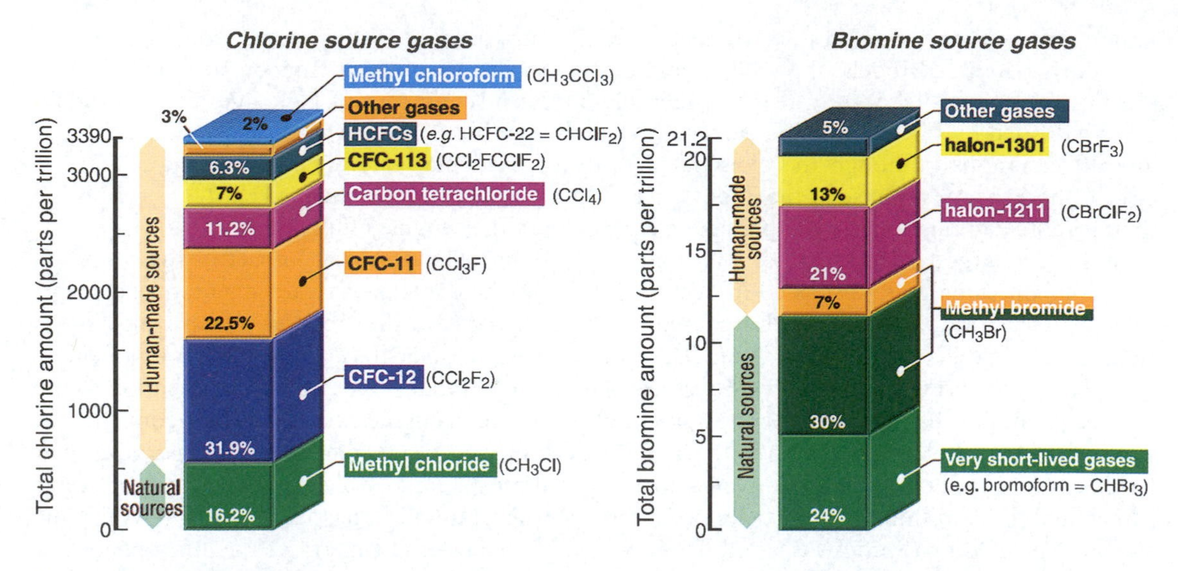

Figure 11.5 Primary sources of chlorine and bromine transported to the stratosphere in 2004. Source: *Scientific Assessment of Ozone Depletion: 2006*, World Meteorological Organization, Global Ozone Research and Monitoring Project, Report No. 50, WMO, 2007.

equal, within experimental error, to the total amount ever manufactured. Tropospheric inertness of the CFCs was thus confirmed, and lifetimes of up to hundreds of years were indicated. Only one escape route is possible for the compounds: transport to the stratosphere followed by ultraviolet photolysis. For CFC-12, the photolytic process is

$$CF_2Cl_2 + h\nu \rightarrow CF_2Cl + Cl. \tag{11.26}$$

Mario Molina and Sherry Rowland noted that the chlorine atoms from reaction (11.26) were a serious threat to the ozone layer, since the CFC source emissions were known to exist, and the Cl they yielded in the stratosphere was known to catalyse ozone destruction (Section 9.3). Tropospheric measurements have since confirmed the presence of all the Man-made CFCs, as well as chlorinated compounds such as carbon tetrachloride (CCl_4) and 1,1,1-trichloroethane (methyl chloroform, CH_3CCl_3) that are almost certainly solely of anthropogenic origin. Figure 11.6 shows how tropospheric concentrations of CFC-11, CFC-12 and CFC-113 have been increasing since 1979. Until about 1993, the increases were inexorable. However, it is interesting (as well as gratifying) to observe the effect that the imposition of successive control measures on production has had. The rate of increase of CFC-12 has become smaller, and the other two CFCs have declined slowly. Estimates of total chlorine from CFC-11, CFC-12, CFC-113, CH_3CCl_3, and CCl_4 suggest a mixing ratio of about 1.6 ppb in 1977 (almost all of which resulted from Man's activities), increasing almost linearly to about 3.1 ppb in 1993, and decreasing slightly to 2.9 ppb by 2006.

From the point of view of ozone depletion, one very significant observation is that the CFCs can be detected in the stratosphere, and the concentration–altitude profiles there are consistent with the photochemical loss mechanism (some additional loss, especially for the hydrogenated species, occurs *via* reaction with OH radicals). Figure 11.7 illustrates for a typical CFC how the concentration remains virtually constant throughout the troposphere, where the compound is inert and well mixed, but drops suddenly beyond the tropopause as photolysis destroys it and releases active chlorine. Measurements of the vertical distribution of CH_3Cl (the most important natural chlorine-bearing species) show that chlorine of anthropogenic origin now predominates in the stratosphere. The threat to the ozone layer is thus real and present.

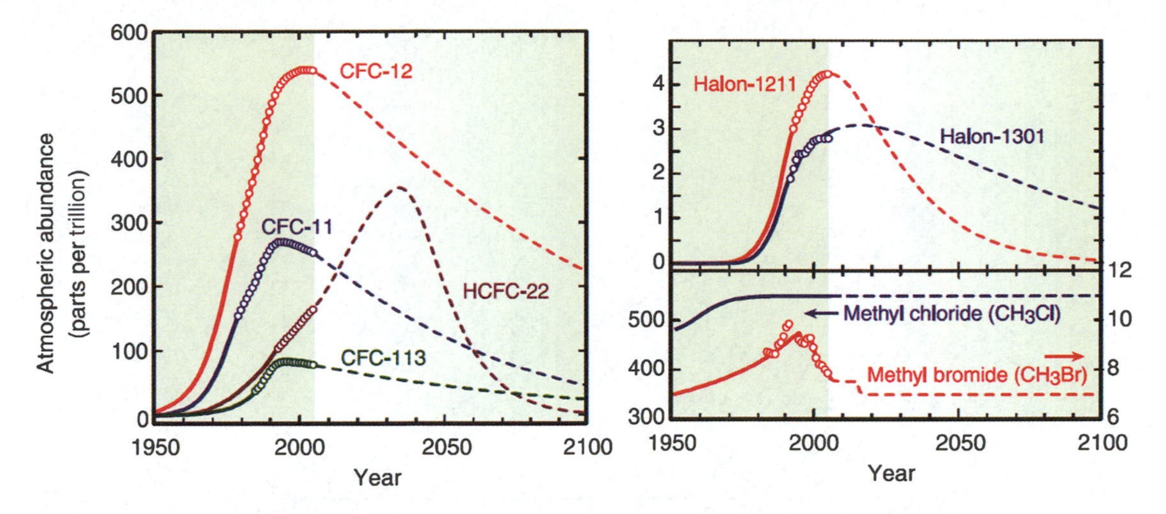

Figure 11.6 Atmospheric abundances of selected halogen source gases in the past, and projected for the future. Source as for Figure 11.5.

Bromine-containing compounds can also interfere with stratospheric ozone. A particular concern with bromine lies in the very high efficiency with which it can destroy ozone in the various catalytic cycles, so that relatively small additional releases can have a disproportionate effect on stratospheric ozone levels. According to one estimate, bromine is 58 times more effective than chlorine at catalysing the destruction of ozone (Section 9.4.5). Source gases are mainly CH_3Br and brominated CFCs (the 'halons'), as indicated in the right-hand panel of Figure 11.5. Methyl bromide is of both natural and anthropogenic origin; we shall return shortly to a fuller discussion of its sources. The halons are of entirely synthetic origin and are used for several purposes, especially as fire extinguishing agents and fire retardants. The compounds halon-1202 (CF_2Br_2), halon-1211 (CF_2BrCl), halon-1301 (CF_3Br), and halon-2402 (CF_2BrCF_2Br) have all been detected in the atmosphere. Of these, the most abundant are halon-1211 and halon-1301. Emissions peaked in 1988 at roughly 43×10^6 kg and 13×10^6 kg for halon-1211 and halon-1301, respectively.

Methyl bromide is a most important potential source of Br atoms in the stratosphere, but there remain several puzzles about its sources and budgets in the troposphere. Perhaps the most striking difficulty is that the budget seems to be out of balance: identified sources provide about 113×10^6 kg yr^{-1} to the atmosphere, but identified sinks remove at least 146×10^6 kg yr^{-1}, so that there is a shortfall of more than 33×10^6 kg yr^{-1}. Natural methyl bromide appears to derive mainly from the oceans, with smaller amounts being emitted from wetlands and salt marshes. One estimate for 2007 puts the total natural source strength at about 64×10^6 kg yr^{-1}. Fumigation (especially of soils) accounted for some 19×10^6 kg yr^{-1} of Man-made CH_3Br; emissions from automobiles that used leaded gasoline may have made a substantial contribution in the past, although emissions are now probably below 5–6×10^6 kg yr^{-1} today. Biomass burning, biofuel production and rapeseed cultivation are more important, and are responsible for the release of about 24×10^6 kg yr^{-1} of CH_3Br to the atmosphere. These numbers thus indicate that the natural and anthropogenic sources of CH_3Br are very roughly comparable in strength, giving the combined emission rate just quoted.

Recent studies indicate a global mean mixing ratio for CH_3Br of 9.2 ppt in the three years before 1999, after which the tropospheric mixing ratio started declining. By mid-2004, the mixing ratio had decreased by approximately 1.3 ppt (14 per cent) from its peak values. CH_3Br measured in ice cores

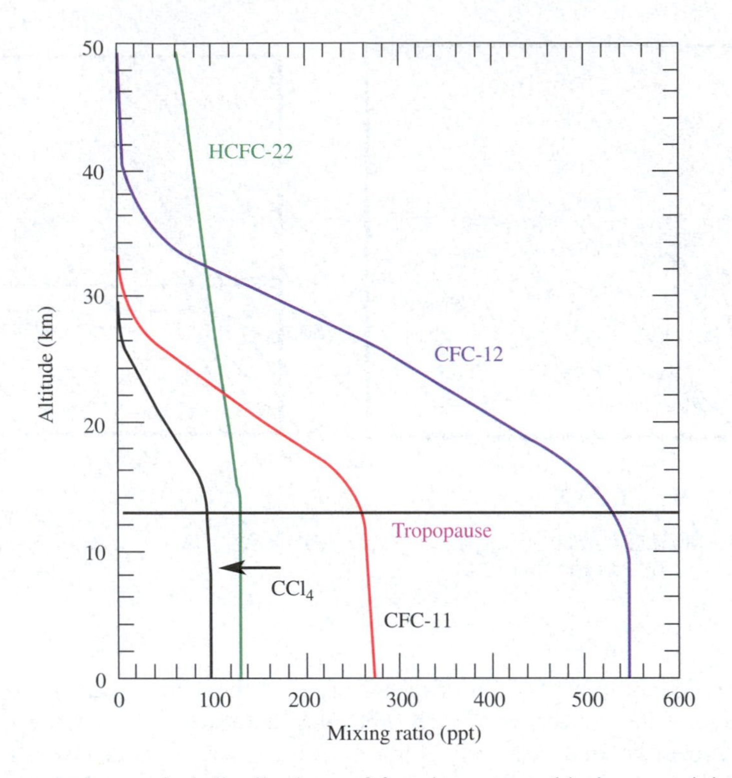

Figure 11.7 Representative vertical distributions of four important chlorine-containing compounds. The mixing ratios of all of the species remain constant in the troposphere, because loss processes are slow. Once in the stratosphere, photolysis decomposes the compounds, at altitudes that depend on the individual absorption spectra, and in so doing liberates atomic chlorine. Source: *Fundamentals of Atmospheric Modeling*, 2nd Ed., M. Z. Jacobson, Cambridge University Press, New York, 2005.

suggests that ambient air mixing ratios over Antarctica during 1700–1900 were 5–6 ppt, and independent measurements provide a value of 5.5 ppt in the 1930s. Thus there seems to be good evidence for an increase in mixing ratios in the latter part of the 20th century, and the recent decrease can be associated with a decline in industrial production.

Besides methyl bromide, there are several other naturally occurring organic bromine compounds, such as CH_2Br_2, $CHBr_3$, CH_2ClBr, and so on, that are found in the atmosphere. Tropospheric mixing ratios are typically about $5–10\times10^{-12}$ (5–10 ppt), but may be higher in the marine environment than elsewhere. Estimates of fluxes to the troposphere suggest a dominant input from natural biogenic sources, and the total bromine flux from such compounds could be comparable to that arising from the combined emissions of CH_3Br and the halons. However, the tropospheric lifetimes are substantially shorter than those of CH_3Br or the halons, so that less reaches the stratosphere.

11.2.4 Halocarbons: Loading and Ozone-Depletion Potentials

Several factors can influence the impact that releases of particular halocarbons might have on stratospheric ozone. They include the detailed chemistry of the halocarbon once in the stratosphere, the fraction of the emission that reaches the stratosphere, and the absolute amount of the material released.

Although there are some remaining uncertainties in stratospheric models, they do indicate an important feature of perturbation of the stratosphere by CFCs. Since the relative importance of different chlorine-related cycles depends on altitude, the vertical distribution of chlorine sources from different halocarbons must be considered. Destruction of relatively small fractions of ozone at around 15–20 km are more important on an absolute scale than larger fractional destructions at higher altitudes. Ozone is also less rapidly replenished at lower altitudes than at higher ones. Some halocarbons are photolysed at shorter wavelengths, and thus at higher altitudes, than others. Other things being equal, substitution of chlorine by fluorine shifts absorptions to shorter wavelengths. For equal concentrations, $CFCl_3$ has a maximum photolysis rate at ~25 km, CF_2Cl_2 at 32 km, while $CClF_2CF_3$ does not produce its maximum contribution until 40 km. It follows that the more heavily chlorinated halocarbons are more active in destroying ozone for two reasons. First, they are photolysed at lower altitudes where their absolute impact is greater. Secondly, they can release more chlorine atoms per molecule to the catalytic cycle.

For the reasons explained at the end of the last section and in Section 9.4.5, bromine is potentially far more destructive towards stratospheric ozone than is chlorine. Estimates of the catalytic activity of Br relative to that of Cl, referred to henceforth as α, vary rather widely. Values of α obtained depend not only on the detailed chemistry of the models employed, but also on exactly what comparison of activity is being made: for example, they are functions of altitude, latitude, and season. Recent consensus indicates that α may be as high as 58, the value we have been adopting in this book.

It is obviously important, for control and legislation if for nothing else, to compare the effects of the release of different natural and anthropogenic halocarbons and other ODSs. One such measure is the *ozone-depletion potential* (ODP). The ODP is defined as the ability of the ODS to deplete stratospheric ozone relative to the depletion caused by the same mass of $CFCl_3$. The ODP will depend on how many atoms of halogen are ultimately released following photolysis, their identity (Cl or Br, or both), how much reaches the stratosphere (related directly to the tropospheric lifetime), the altitude at which the atoms are released, and so on. Calculations of ODPs thus necessarily use a suitable chemical-transport model (CTM).

Table 11.4 shows the lifetimes and ODPs for some of the most important ODSs. Note the similar ODPs for the three CFCs, and the large ODPs (>1) of the bromine-containing halons. Methyl bromide shows an interesting competition between a short lifetime (small ODP) and its bromine content (large ODP). The ODP is only a factor of roughly two smaller than that of the standard,

Table 11.4 Lifetimes, ODPs and emission rates of some ozone-depleting substances (ODSs).[a]

Halocarbon	*Atmospheric lifetime (yr)*[a]	*Ozone-depletion potential*[a]	*Release rate*[b] *(10^6 kg yr^{-1})*	
			1990	*2003*
$CFCl_3$ (CFC-11)	45	[1.0]	250	93
CF_2Cl_2 (CFC-12)	100	0.9	371	120
$CF_2ClCFCl_2$ (CFC-113)	85	0.9	236	8
CHF_2Cl (HCFC-22)	11.8	0.04	178	350
CF_3Br (halon-1301)	65	16	4	3
CF_2BrCl (halon-1211)	16	7.1	10	7
CH_3CCl_3	4.8	0.12	718	20
CCl_4	26	0.73	63	95
CH_3Br	0.7	0.51	206	180

[a]Source: *Scientific Assessment of Ozone Depletion: 2006*, World Meteorological Organization, Global Ozone Research and Monitoring Project, Report No. 50, WMO, 2007.
[b]1990 emissions taken from *Scientific Assessment of Stratospheric Ozone: 1998*, World Meterological Organization, Geneva, 1999.

$CFCl_3$, yet the lifetime of CH_3Br is more than 60 times less and the molecule contains only one halogen atom instead of three: but that halogen atom is Br and not Cl.

It must be emphasized that exact values of these ODPs (as well as the atmospheric lifetimes) depend on the particular model used for their assessment. In addition, there is a potentially serious problem in that the ODP is based on steady-state concepts, comparison being made with the long-lived $CFCl_3$, while short-term releases of highly reactive compounds put a transient burden on the atmosphere. Improvements have been proposed that attempt to deal with the problem of timescale by defining a *time-dependent ozone-depletion potential*.

ODPs are essentially concerned with quantifying the impacts on stratospheric ozone of individual halocarbon compounds. Another way of viewing the impacts of the *mixture* of halocarbons released by nature and by Man has become increasingly popular. In this approach, the starting point is the evaluation of an *equivalent effective stratospheric chlorine* (EESC) loading of all the compounds of interest. The EESC is the sum of atmospheric mixing ratios of the constituent compounds, each suitably weighted by the number of halogen atoms it contains, by a factor that describes the halogen release, and, in the case of bromine compounds, by α. Figure 11.8 shows a typical set of calculations of EESCs, with the future release rates reflecting one of the 'protocol scenarios' that will be described in the next section. The figure indicates the contribution made by each of a number of individual compounds or classes of compounds. Of these, the *hydrochlorofluorocarbons* (HCFCs) are intended as (temporary) alternatives to CFCs (as also described in the next section): Table 11.4 shows the very low OPD of one such HCFC. However, it will be seen immediately that by far the largest contribution to the EESC is that made by CFCs, and that will remain true until the end of the 21st century.

The next step is to relate the EESC to the resultant ozone depletions. One approach is to use observations of trends in both EESC and O_3 (Section 9.6) to predict changes of O_3 in the future. For simplicity, it may be assumed that there is a linear relation between changes in ozone-column amount and EESC. The year 1980 has been used as the reference base in many of these calculations

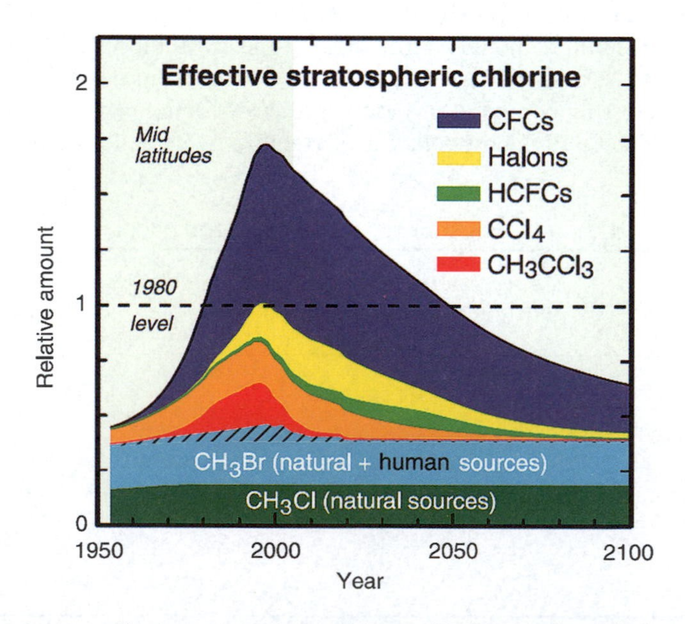

Figure 11.8 The *equivalent effective stratospheric chlorine* (EESC) loading of the atmosphere, past and future. The EESC measure takes into account the numbers of halogen atoms released by each compound, the altitude distribution of photolysis, and the catalytic activity of bromine relative to chlorine. Source as for Figure 11.5.

because up to that date the chlorine and bromine in the atmosphere had not been sufficient to significantly deplete stratospheric ozone. The EESC in 1980 is taken to be 1.895 ppb, with an average rate of increase of 0.860 ppb per decade. This procedure then allows conversion of projected EESC loadings, such as those shown in Figure 11.8 or those from any other chosen scenario, to predicted amounts of stratospheric ozone in the future. Such predictions form the basis for legislation for control of emissions, and it is to that subject that we turn next.

11.2.5 Halocarbons: Control, Legislation, and Alternatives

So long as the release rates are unaltered, a steady-state concentration of halocarbon must ultimately be reached. Both the time taken to reach the steady state and the value of the limiting concentration depend on the lifetime. With the 110-year lifetime of Table 11.4, a steady-state mixing ratio for CF_2Cl_2 of 1.8×10^{-9} would be reached if 1990 release rates were maintained, a value more than three times the 1990 atmospheric burden shown in Figure 11.6.

Constant release, leading to steady-state atmospheric concentrations, is only one of many scenarios that can be imagined for the CFCs. For a start, historical release rates have been anything but constant. From a few hundred thousand kilograms of CF_2Cl_2 produced in 1931, an all-time high of 4.74×10^8 kg were produced in 1974. (The cumulative production up to 1990 was about 10^{10} kg.) Similar figures apply to $CFCl_3$. Some legislation was passed in the 1970s forbidding certain uses of the CFCs, especially in the US, and production in 1981 was more than 20 per cent less than it had been in 1974. By the late 1980s, however, production had begun to creep up again, to reach the rates given in Table 11.4. A much more wide-ranging control is embodied in the 'Montreal Protocol on Substances that Deplete the Ozone Layer' that was agreed in September 1987 and that entered into force in January 1989. Each party to the Protocol was to freeze, and then reduce according to an agreed timetable, its production and consumption of five specified CFCs and the halons. However, it soon became apparent that, even with the 'Montreal' controls, concentrations of the CFCs would continue to increase, albeit perhaps only half as fast as without the controls. A series of amendments (London, 1990; Copenhagen, 1992; Vienna, 1995; Montreal, 1997; Beijing 1999) have therefore been proposed, and ratified by many, but not all, of the producing countries. These controls called for the 'developed' countries to have phased out production of the halons by 1994 and the five CFCs by 1996; 'developing' countries have until 2010 to completely phase out their production of these compounds. Similar requirements are placed on most other halogen-containing anthropogenic ozone-depleting substances, including methyl bromide. The projections for future EESC loadings shown in Figure 11.8 are, in fact, based on these requirements of the Protocol incorporating the 1997 and 1999 amendments. A comparison of the release rates of several ODSs given in Table 11.4 for 1990 and 2003 also shows the decreases and shifts in production brought about by the protocols.

Reductions in the use of CFCs can be achieved by avoiding their use altogether where possible, or by improving recovery and recycling of them. However, there are instances where halocarbons offer outstanding advantages in particular applications, such as use as refrigerants. There has been an intensive search, therefore, for halocarbons that are 'alternatives' to the conventional ones. The requirements are that the alternative compound should have a low ODP (which means, in essence, that it should have a short tropospheric lifetime or contain no chlorine or bromine) while retaining the desired physical and thermodynamic properties (such as boiling point or heat of vaporization). Two classes of compound that have received particular attention are the *hydrochlorofluorocarbons* (HCFCs) and the *hydrofluorocarbons* (HFCs). Example replacements are CF_3CHCl_2 (ODP $\cong$ 0.013) for $CFCl_3$ (ODP $= 1$) as a foam-blowing agent; CF_3CH_2F (ODP $\cong 0$) for CF_2Cl_2 (ODP $= 0.9$) as a refrigerant, and CH_2ClCF_2Cl (ODP < 0.05) for $CF_2ClCFCl_2$ (ODP $= 0.9$) as a solvent in the electronics industry. Because both HCFCs and HFCs contain hydrogen in place of one or more of the chlorine atoms in the CFCs, they are subject to attack (hydrogen abstraction) by OH in the troposphere (see Chapter 8). The lifetimes are consequently much shorter than those of the CFCs, so that much smaller amounts reach the stratosphere. Furthermore, the HFCs contain no chlorine at all, so that chlorine-based catalytic cycles for the removal of ozone cannot operate. Although there has been some discussion about catalytic chains involving F- or CF_3-based radicals, it seems that they cannot give rise to much removal of stratospheric ozone. Aspects of the use of HCFCs and HFCs that have given rise to concern are their toxicities and the possible adverse impact of their degradation products in the troposphere. For example, CF_3COOH, a known neurotoxin, might well be formed in the degradation of compounds containing the CF_3 moiety. Of the 'alternative' compounds, the HCFCs, in particular, are seen only as *transitional substances*, in the sense that they, too, will be phased out ultimately.

11.2.6 Halocarbons: Future Ozone Depletions

Projected changes in atmospheric loadings of various halocarbons, such as those described in the last section, can be used in conjunction with numerical models of the atmosphere to predict how atmospheric ozone concentrations might alter in the future. Different models, possessing different physical formulations and making different assumptions, produce different outcomes. The predicted extent of ozone depletion caused by CFCs has, of course, changed as the chemical input data to the models have been refined, and it will doubtless continue to change in the future. Although the predicted magnitude of the effect has altered, the sense has always been the same: ozone is lost.

The observational database is now sufficiently extensive for historical changes in stratospheric ozone concentrations to be linked to changes in stratospheric halogen loading (EESC). Figure 11.8 has already been introduced to indicate the EESCs expected from one specific scenario, and it is then a simple matter to obtain a rough estimate of ozone levels that might be found at any time in the future, assuming of course that changes in stratospheric ozone levels will continue to be dominated by changes in the halogen content of the stratosphere.

Depletions of ozone, whether predicted by models or by the pseudo-experimental method, depend on altitude, season, and latitude. Integrated column-ozone concentrations are generally depleted more at high latitudes than at the equator, and the high-latitude depletion is greatest during winter. However, the changes in column abundances are all generally much smaller than the local changes at high altitudes. Some models show depletions for high latitudes of more than 50 per cent at altitudes of 45 km, where the effects of CFC release are at a maximum. Depletion of ozone in the upper stratosphere appears to be partially compensated by increased penetration of solar ultraviolet radiation, which leads to an increased rate of generation of ozone in the lower stratosphere. This compensation is sometimes referred to as 'self-healing'. A further complication arises because changes in concentrations of other atmospheric gases, such as N_2O, CH_4, and CO_2, also affect both the distribution and the total column density of ozone.

Regardless of the exact quantitative link between ozone and halogen levels in the stratosphere, it is abundantly obvious from Figure 11.8 that, without the controls imposed in 1996, ozone depletions could have reached catastrophic proportions in the 21st century. A simple extrapolation of the EESC values shows that the Montreal Protocol of 1987 would apparently have permitted a doubling of halogen loading in each decade. As discussed in Section 11.2.1, one of the main concerns about reductions in stratospheric ozone has been the increase in UV-B radiation reaching the Earth's surface, and the consequential increase in the incidence of skin cancer. It is therefore interesting to end the present examination of the impact of the halocarbons on atmospheric ozone by looking at how the EESCs might change for various emission and control scenarios, and extend the calculations to the projected number of excess cases of skin cancer. Figure 11.9 indicates the

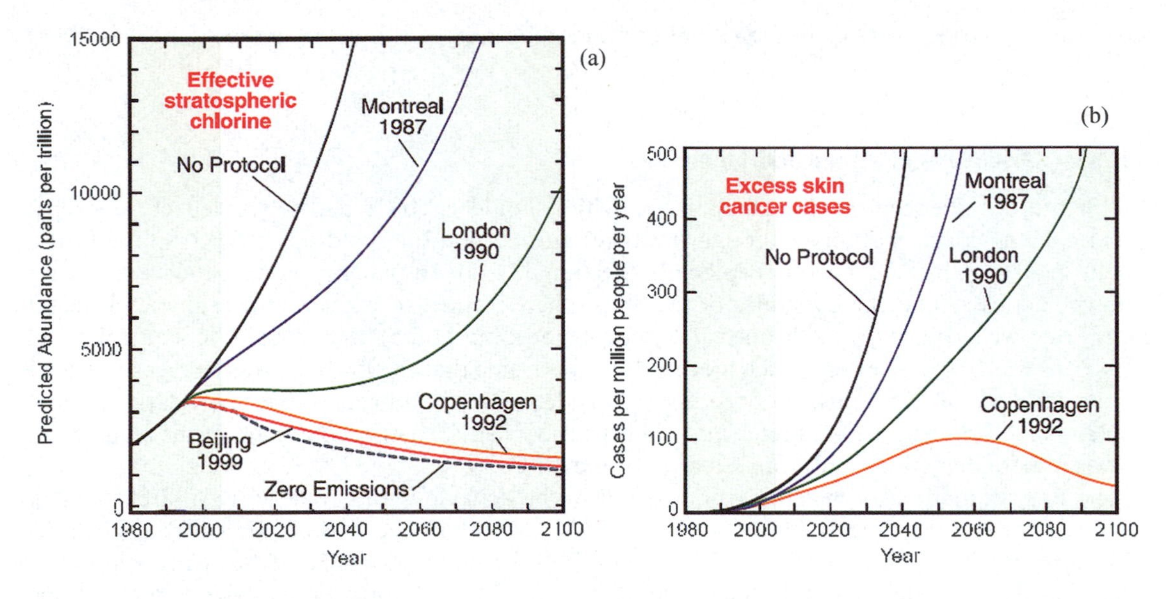

Figure 11.9 (a) Effects of the Montreal Protocol and later amendements on the modelled evolution of EESC into the future. (b) The estimated accompanying increased incidence of skin cancer for the four main scenarios. Source as for Figure 11.5.

results of one such set of estimations. Adopting the 1987 Montreal protocol was better than doing nothing, but not much. Only after the much more draconian demands of the later protocols, starting with the 1982 amendments from Copenhagen, did the EESCs show a sustained decrease from their peak levels, and skin cancers even then continue to increase until the middle of this century. If the projections are correct, and there is really no reason to doubt them, this is the situation with which most of the readers of this book will have to contend.

11.3 POLAR OZONE 'HOLES'

Section 9.4 described some special features of the ozone chemistry that may occur in polar regions. The meteorology of these regions may allow the formation both of a vortex that essentially isolates the gases within during the long polar night and of polar stratospheric clouds (PSCs) on the particles of which unusual heterogeneous chemical reactions can occur. The end result is that, when the Sun rises again in the early Spring, very large depletions of ozone may occur that have popularly become known as ozone 'holes'. The phenomenon is more marked and more frequent in the Antarctic than the Arctic because temperatures are lower, so that both vortex generation and PSC formation are more common in the south than in the north. The ozone-depletion chemistry presented in Sections 9.4.3–9.4.5 all turns on ozone destruction catalysed by chlorine and bromine. We now know from Sections 11.2.3–11.2.6 that almost all of the halogen precursors are substances manufactured by Man (and most of them do not exist in nature at all). Released in highly populated and reasonably temperate parts of the Earth, they exert their most dramatic damaging effects in remote polar regions of the planet. We have thought that the story of the ozone holes will be less confusing and convoluted if it is told in two parts. The first part is about the chemistry itself, and is an extension of the more general presentation of stratospheric ozone; the material clearly all belongs together in Chapter 9, and we suggest that the reader looks over the relevant sections before continuing here. In the present chapter, the emphasis switches to the discovery of the unexpectedly large polar-ozone depletions and to establishing a link between the ozone losses and Man.

11.3.1 Discovery of Abnormal Depletion

In 1985, it did not seem as though Man's activities should *yet* have had a large effect on stratospheric ozone concentrations, although it was recognized that the build-up of CFCs in the future might lead to substantial ozone depletions (Section 11.2.6). In general terms, models based on known chemistry and the dynamics of the atmosphere seemed to explain stratospheric chemical behaviour well. Indeed, the 'Montreal Protocol' (Section 11.2.5) that sought to control CFC emissions was based on exactly such models. It is thus ironic that the Protocol was being established in the autumn of 1987, just as experimental evidence was confirming extremely large annual depletions of ozone in the Antarctic, and atmospheric scientists had become sure of the connection between these depletions and Man's release of the CFCs.

The first intimation of something unexpected in the behaviour of ozone over Antarctica came from scientists of the British Antarctic Survey (BAS), who had been measuring ozone concentrations regularly from their base at Halley Bay at 76°S for many years. The BAS team believed in 1982 that they had detected a decline in (Southern Hemisphere) springtime ozone concentrations since 1977, and by October 1984 they were sure, with something like 30 per cent total ozone loss, and the depletion apparently increasing over the years 1982, 1983, and 1984. It is now apparent that this thinning of the ozone layer in the Antarctic spring had already been going on for more than a decade. Figure 11.10 shows how the mean October ozone concentrations measured from Halley Bay have evolved. There is no room to doubt that something quite dramatic has happened since the mid-1970s, and it is the absence of much of the normal stratospheric ozone that has been called the

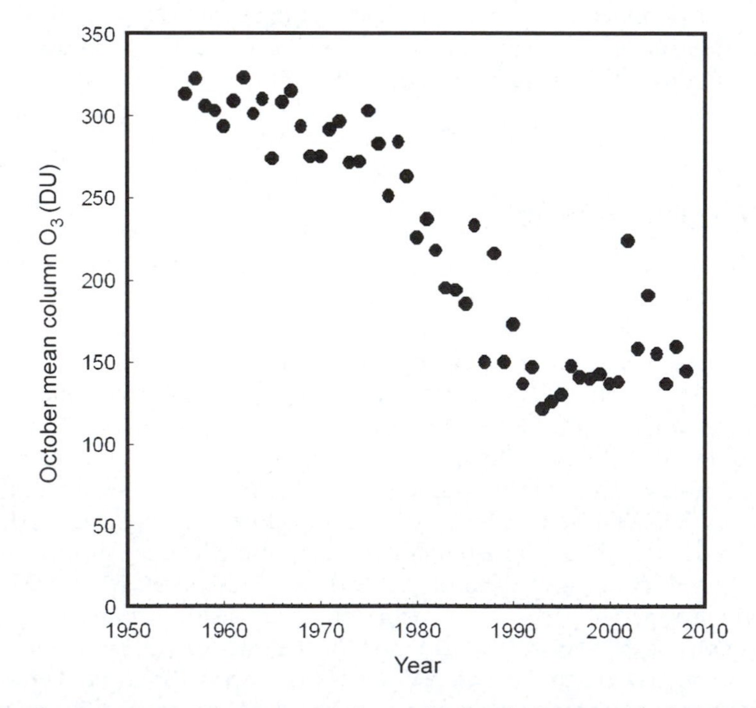

Figure 11.10 Decline in mean October ozone levels over Halley during the period 1957–2008. Data from the British Antarctic Survey (http://www.antarctica.ac.uk/met/jds/ozone/data/ZOZ5699.DAT), July 2009. Halley station was formerly referred to as Halley Bay.

'ozone hole'. Subsequent examination of records from satellite-borne ozone instruments (see Section 4.3 and Figure 4.2) such as the Total Ozone Mapping Spectrometer (TOMS) and the Solar Backscattered Ultraviolet (SBUV) instrument on the Nimbus 7 satellite showed that these more sophisticated devices also gave evidence for the ozone depletion. It is, however, instructive that the depletion of ozone was recognized first by the British scientists working with relatively simple equipment from the ground. The magnitudes of the changes were not predicted in 1985 by any model of future stratospheric composition, even for 50 or 100 years ahead. The depletions were the more startling because they occurred in the present-day atmosphere.

Satellite observations (TOMS, GOME and SCIAMACHY) show that the depleted region has grown much deeper, and covered greater areas of the Earth, since 1979. The hole has been defined as an ozone column of 220 Dobson units (DU) or less: bear in mind that the pre-hole column amounts were typically 300–350 DU at Halley Bay during October. Defined in this way, the hole was first larger in footprint than the Antarctic continent in 1987. One of the lowest column amounts ever measured was 88 DU on 28 September 1994. Figure 11.11 shows a series of contour maps of mean October ozone-column amounts for the years 1995–2008. In every one of those years, an ozone hole developed (dark blue); in most of the years, some areas experienced an overhead column of <150 DU. October 2006 seems to have seen the current 'record largest' ozone hole ever in terms of low ozone (as low as 93 DU) over a large area (the hole covered 11.4 million km^2).

The Antarctic ozone hole is very clearly a seasonal phenomenon, with maximum depletions in late September or early October. The monthly total ozone has recently been between 40 and 55 per cent below the pre-hole values, with up to 70 per cent deficiencies in short periods of a week or so. It has generally been assumed that the hole starts to develop in the austral spring, when the Sun returns to the main body of Antarctica, but there is some evidence to suggest that the hole really starts each year in mid-winter at the sunlit edge of Antarctica. This result may be important,

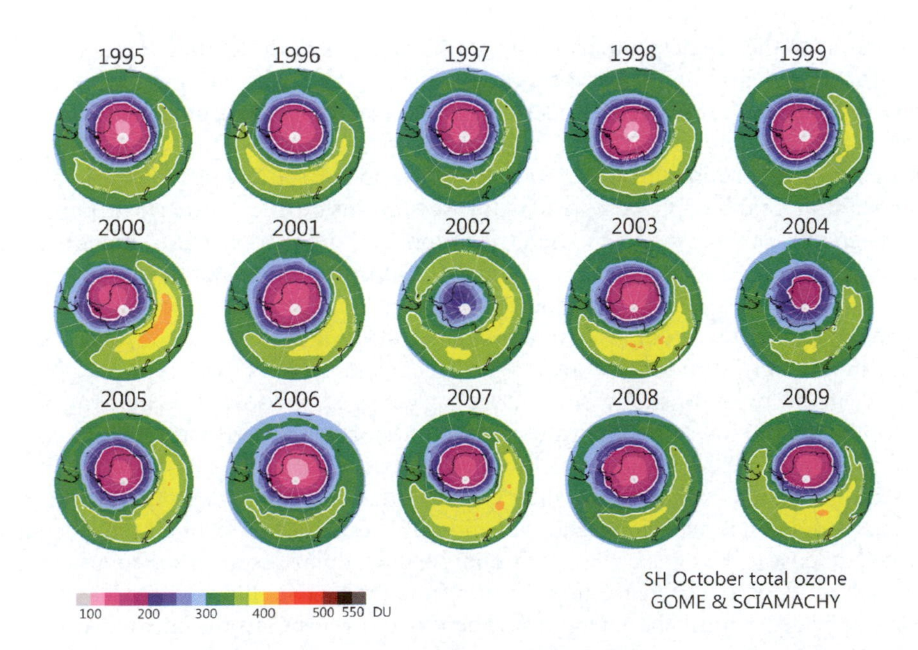

Figure 11.11 Mean October ozone columns in the Southern Hemisphere for the years 1995–2009. Data retrieved from GOME and SCIAMACHY satellite observations. Plot kindly provided by J. P. Burrows and M. Weber, Institute for Environmental Physics and Remote Sensing, University of Bremen, December 2009.

because ozone-poor air from the edge of the ozone hole regularly passes over southern South America, thus exposing populated regions to elevated doses of UV radiation.

The use of the word 'hole' to describe what happens to ozone in the Antarctic hardly seems an exaggeration. Up to 80 per cent of the ozone in the lower stratosphere (12–20 km) has been lost in recent years in October, and at some altitudes virtually all the ozone is lost, as Figure 9.5 illustrates (see Section 9.4.1).

11.3.2 Origin of Chlorine Compounds; Dynamics

Section 9.4 sets out the unusual chemistry that leads to polar ozone holes. This anomalous chemistry depends on the presence of chlorine and bromine compounds, and most of those halogen compounds have been released by Man in the form of the chlorofluorocarbons. The concentration of the CFCs in the lower atmosphere near the South Pole is almost exactly the same as it is in rural areas of Britain or the United States. In the Antarctic, the compounds have been transported in from populated regions of the Earth. However, at higher altitudes over Antarctica, as the stratosphere is reached, the concentrations of CFCs drop very abruptly, as shown already in Figure 11.7. The absence of CFCs in the stratosphere means, of course, that the chlorine atoms have already almost all been released by photochemical decomposition, and are available to destroy ozone. This release may occur largely over mid-latitudes and the tropics, the chlorine being transported (mainly in the form of the reservoir gases) to polar regions.

The records of polar stratospheric ozone show a considerable variability in the losses from year to year, as even a cursory glance at Figures 11.10 or 11.11 will show. Within any year, too, there is much variability from day-to-day in extent and depth of the ozone hole, even though the progression through formation to recovery follows the trends that we have described. Variability is largely consequent on the dynamical features of the vortex, and the random nature of the forces that act on it.

Features that have the greatest impact on ozone are changes in the spatial and temporal extent of low temperatures, and how long the vortex endures into the spring. For example, the *relatively* weak hole seen in 2002 (Figures 11.10 and 11.11) was a result of meteorological factors: the vortex (see Section 9.4.2) was unusual in form, and broke up very early. Formation of PSCs requires temperatures below a certain threshold, so that fluctuations of a few degrees in temperature can produce substantial changes in the extent of processing inside the vortex. Furthermore, similar small changes greatly affect the extent of denitrification, and thus, in turn, how quickly the reservoir $ClONO_2$ can be reformed after the Sun returns. If chlorine remains activated when sunlight is present, ozone loss rates increase substantially. A similar influence can be seen at work when wave activity distorts the vortex from a symmetric polar flow into one that transports chemically processed air into lower latitudes that are sunlit. Volcanic eruptions (Section 9.5.5) further add to the variability of the polar ozone in several ways. Increased stratospheric aerosol, for example, leads to a cooling in the lower polar stratosphere, as well as enhanced background chemical processing and additional condensation nuclei for the formation of PSCs.

Superposed on the factors that lead to interannual and intra-annual variability seem to be others that provide a trend for a deepening and widening of the Antarctic ozone hole since its identification in 1984, at least if the evidence of Figure 11.10 and similar results is taken at face value. One major influence must be the increase in anthropogenic inorganic chlorine that has been reaching the stratosphere at least until the late 1990s. There may even be a non-linear effect, because the concentration of the chlorine monoxide dimer, a key species in ozone destruction, depends on the square of the active chlorine concentration. A positive feedback between ozone concentrations and temperatures may also have an amplifying influence. Reduced ozone in the stratosphere implies less solar heating, so that the vortex remains cold, and it and PSC processing persist longer. In turn, these factors will lead to increased ozone depletion. Long-term declines in ozone are thus expected to cause larger ozone holes each year, and there is certainly observational evidence that the vortices in both hemispheres have been lasting longer in recent years. Changes in tropospheric source gas concentrations may have similar effects. Increased CO_2 is expected to *reduce* lower-stratospheric temperatures and thus increase the frequency and extent of PSC formation; similarly, increased CH_4 burdens are likely to have the same effect.

11.3.3 The Arctic Stratosphere

It is of obvious interest to discover if the anomalous chemistry seen in the Antarctic atmosphere could also be of significance in the Arctic. Because the winter stratosphere is generally warmer over the Arctic than over Antarctic regions (the temperatures are roughly 10 °C higher), polar stratospheric clouds are far less abundant and persistent in the north than in the south polar regions. This situation is illustrated in Figure 9.6. Temperatures can be low enough for PSC formation (Section 9.4.2) for months on end in the Antarctic, and even at the upper range of temperatures experienced are always below the threshold for PSC formation in mid-winter. On the other hand, sufficiently low temperatures are experienced in the 'average' Arctic stratosphere for just a few weeks, and in many winters are not reached at all. The Arctic vortex is generally smaller, less stable, and shorter lived than the Antarctic vortex. A further difference between the two hemispheres lies in how the poleward and downward transport of ozone-rich stratospheric air from lower latitudes during the autumn and winter is stronger in the north than in the south. As a result, total ozone values in the Arctic are considerably higher than in the Antarctic at the beginning of each winter season (see Figure 5.4). Finally, the variabilities of stratospheric ozone, within one winter–spring period and from year-to-year, are particularly large in the Arctic, and it has proved difficult to establish clearly the signal of chemically induced seasonal ozone loss.

Despite the formidable difficulties, ozone losses have now been detected that are comparable with those seen in the Antarctic, and that can thus properly be regarded as 'Arctic ozone holes'. Figure 11.12 is the Northern-Hemisphere analogue of Figure 11.11, being the March-average column-ozone amounts for 1996–2009. Normal monthly averaged values for March are 460–500 DU (red and brown colour codes), so that there are significant ozone depletions in many of the years represented here, and especially in 1997 and 2000. The evidence for a 'hole' is most apparent for 1997. Satellite instruments observed a record ozone low of 219 DU as the spring Sun dawned over the North Pole on 24 March 1997, and the average 1997 levels were 40 per cent lower than the March averages for the years 1979–82. Beginning in early March, sunlit regions within the Arctic circle (*e.g.* central and eastern Siberia) experienced ozone thicknesses in the range 240–260 DU.

The winters of 1995–1996, 1996–1997, and 1999–2000 were characterized by particularly low polar stratospheric temperatures, and by low spring ozone concentrations. The winter polar vortexes of 1996–1997 and 1999–2000 were unusually strong and persisted into March. It is thus highly probable that the signature of Arctic ozone depletion has become sharper as polar stratospheric temperatures have dropped in recent years, and as the anthropogenic halogen burden has built up to its maximum level. In contrast, losses were relatively small in the winter of 1998–1999 (less than five per cent loss in the vortex), and this was a winter in which there was a warming in December, the vortex was perturbed, and no PSCs were seen after mid-January 1999.

Figure 11.13 brings together for comparison the observational data for polar ozone abundances in the Arctic and Antarctic stratospheres over a period extending to March 2007. By continuing the line for the pre-1982 average, the decreases in measured column O_3 amounts in more recent years can be seen clearly as the deviation from the line (coloured zone). Essentially all of the decrease in the Antarctic, and usually most of the decrease in the Arctic, each year is attributable to chemical

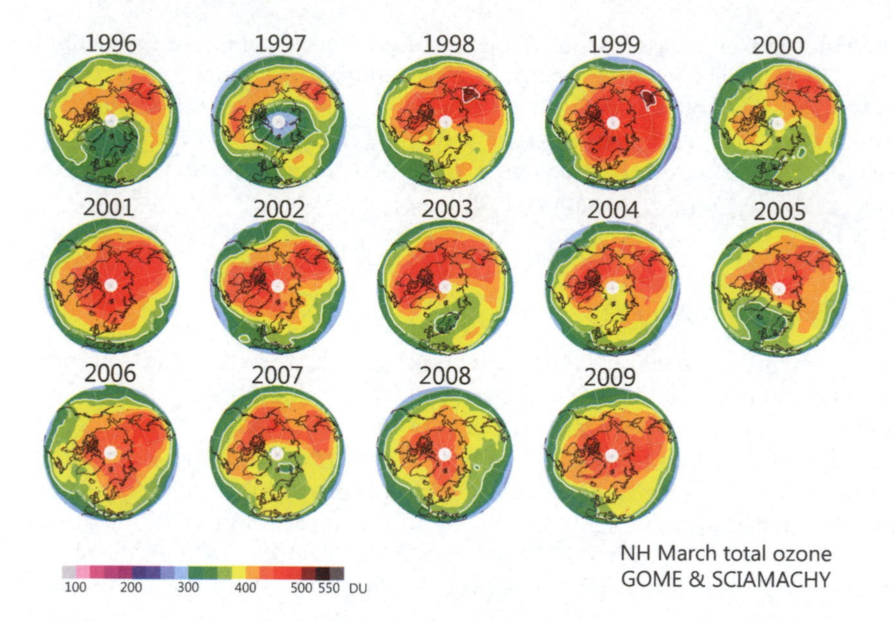

Figure 11.12 Mean March ozone columns in the Northern Hemisphere for the years 1996–2009. Data retrieved from GOME and SCIAMACHY satellite observations. Plot kindly provided by J. P. Burrows and M. Weber, Institute for Environmental Physics and Remote Sensing, University of Bremen, July 2009.

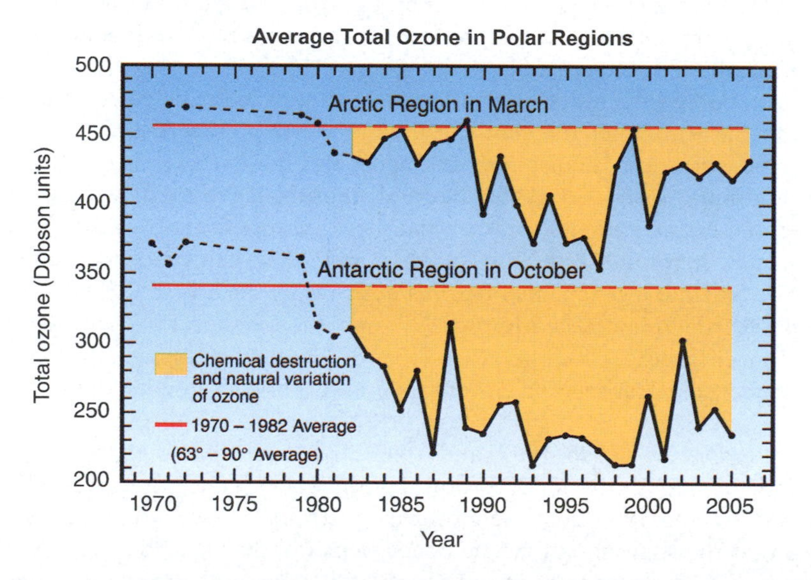

Figure 11.13 Ozone in polar regions. Average springtime total-ozone values measured between 1970 and 1982 (red lines) are compared with those in later years. Each point represents a monthly average in October in the Antarctic or in March in the Arctic. Source as for Figure 11.5.

destruction by reactive halogen gases. Models of Arctic ozone chemistry and transport can now simulate quite well the observations for individual years. One dramatic conclusion from the model studies is that, had the halogen loadings not been limited by the control protocols, we should already have been anticipating Arctic ozone losses as large as those experienced in the Antarctic.

Given the distribution of the world's human population, very serious consequences might have followed.

11.3.4 Implications of the Polar Phenomena

There are two questions to be faced in assessing what the effects of polar-ozone depletion might be. First, there are the direct consequences of reduced ozone in those geographical regions where the ozone is lost. Such consequences range from meteorological impacts, including changes of temperature and weather patterns, to the results of increased exposure of living organisms to UV radiation. Secondly, there might be remote consequences on parts of the environment not immediately in or under the ozone-hole region. It is, perhaps, worth noting immediately that, although ozone depletions in the Arctic are much less than those in the Antarctic *so far*, the Arctic is much closer to high densities of population, of humans at least, and the off-pole centreing of the Arctic vortex brings the Northern Hemisphere ozone hole even closer to many of us. For those living in the United Kingdom, March 1996 brought record low values of total ozone. On 3 March, the total ozone over Camborne (50°N) was 206 DU, and two days later, Lerwick (60°N) recorded 195 DU. This was the first time that a daily value below 200 DU had been observed over the UK. While meteorological factors contributed to the low ozone values, it is clear that Arctic air in which photochemical ozone loss had *already* occurred was lying over the UK at the time.

A major reason for concern about stratospheric ozone depletion is the biological danger posed by increased solar ultraviolet radiation reaching the Earth's surface (Section 11.2.1). In that context, the creation of polar ozone holes must increase the amount of UV that can penetrate to the ground, especially in the significant UV-B region (λ = 280–320 nm). Spring-time biologically active irradiances in the Antarctic can exceed those experienced in the (unperturbed) midsummer. For example, even by as early as 5 November 1987, the calculated UV-B irradiance averaged over the day exceeded the solstice value by over 50 per cent, and the irradiance at noon was relatively even more enhanced. For the same overhead ozone column, the UV exposure in the Antarctic spring is comparable with typical values at a mid- to low-latitude location such as Miami. A particular threat is posed to indigenous organisms rather than to human visitors to the Antarctic. One area of concern is the possible sensitivity of the spring bloom of phytoplankton living in the surface waters around Antarctica. The UV dose is already greatest in spring, when the ice is much more transparent than in the summer.

Quite apart from the local effects of polar-ozone depletion, it is important to know if the processes occurring within the chemically perturbed region can influence ozone concentrations at lower latitudes. Although the vortex is conveniently pictured as a 'containment vessel' that isolates the chemically perturbed region, there is no seal around the polar air, and the containment provided by the vortex is somewhat leaky. Ozone-deficient or chemically processed air can thus be transported to lower latitudes. Later in spring, when the vortex breaks up, more extensive lateral mixing takes place. Export of ozone-poor air from polar regions to middle latitudes might lead to a dilution of ozone that could persist for up to a year because of the relatively slow photochemical replacement of ozone, and transport of processed air rich in active chlorine could enhance the effect. If the deficit lasts from one spring to another, there could be a cumulative, permanent depletion of ozone. There is now clear evidence for a reduction in stratospheric ozone at mid-latitudes (see Section 9.6), although the causes remain an open question. One potential explanation is erosion of ozone-poor air from the vortices. In the Northern Hemisphere, frequent extravortical PSCs, probably enhanced by lee-wave activity above the Arctic mountain ridges, may also play a role akin to that of PSCs in the vortex itself.

The understanding that heterogeneous chemistry is central to Antarctic and Arctic ozone depletion has brought a heightened awareness of the importance of aerosol processes in the chemistry as well as the physics of the stratosphere. Laboratory experiments have shown that reactions can proceed on sulfuric acid particles that are similar to those taking place on the surfaces of PSCs. Sulfuric acid aerosol is widely distributed in the stratosphere, and might thus be responsible for ozone depletion in non-polar regions. Such activity is obviously enhanced in periods following volcanic eruptions that deposit sulfate aerosol in the stratosphere. Clear signatures are apparent in the observations of chemical processing following the El Chichón and, especially, the Mount Pinatubo explosions (see Section 9.5.5) and show that the aerosol was responsible for heterogeneous loss. Ozone measurements exhibit record lows following the eruptions. Total column NO_2 abundances also show a marked decrease, in the same way as in the Antarctic perturbed chemistry. This circumstantial evidence is indicative of chemical processing on aerosol surfaces.

11.4 THE 'GREENHOUSE EFFECT', GLOBAL WARMING AND CLIMATE CHANGE

Of all the impacts that Man has on the atmosphere, potentially the most damaging of all is brought about by human activity adding to the atmospheric burden of greenhouse gases (GHGs), whose properties were introduced in Section 2.3.2. Increased radiation trapping (*radiative forcing*) will bring about global warming of an extent that constitutes *climate change*. Carbon dioxide from the burning of fossil fuel is the major contributor; but gases ranging from CH_4 and N_2O to the HFCs introduced as replacements for the CFCs (Section 11.2.5) are radiatively active, so that changes in agricultural practices or shifts in manufacturing emphasis can further enhance the anthropogenic greenhouse warming. Since each GHG possesses its own individual absorption spectrum and absorption intensities, each gas is assigned a *greenhouse warming potential* (GWP) in order to compare its greenhouse activity relative to that of CO_2. The precise definition is complex, but essentially it is the ratio of the warming produced by injection of 1 kg of the target gas to that produced by the same mass of CO_2 and measured over a specified *time horizon* (usually 100 yr).

Carbon dioxide concentrations are increasing at an unprecedented rate, and the increases are largely of Man's making (see Figure 6.3 and the associated discussion). In response to the perceived impacts of global increases in temperature, a concerted international effort has been made to assess the current state of knowledge by the establishment of an *Intergovernmental Panel on Climate Change* (IPCC). The present section of the book provides a brief introduction to the problem, and largely—but, as will appear later, not entirely—reflects views put forward in the panel's *Fourth Assessment Report* (*Climate Change 2007*), abbreviated here to AR4. Involving over 3800 scientists from over 150 countries and six years of work, AR4 reviewed and analysed scientific studies published up to the end of 2006, and in a few cases, to early 2007. The report states that warming of the climate system is no longer open to question, as is now evident from observations of increases in global average air and ocean temperatures, widespread melting of snow and ice and rising global average sea level. Figure 11.14 displays some of the evidence for these statements.

It is very unlikely that the 20th-century warming can be explained by natural causes. The second half of the 20th century was unusually warm, and was probably the warmest 50-year period in the Northern Hemisphere in the last 1300 years. This rapid warming is consistent with how the climate should respond to a rapid increase in greenhouse gases, but inconsistent with the way the climate should respond to natural external factors such as variability in solar output and volcanic activity. When the effects of increasing levels of greenhouse gases as well as natural external factors are included in climate models, the models simulate well the warming that has occurred over the past century. The models fail to reproduce the observed warming when run using only natural factors. When human factors are included, the models also simulate a geographic pattern of temperature change around the globe similar to that which has occurred in recent decades. This spatial pattern, which has features such as a greater warming at high northern latitudes, differs from the most

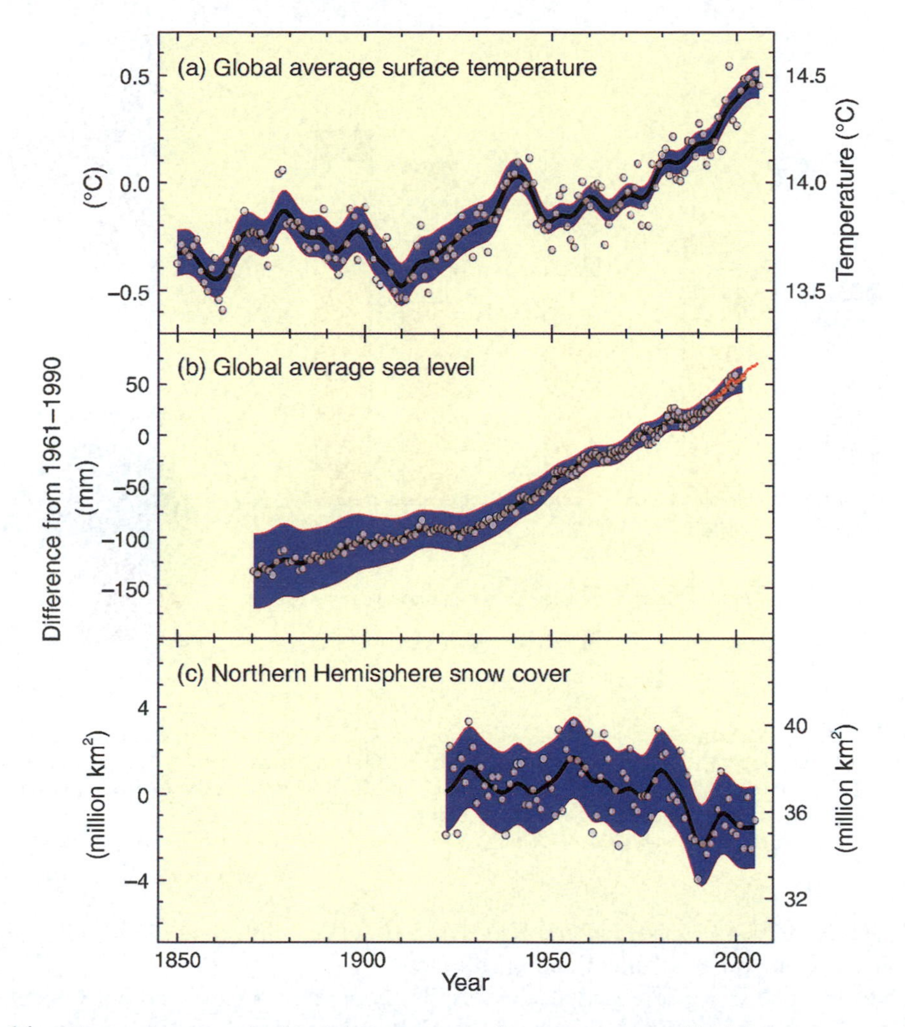

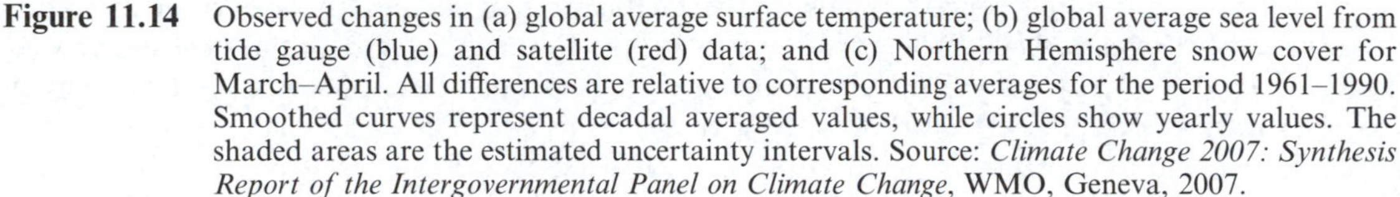
Figure 11.14 Observed changes in (a) global average surface temperature; (b) global average sea level from tide gauge (blue) and satellite (red) data; and (c) Northern Hemisphere snow cover for March–April. All differences are relative to corresponding averages for the period 1961–1990. Smoothed curves represent decadal averaged values, while circles show yearly values. The shaded areas are the estimated uncertainty intervals. Source: *Climate Change 2007: Synthesis Report of the Intergovernmental Panel on Climate Change*, WMO, Geneva, 2007.

important patterns of natural climate variability that are associated with internal climate processes, such as El Niño (see Section 9.5.2).

11.4.1 Radiatively Active Gases and Particles in the Atmosphere

Figure 11.15 shows the emission rates for the most important greenhouse gases as presented in the 2007 IPCC report (AR4) for the years 1970–2004. The rates have been scaled in panel (a) by the greenhouse warming potentials: the resulting CO_2-equivalent values then facilitate comparison of the impact of each of the GHGs. A substantial increase has been experienced over the 34-year period in fuel- and industry-derived CO_2, but CO_2 from deforestation, CH_4 and N_2O emission rates have all increased during the period. Panel (b) of the figure shows the scaled contribution of the

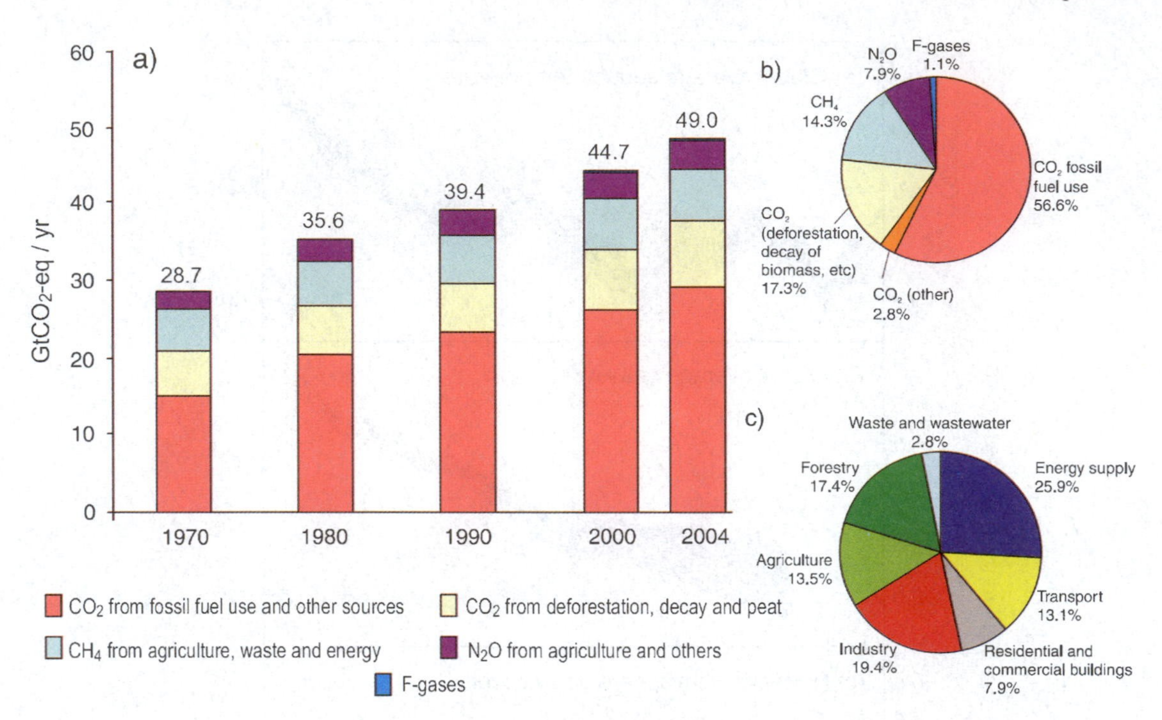

Figure 11.15 (a) Global annual emissions of anthropogenic GHGs from 1970 to 2004. (b) Share of different anthropogenic GHGs in total emissions in 2004 in terms of CO_2-eq. (c) Share of different sectors in total anthropogenic GHG emissions in 2004 in terms of CO_2-eq. (Forestry includes deforestation.) Source as for Figure 11.14.

individual GHGs in 2004 (F-gases include HFCs and HCFCs), while panel (c) gives a clear idea of the contribution from various human activities.

Air bubbles trapped in ice-core samples, especially those from Antarctica and Greenland, have been used to discover the concentrations of the GHGs present in the atmosphere in the more distant past. In the period 800–2500 yr before present, the CO_2 mixing ratio in the atmosphere was 2.6×10^{-4} (260 ppm). By early 2009, the mean mixing ratio was about 386 ppm, and showing a yearly growth of 1.58 ppm. The carbon dioxide increase has come about almost entirely since the Industrial Revolution, and mainly within the last 90 years, as a direct result of Man's activities. Increased CO_2 emission is the primary cause, with burning of fossil fuels (98 per cent) and cement manufacture (2 per cent) as the major contributors. Deforestation in the tropics and changing agricultural practices reduce the efficiency of CO_2 recycling, thus providing a secondary aggravation of the CO_2 increase.

Methane is another important contributor to the greenhouse effect whose concentrations have been rising. Greenland polar ice cores show that [CH_4] is constant from depths of 250 m to 1950 m, corresponding to the period 1580 AD to about 20 000 BC! In 1580, methane levels began to rise from their mixing ratio of 715 ppb, at first slowly, but since 1918 much more rapidly, to reach their mixing ratio of 1774 ppb in 2005. What has caused an accelerating increase in methane concentrations since the 17th century, after they had remained constant for 20 000 years? Man seems implicated because of the timescale. Increasing [CH_4] could result from increasing production of the gas. The growth in rice-paddy cultivation and in cattle farming, which are a response to population growth, could account for increased methane production. If this explanation is correct, then it seems probable that methane levels will continue to rise in the decades ahead.

Input of nitrogen into cropland is an additional way in which average world temperatures might be raised in the future. Commercial fertilizers and nitrogen-fixing leguminous crops (Section 6.4) both have the potentiality of increasing emissions from soils of N_2O, a greenhouse gas. The nitrogen fixed annually by combustion and in manufacturing fertilizers has now reached almost half of the amount of N_2O that plants produce naturally. Atmospheric N_2O concentrations, which had hardly risen at all in the preceding 10 000 yr, increased from their value in 1900 of 270 ppb to 319 ppb in 2005. Once again, there is a coupling with chemistry, this time stratospheric, because N_2O is an important 'natural' source of the NO_x that controls ambient O_3 concentrations (Section 9.3.1).

Release by Man to the atmosphere of halocarbons (CFCs, HCFCs, HFCs, and halons: see Sections 11.2.3–11.2.6) may represent another anthropogenic influence on climate. As described in Section 11.2.3, the concentration of these synthetic compounds has built up in the troposphere (see Figure 11.6). The CFCs themselves possess strong absorption bands in the infrared region that, by chance, coincide with regions where CO_2 itself has relatively weak absorption. They thus have the potential of closing the atmospheric 'windows' through which radiation could escape to space, and the contributions of such compounds to greenhouse warming may consequently be much greater than the simple additive effect of the radiation trapped by the CFCs on their own (see Section 2.3.2). Section 11.2 concentrates on the impact that CFCs may have on depletion of ozone in the stratosphere, but it is evident that CFCs may have an additional adverse environmental impact by contributing to global warming. In general, the control strategies discussed in Section 11.2.5 are those needed to counter both threats, and reduced tropospheric lifetime is clearly one desirable quality in an 'alternative' halocarbon designed to replace the CFCs. However, it must be recognized that the compounds selected to reduce ozone-depletion potentials (ODPs: Section 11.2.4) may not be ideal in terms of their infrared absorption intensities and wavelengths, and due consideration must be given to these factors in assessing overall environmental impact. In particular, the HFCs possess very large GWPs, yet their control by the ozone protocols is very limited because their ODPs are zero and they are regarded as highly attractive replacements for the banned CFCs and the transitional HCFCs.

The complexity of the climate system is nicely exemplified by a further influence of the CFCs and other halocarbons. These compounds may certainly produce direct heating through the trapping of radiation, but those that deplete stratospheric ozone have a secondary effect. Ozone itself is an important greenhouse gas in both the stratosphere and the troposphere. Reductions of ozone concentrations in the lower stratosphere produce a net *cooling* of the atmosphere, which somewhat offsets the direct radiative heating. This process is an example of a *negative feedback* in the climate–chemistry system, and we shall return to the subject in Section 11.4.3.

Other factors are believed to have a cooling rather than a heating effect. For example, both sulfate aerosol (Sections 8.4, 8.6 and 11.1.3) and the aerosol from biomass burning (Section 11.1.8) affect the energy balance. They absorb some radiation from the Sun, and reflect some back to space as well. Averaged over the planet, anthropogenic particles seem to reflect more than they absorb, and are thus direct contributors to some cooling. In a more complicated indirect mechanism, such particles also act as cloud-condensation nuclei and thus promote the formation of clouds and fogs. We discuss clouds further in Section 11.4.3, but, once again *on average*, the clouds nucleated by anthropogenic particles appear to exert a net cooling effect.

11.4.2 Radiative Forcing

The influence of a factor that can cause climate change, such as a greenhouse gas, is often evaluated in terms of its *radiative forcing*. This quantity is defined as the perturbation to the energy balance between incoming solar radiation and outgoing infrared radiation of the Earth–atmosphere system,

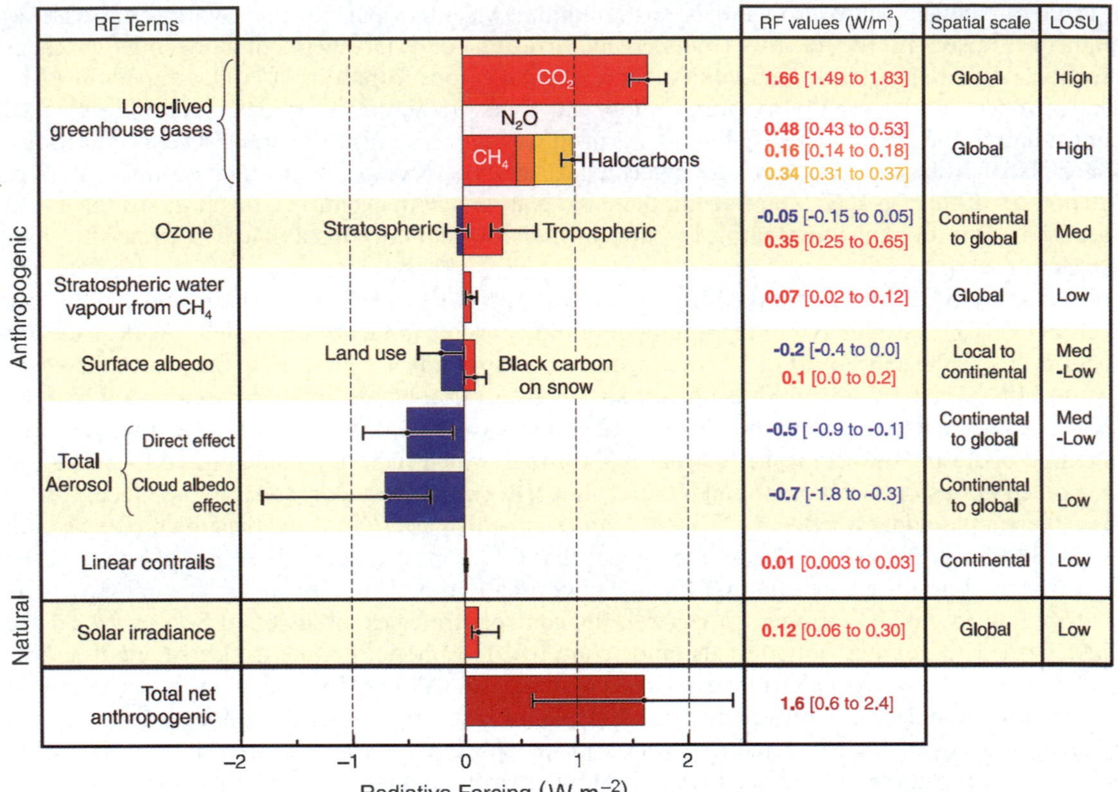

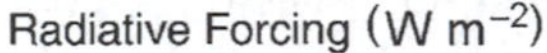

Figure 11.16 Global average radiative forcing (RF) in 2005 (best estimates and 5–95 per cent uncertainty ranges) with respect to 1750 for CO_2, CH_4, N_2O and other important agents and mechanisms, together with the typical geographical extent (spatial scale) of the forcing and the assessed level of scientific understanding (LOSU). Aerosols from explosive volcanic eruptions contribute an additional episodic cooling term for a few years following an eruption. The range for linear contrails does not include other possible effects of aviation on cloudiness. Source as for Figure 11.14.

and is measured in units of power per unit area (watts per square metre: $W\,m^{-2}$). The term forcing is used to indicate that Earth's radiative balance is being pushed away from its normal state. Important challenges for climate scientists are to identify all the factors that affect climate and the mechanisms by which they exert a forcing, to quantify the radiative forcing of each factor and to evaluate the total radiative forcing from the group of factors.

Figure 11.16 shows estimates for the radiative forcing due to different atmospheric changes that are thought to have occurred between the pre-industrial period and 2005. The various anthropogenic components are those introduced qualitatively in the previous section. The radiative forcing from CO_2 increased by 20 per cent from 1995 to 2005, the largest change for any decade in at least the last 200 years. The combined radiative forcing due to increases in CO_2, CH_4 and N_2O according to the values given in the figure has a most probable value of $+2.6\,W\,m^{-2}$, and its rate of increase during the industrial era is very likely to have been unprecedented in more than 10 000 years.

The values for radiative forcing can be compared with the total amount of incoming radiation that heats[iii] the Earth, which is roughly $235\,W\,m^{-2}$. Taking, as an example, the greenhouse gases

[iii] The Sun delivers $342\,W\,m^{-2}$, but of this amount, $107\,W\,m^{-2}$ is reflected back to space.

alone, something like 2.6 W m^{-2} less power would radiate back to space now than before the Industrial Revolution *if the surface temperatures had remained constant*. Since the radiation arriving and returning to space must be in balance, the conclusion has to be that temperatures must, in reality, rise if there is net positive radiative forcing. It is the correct calculation of this temperature rise that is the subject of Section 11.4.3.

Natural forcings arise that are due to changes in solar radiation and explosive volcanic eruptions. Solar output has increased gradually in the industrial era, causing a small positive radiative forcing (see Figure 11.16). This is in addition to the 11-year cyclic changes in solar radiation. However, these changes are both very small compared to the differences in radiative forcing that have resulted from human activities.

11.4.3 Feedbacks and Models

The radiative forcing accompanying a doubling of atmospheric CO_2 concentration from its preindustrial value would be very roughly 4–5 W m^{-2}. A simplified calculation shows that, in order to restore the energy balance, the surface and lower atmosphere would have to warm up by about 1.2 °C. One aspect of the simplifications made in reaching this conclusion is that the *only* change is in temperature. Cloudiness, water-vapour content, snow cover, and many other factors are assumed to be unaffected. In reality, of course, these assumptions will not be justified. According to current best estimates, the true average temperature change following a doubling of CO_2 concentrations would be nearer 3.0 °C, the likely range being 2.0–4.5 °C. This change for doubled CO_2 is termed the *climate sensitivity*.

The difference between the results of the simple and the sophisticated calculations comes about because of the intervention of a variety of meteorological and biological feedbacks, both positive and negative in effect, which working together have a net positive amplification factor that more than doubles the temperature rise predicted without feedback.

The meteorological feedbacks include (i) *water-vapour feedback*; (ii) *cloud–radiation feedback*; (iii) *ocean–circulation feedback*; and (iv) *ice–albedo feedback*. Of these, the first is the most important, and in many ways the simplest. If the temperature rises, the vapour pressure of water is increased, and more water evaporates from the oceans and surface waters. Water is itself a major greenhouse gas, and the increased atmospheric concentrations produce a further increase in temperature. There is thus a strong positive feedback operating. Just this single influence increases the effect of doubling CO_2 concentrations by a factor of at least two. Cloud–radiation feedback is more complex, because clouds affect atmospheric radiation in two opposing ways. They reflect some radiation back to space, thus reducing the solar energy available. On the other hand, they also absorb thermal energy emitted from the surface, and themselves emit thermal radiation, thus acting as a blanket to increase surface temperatures. The ocean feedbacks arise in three ways: oceans are the source of most of the atmospheric water vapour; they redistribute heat throughout the climate system as a result of their circulation; and they act as a buffer against rapid change because their heat capacity is much greater than that of the atmosphere. Finally, the albedo (reflectivity) of ice and snow provides a positive-feedback mechanism. The surfaces are strong reflectors of radiation. Melting of the cover as a result of warming permits the previously reflected radiation to be absorbed by the surface, and bring about yet further warming.

Biospheric feedbacks may also be of importance in the climate system. One of the most fascinating is the *plankton multiplier* effect. During cold periods (presumably associated with low atmospheric CO_2), convection in the upper layers of the oceans is enhanced, and the resulting thicker mixed layer near the surface promotes biological activity. Growth of marine plankton results in some long-term removal of CO_2 to the deep ocean. There is thus a positive feedback present that amplifies changes in atmospheric CO_2 levels. Several other biological positive feedbacks have been identified.

Methane may be involved in a positive-feedback process that will amplify temperature changes. Enormous quantities of methane are locked up as the clathrate hydrates such as $CH_4.6H_2O$, which are inclusion compounds of methane trapped in the interstices of ice crystals. Gas hydrates below the permafrost of polar regions may contain 2×10^{15} kg of carbon (2.6 times the carbon content of the atmosphere), while estimates for the content of sea-floor sediments are even larger (up to 10^{17} kg). Phase diagrams for methane hydrate can be used to discover at what depths (equivalent to pressures) and temperatures the compound is only marginally stable and might therefore be released in a warmer climate. Arctic Ocean sediments seem the most likely to be labile and to be exposed to higher temperatures. Decomposition over a 100-year period from a layer 40 m thick at ~300 m depth and half-way round the Arctic Ocean could release $\sim8\times10^{12}$ kg yr^{-1} of CH_4, enough to produce a significant positive-feedback effect on CO_2-induced warming. Because of the depth of the sediments, the feedback is unlikely to become important during the next century, but if global warming were to continue for longer, releases from hydrates might become the major contributor to atmospheric CH_4.

Numerical models are clearly the only available tool with which quantitative expression can be given to the climate system, involving as it does interacting, non-linear, positive- and negative-feedback components. The essence of the methods is that outlined in Section 4.4, and discussed several times elsewhere, especially in Chapters 8 and 9. Many of the feedback processes are already properly treated in weather-forecasting models. Others, such as the cloud–radiation and ocean–circulation feedbacks have had to be incorporated in new approaches. Increasing computational power is making such modelling feasible, and the results are apparently reliable, as we shall see later. Our discussion has so far considered only radiation trapping by the greenhouse effect, together with direct chemical intervention. There follow a multitude of additional feedbacks and couplings. For example, CO_2 and the other GHGs can *cool* the stratosphere, because when their concentration is low, they no longer trap radiation, but emit it to space (see Section 2.3.2). At these lower temperatures, the rates of loss of O_3 are decreased (Sections 9.2–9.3), so that O_3 levels tend to increase. But O_3 is the very agent that warms the stratosphere, so there is evidently a *chemical* (negative) feedback loop in operation. Models are now beginning to treat climate and chemistry together in an attempt to simulate the 'real-life' situation.

11.4.4 Projected Changes in Concentrations, Forcing, and Climate

Regardless of the quantitative connection between concentrations of CO_2 in the atmosphere and climate, it seems certain that concentrations of greenhouse gases will continue to increase throughout this century and beyond. As a consequence, radiative forcing will also increase. In 2000, the IPCC published its Special Report on Emissions Scenarios (SRES), and these scenarios are currently the basis for most discussions of projected future changes in GHG concentrations. Figure 11.17(a) shows the CO_2-equivalent emissions in some of the scenarios, as well as for two extreme scenarios suggested after the SRES. For our purposes, we shall take just one illustrative scenario, the mid-range A1B, which assumes a world of very rapid economic growth, a global population that peaks in mid-century, and the rapid introduction of new and more efficient technologies balanced between fossil-intensive and non-fossil energy resources. In this scenario, the best estimate of temperature rises for 2090–2099 relative to 1980–1999 is 2.8 °C, the likely range being 1.7–4.4 °C. Figure 11.17(b) shows the model-derived temperature changes for the main GHG emission scenarios of Figure 11.17(a), as well as the historical record for the preceding century. At first sight, 2.8 °C does not seem a very large increase. Normal temperature variations in one location, between day and night, between one day and the next, and between one season and another can be greatly in excess of one or two degrees Celsius. But the warming of 2.8 °C is a *global* average for all seasons. This temperature change within 100 years probably represents a greater rate

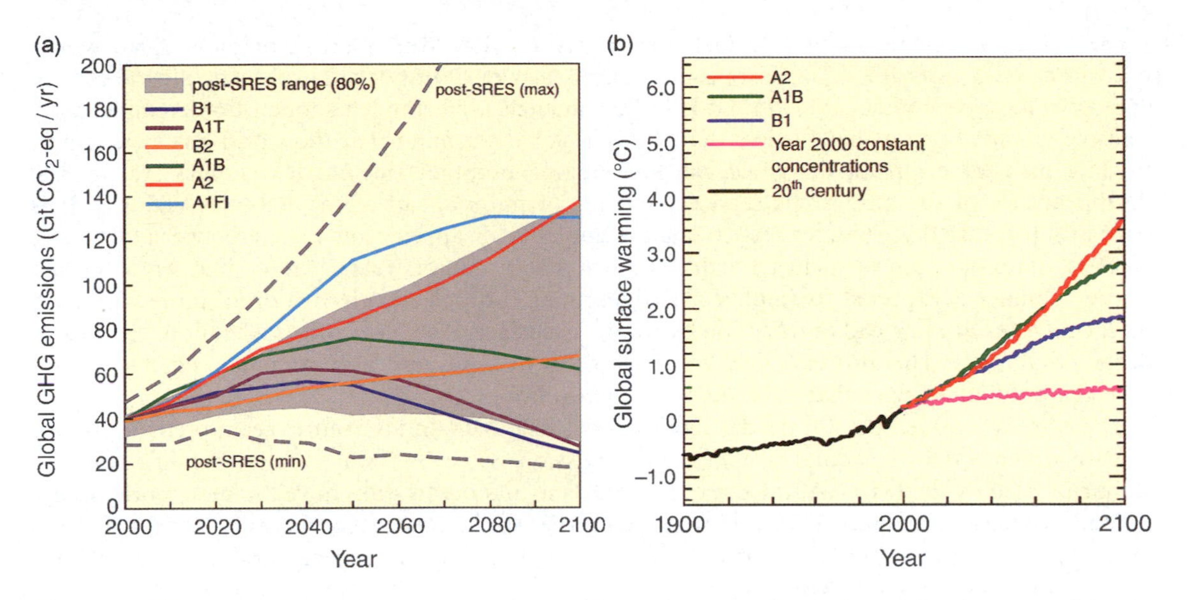

Figure 11.17 (a) Global GHG emissions for six illustrative scenarios (coloured lines) proposed in the IPCC's Special Report on Emissions Scenarios (SRES), together with the 80th percentile range of recent scenarios published since SRES (grey-shaded area). Dashed lines show the full range of post-SRES scenarios. The emissions include CH_4, N_2O and F gases as well as CO_2, and are assimilated on the figure as CO_2-equivalents in Gtonne (10^{12} kg) yr^{-1}. (b) Surface warming (global averages) relative to 1980–1999 for three of the SRES scenarios, shown as continuations of the 20th-century simulations. Also shown is the modelled warming if concentrations are held constant at year-2000 values. Source as for Figure 11.14.

of increase than any seen since the end of the last ice age. Global average temperatures change 5–6 °C between the middle of an ice age and the warm periods in between, so that 2.8 °C is the climatic equivalent of one-third to one-half of an ice age, a very significant change. From an ecological point of view, one of the major problems concerns the *rate* of change. Systems can adapt to a few degrees Celsius of temperature change, but over tens of thousands of years, not hundreds.

One of the most significant impacts that a warmer climate would have would be to increase the level of the oceans. Large changes in sea level have indubitably occurred in the past. Melting or growth of large ice sheets that cover the polar regions could be responsible for the long-term changes: for example, in bringing about the 100 m increase between 18 000 years ago and today that separated Britain from the rest of Europe. On a shorter timescale, the most important contributor to a rise in sea level is thermal expansion of the water. Next in order of importance is the melting of glaciers and ice caps. In the worst-case scenario of AR4, the maximum sea-level rise by 2100 was given as 59 cm. However, since the publication of this key report in 2007, scientific research on climate change and its impacts has continued and new studies are revealing that global warming is accelerating, at times far beyond forecasts outlined in earlier studies included in AR4. By 2009, it seemed as though the AR4 predictions had drastically underestimated the problems. New calculations[iv] suggested that *global* temperatures would increase by 4 °C in as little as 50 years if GHG release remains unchecked. This increase translates to an average on land of 5.5 °C, with astonishing rises of 12 °C or more in Siberia and northern Canada.

[iv] On 21 October, the UK Government released a poster based mainly on work of the Hadley Centre of the Meteorological Office, showing in the form of a global map the temperature changes now predicted by 2060. A version of the map may be accessed online from links provided at: http://www.metoffice.gov.uk/corporate/pressoffice/2009/pr20091022.html. The map is interactive, and allows the user to view the geographical distrubution of impacts on different systems of the predicted warming.

Oceans seem likely to rise twice as fast as suggested in AR4. More than 1.2 m sea-level rise would put vast coastal areas at risk, in Europe and around the world. Important aspects of climate change appear to have been under-rated in AR4, and the impacts are being felt sooner. For example, early signs of change suggest that the less than 1 °C of global warming that the world has experienced to date may have already triggered the first tipping point of the Earth's climate system, the disappearance of summer Arctic sea-ice (30 years or more ahead of the AR4 projections). It is currently forecast that summer sea-ice could completely disappear somewhere between 2013 and 2040—a state not seen on planet Earth for more than a million years. An ice-free Arctic Ocean during summer is expected to amplify global warming through the absorption of more heat by a dark ocean surface (instead of reflection by the white surface of sea-ice) and as a result of changes in ocean circulation. This process could open the gates to rapid and abrupt climate change, rather than the gradual changes that have been forecast so far.

There are several reasons for the discrepancies between AR4 and the more recent predictions, not least of which is that the actual growth rate of emissions since 2000 was greater than in any of the scenarios used by the IPCC. At the same time, the land and ocean sinks have declined more rapidly in their capacity to remove CO_2 than forecast in earlier studies. A further factor is that some of the feedbacks, such as the one described at the end of the previous paragraph, seem to have been severely underestimated in AR4.

The end of the 20th century already saw evidence for some unprecedented melting of glaciers. Glaciers in the Himalayas make up the largest body of ice outside the polar caps, and some (disputed) reports suggest that they are retreating faster than those anywhere else on Earth. The Alps have lost about 50 per cent of their ice in the past century, and 14 of the 27 glaciers that existed in Spain in 1980 have now disappeared. Glaciers on Mount Kilimanjaro, in Kenya, are only one-quarter as big as they were a century ago. Global warming or not, something is causing the ice to melt! Most recent satellite and *in-situ* data show that seas are now rising by more than 3 mm a year—more than 50 per cent faster than the average for the 20th century.

The ice sheets of the Antarctic and Greenland are special cases. Over the relatively short term, they provide two competing influences. A warmer climate is damper, and there is more snowfall that builds up the ice sheets; but there is also more melting into the oceans near the boundaries of the sheets. For Greenland, the losses are probably greater than the ice accumulation, but the reverse is true in Antarctica. The ice sheets are more like geological formations than components of the hydrologic system. Since they rest on solid bedrock, their melting, or sliding into the ocean, would cause a rise in sea-level. Straightforward melting of the enormous masses of ice would probably take many thousands of years, but another mechanism could accelerate the process. Much of the West Antarctic ice sheet (and some of the East Antarctic ice sheet) rests on bedrock that is below sea-level. Warmer ocean water could work its way under the ice sheet, separating it from the bedrock and causing it to slide towards the ocean. Disintegration and melting of the ice would then be relatively rapid because of the more intimate contact with the water and because of the diminished thermal insulation. It is the accelerated melting of the vast, land-based ice sheets in Greenland and Antarctica, caused by rapidly rising temperatures at high latitudes, that is now speeding up the increase in sea level beyond anything previously forecast. The Greenland ice sheet, in particular, is not simply melting, but is melting 'dynamically': that is, it is collapsing in parts as meltwater seeps down through crevices and speeds up its disintegration. Critically, this process was not taken into account in the AR4 report, leading to estimates of sea-level increase that were far too low. Greenland is losing 200 to 300 km^3 of ice into the sea each year, about the same amount as all the ice in Arctic Europe. This process on its own is causing the global sea level to rise by more than a millimetre a year, whereas a decade ago Greenland's contribution to sea-level rise was non-existent.

Temperatures will rise more over continental areas than over the oceans, and the Northern Hemisphere will warm more quickly than the Southern. The area covered by snow will contract, and widespread increases in thaw depth are expected over most permafrost regions. Sea-ice is projected to shrink in the Antarctic as well as the Arctic. It is very likely that hot extremes, heat waves and heavy precipitation events will become more frequent, and that future tropical cyclones (typhoons and hurricanes) will become more intense, with larger peak wind speeds and more heavy precipitation associated with ongoing increases of tropical sea-surface temperatures. As the CO_2 levels progress to a doubling of pre-industrial values, the southern parts of Europe and the United States would become truly tropical. Mid- to low-latitude land masses seem likely to experience a wetter climate than at present, with the deserts of North Africa, north-west India, and the south-west USA becoming prairie-like. Conversely, northern Europe and most of the central parts of North America and the Soviet Union would become drier. The climate would be getting close to that which existed at the beginning of recorded history, some 4500 to 8000 years ago (the *Altithermal period*), when the world was definitely warmer.

Warming in the polar regions could approach as much as 10 °C by the middle of the next century. Formation of sea-ice will be reduced in both polar regions, and climatologists are tantalized by the possibility that Arctic Ocean ice might disappear and not return. Modification of the entire climate of the Arctic Basin would ensue, with profound ecological consequences on land and in the sea, and with possible release of much CH_4 from methane hydrate. Permafrost under the Arctic Sea north of Siberia has started to melt as average temperatures in the area have increased almost 10 °C in 30 years. The release of CH_4 caused by the melting is now believed to be causing methane levels to rise after being stable for a decade. This permafrost was left over from the last ice age when sea levels were much lower. The area at risk is thought by some to contain as much carbon dioxide and methane as all of the coal, oil and natural gas ever burned and all that still remains in the ground.

Beyond the 21st century, anthropogenic warming and sea-level rise would continue for centuries even if GHG concentrations were to be stabilized, with a further increase in global average temperature of about 0.5 °C expected by 2200. Thermal expansion alone would lead to 0.3 to 0.8 m of sea-level rise by 2300 (relative to 1980–1999), and expansion would persist for many centuries. Contraction of the Greenland ice sheet is expected to continue to contribute to sea-level rise after 2100. Current models suggest that ice-mass losses increase with temperature more rapidly than gains due to increased precipitation and that the surface mass balance becomes negative (net ice loss) at a global average warming (relative to pre-industrial values) in excess of 1.9 to 4.6 °C. If such a negative surface mass balance were sustained for millennia, that would lead to virtually complete elimination of the Greenland ice sheet and a resulting contribution to sea-level rise of about 7 m. The corresponding future temperatures in Greenland (1.9–4.6 °C) are comparable to those inferred for the last interglacial period 125 000 years ago, when palaeoclimatic information suggests reductions of polar land-ice extent and 4–6 m of sea-level rise. Dynamical processes related to ice flow—which are not included in current models but are suggested by recent observations—could increase the vulnerability of the ice sheets to warming, increasing future sea-level rise. Understanding of these processes is limited and there is no consensus on their magnitude. Current global model projections are for the Antarctic ice sheet to remain too cold for widespread surface melting and to gain mass due to increased snowfall. However, net loss of ice mass could occur if dynamical ice discharge dominates the ice-sheet mass balance. Both past and future anthropogenic CO_2 emissions will continue to contribute to warming and sea-level rise for more than a millennium, because of the timescales on which this gas is removed from the atmosphere.

11.4.5 Aircraft

The potential impact of aircraft-related climate change deserves special mention because some climate impacts are unique to aviation: they include the formation of condensation trails (contrails), their impact on cirrus cloudiness, and the high-altitude atmospheric injection of aerosols. Persistent contrail formation and induced cloudiness are indirect effects from aircraft operations because they depend on variable humidity and temperature conditions along aircraft flight tracks. Future changes in atmospheric humidity and temperature distributions in the upper troposphere will thus have consequences for aviation-induced cloudiness. Aviation aerosols may also alter the properties of clouds that form later in air containing aircraft emissions.

Section 11.2.2 discusses aircraft operation in relation to stratospheric ozone losses, but we documented there both the expansion of aviation over recent decades and the several ways in which aircraft emissions might affect the atmosphere. Even in relation to the contribution of aircraft to the general GHG burden, emission of CO_2 alone is not the end of the story. Subsonic aircraft emit NO_x, which generates ozone; ozone increases will, on average, tend to warm the Earth's surface. Aircraft operations are estimated to have increased O_3 in the 1990s by up to six per cent in northern mid-latitudes at cruise altitudes. On the other hand, aircraft NO_x tends to decrease concentrations of CH_4, thus reducing the greenhouse heating from this gas. However, the changes in tropospheric O_3 occur mainly in the Northern Hemisphere, while the CH_4 changes are global, so that they do not offset one another.

Water vapour emitted in the troposphere is fairly rapidly released by precipitation; that released in the lower stratosphere can build up to higher concentrations. Aircraft produce persistent contrails in the upper troposphere in ice-supersaturated air masses that are probably more important than the vapour-phase water. Contrails are thin cirrus clouds, which reflect solar radiation and trap outgoing longwave radiation. The latter effect is expected to dominate for thin cirrus resulting in a net positive radiative forcing. The current best estimate for the forcing produced by persistent linear contrails for aircraft operations in 2005 is $+0.010\,W\,m^{-2}$, but the uncertainty range is conservatively estimated to be a factor of three because of the poor level of scientific understanding of contrail radiative forcing.

Individual persistent contrails are routinely observed to shear and spread, covering large additional areas with cirrus cloud. Aviation aerosol could also lead to changes in cirrus cloud. *Aviation-induced cloudiness* (AIC) is defined to be the sum of all changes in cloudiness associated with aviation operations. An estimate of AIC thus includes persistent contrail cover. Because spreading contrails lose their characteristic linear shape, a component of AIC is indistinguishable from background cirrus. This basic ambiguity prevents the formulation of a best estimate of the extent of AIC and the associated radiative forcing. The ratio of induced cloudiness cover to that of persistent linear contrails ranges from 1.8 to 10, indicating the uncertainty in estimating AIC. Some studies have found significant positive trends in cirrus cloudiness in regions of high air traffic and lower-to-negative trends outside air-traffic regions, and suggest cirrus-cover increases for Europe of 1 to 2 per cent per decade over the last one to two decades.

Aircraft engines emit soot and species that produce sulfate aerosol. Although aerosol accumulation is expected to grow with increased burning of aviation fuel, the mass concentrations are likely to remain small compared to surface sources, and direct radiative forcing by the aerosols can probably be neglected. However, aerosols from aircraft may play a role in enhanced cloud formation, which might have a more important effect on the radiation balance. An important concern is that aviation-generated aerosol particles can act as nuclei in ice-cloud formation, thus altering the microphysical properties of clouds, and perhaps cloud cover. At present, these factors are not understood in any quantitative form, which does not mean, of course, that the contribution is not significant!

11.4.6 Impacts of Climate Change

The socio-economic ramifications of climate change lie outside the scope of this book. Figure 11.18 presents some of the impacts identified by the IPCC as being associated with global average

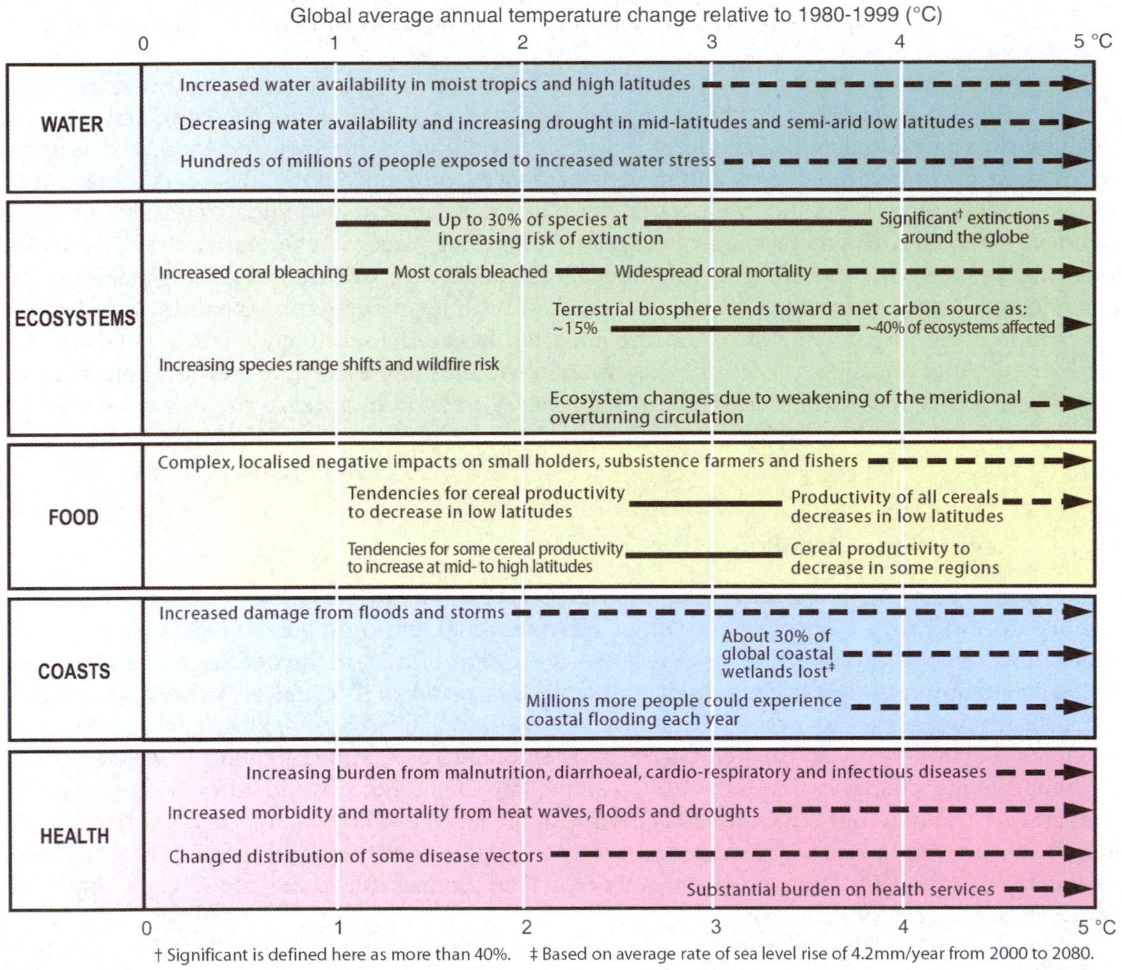

Global average annual temperature change relative to 1980-1999 (°C)

0 1 2 3 4 5 °C

WATER
Increased water availability in moist tropics and high latitudes
Decreasing water availability and increasing drought in mid-latitudes and semi-arid low latitudes
Hundreds of millions of people exposed to increased water stress

ECOSYSTEMS
Up to 30% of species at increasing risk of extinction
Significant† extinctions around the globe
Increased coral bleaching
Most corals bleached
Widespread coral mortality
Terrestrial biosphere tends toward a net carbon source as:
~15%
~40% of ecosystems affected
Increasing species range shifts and wildfire risk
Ecosystem changes due to weakening of the meridional overturning circulation

FOOD
Complex, localised negative impacts on small holders, subsistence farmers and fishers
Tendencies for cereal productivity to decrease in low latitudes
Productivity of all cereals decreases in low latitudes
Tendencies for some cereal productivity to increase at mid- to high latitudes
Cereal productivity to decrease in some regions

COASTS
Increased damage from floods and storms
About 30% of global coastal wetlands lost‡
Millions more people could experience coastal flooding each year

HEALTH
Increasing burden from malnutrition, diarrhoeal, cardio-respiratory and infectious diseases
Increased morbidity and mortality from heat waves, floods and droughts
Changed distribution of some disease vectors
Substantial burden on health services

0 1 2 3 4 5 °C

† Significant is defined here as more than 40%. ‡ Based on average rate of sea level rise of 4.2mm/year from 2000 to 2080.

Figure 11.18 Examples of impacts projected for climate changes (and sea level and atmospheric CO_2 where relevant) associated with different extents of increase in global average surface temperature in the 21st century. Entries are placed so that the left-hand side of text indicates the approximate level of warming that is associated with the onset of a given impact. Adaptation to climate change is not included in these projections, and impacts will vary according to the extent of adaptation, rate of temperature change and socio-economic pathway. Source as for Figure 11.14.

temperature change. Many practical issues will have to be faced, with some people being better off, and some faring worse, on a warmer Earth. As stated earlier, it is the *rate* of change that presents the greatest challenge. Agriculture will be much affected, with mid-latitude farmers expecting about ten days longer growing season for every degree Celsius rise in temperature. Large areas of Africa, the Middle East, India, and central China might cease to be water deficient. On the other hand, the 'food basket' areas of North America and the Soviet Union would become much drier, and it would be harder to grow grain and other major food crops. A higher level of CO_2 in the atmosphere and more rainfall in the subtropics would both enhance forest growth and allow it to spread to places that are now too arid. Present dangerous trends in deforestation might therefore be arrested. Costs of space heating in winter would decrease, but air conditioning, where used, would become more expensive. Heat stress is expected to have an adverse effect on human health, and a warmer climate might encourage the spread of certain tropical diseases, such as malaria, to regions that were previously free of these problems.

Two of the most important impacts of a rise in temperature follow from the changes in the water cycle. Cycling of water between the oceans and the land masses provides the fresh water that is essential for many forms of life, and most especially to fulfil the needs and expectations of humans. An increase in temperature means that more water will evaporate from the surface. Although most scenarios show more rainfall in some parts of the world, other parts would have less, especially in summer. Coupled with greater rates of evaporation in these places, the potential clearly exists for substantial exacerbation of drought conditions. A rise in sea-level of 50 cm might not bother people living inland, except indirectly. Unfortunately, half of the human population inhabits coastal zones, and some of the lowest lying areas are the most fertile and densely populated. Bangladesh, the Netherlands, and certain small oceanic islands are obvious examples. In Bangladesh alone, roughly six million people live within one metre of sea-level. A fraction of a metre rise in sea-level would obviously jeopardize their livelihoods and even existence.

11.4.7 Legislation and Policy

Sections 11.4.4. and 11.4.5 make it evident that forecasts of the extent of global warming in the next century do not provide absolute certainties for policy-making, only 'likelihoods'. Furthermore, projections into the future about fossil-fuel use, and hence of carbon dioxide release, also depend on many assumptions. Yet it is now that action needs to be taken if the untoward consequences of global warming are to be averted or alleviated. We have already pointed out that there is an inevitable delay between cutting emissions of greenhouse gases and the point at which anthropogenic climate change will cease. For example, by 2030 there would already have been a temperature rise of about 1 °C since 2000 even with the modest A1B scenario, and sea levels could have risen in some places by as much as 90 cm if the current growth-rate of $3\,mm\,yr^{-1}$ were sustained. Even if CO_2 emissions were stabilized then, temperatures and sea levels would both continue to rise, but more slowly.

A re-examination of the climate impacts reported in the AR4 indicates that 80 per cent cuts in global greenhouse gas emissions are needed by 2050 to keep the global average temperature rise below 2 °C, and to limit climate impacts to 'acceptable' levels. Such a cut would stabilize atmospheric greenhouse gas concentration at 400–470 ppm CO_2 equivalents. However, even with an 80 per cent cut in emissions, damage will be significant, and much more substantial efforts at adaptation will be required than those currently planned to avoid such damage. Clearly a decrease of this magnitude will require the developed nations to make enormous cuts in emissions of GHGs, since developing countries still have basic energy needs that are likely to mean some growth of emissions over the next decades.

International action is beginning to be taken to control the emission of greenhouse gases, rather in the same manner as legislation has been enacted to reduce emissions of gases that deplete the ozone layer (Section 11.2.5). In December 1997, the *Convention on Climate Change* was held in Kyoto, Japan, and in the ensuing *Kyoto Protocol* some consensus on reductions was agreed. Developed countries agreed to reduce their overall emissions of a basket of six greenhouse gases by 5.2 per cent below 1990 levels over the period 2008–2012, with differentiated, legally binding targets. The then 15 EU Member States adopted a collective target to reduce EU emissions by eight per cent, the US by seven per cent, and Japan by six per cent. Under the 'bubble' arrangement, the EU's target is distributed between Member States to reflect their national circumstances, requirements for economic growth, and scope for further reduction in emissions. Each Member State has a legally binding target, with the UK undertaking to reduce its emissions by 12.5 per cent. The Protocol also established mechanisms to assist in meeting Kyoto targets in the most efficient and cost-effective manner. These mechanisms recognize that greenhouse gases will contribute to global increases in temperature, regardless of the source, and that equally it does not matter where reductions are made provided that they are real reductions. The Kyoto mechanisms include international emissions trading and the use of credits from emission-reducing projects in one country to meet the Kyoto target of another country. The Kyoto Protocol entered into force in February 2005 and to date has been ratified by 162 countries. However, a number of battles remain to be won. Underlying all the discussions are clear divisions between those who believe the threat of global warming to be real and those who do not, between environmentalists and business, and, in the last analysis, between the developed and the developing nations. Uncertainty about entry into force of the Kyoto Protocol led to a loss of momentum in international action and a slow start to the carbon market. Crucially, it also undermined international political will to address the increasingly urgent question of what action would be taken by countries beyond the first set of Kyoto targets that expire at the end of 2012.

Some of the momentum now seems to have been regained. At a meeting of leaders of the G8 nations at Gleneagles in July 2005, they agreed that climate change was a serious and long-term challenge, that it was caused by human activity, and that urgent action should be taken to make significant reductions in greenhouse-gas emissions. This statement was a significant one in that it came from countries that had not previously committed themselves to the case for urgency so publicly. In October 2005, the European Commission launched a new phase of the European Climate Change Programme (ECCP) to consider further measures to contribute towards the EU's Kyoto Protocol target and beyond. The measures include geological carbon capture and storage, and the programme targets passenger road transport, aviation, and non-carbon dioxide emissions. For the first time, the ECCP will also consider adaptation to the impacts of climate change.

The year 2009 was a crucial one in the international effort to address climate change, culminating in the United Nations Climate Change Conference COP15 (*15th Conference of the Parties*) held in Copenhagen in December. Five meetings held earlier in the year had seen delegates discuss key negotiating texts intended to serve as the basis for an ambitious and effective international climate change deal to be clinched at the Copenhagen conference. Many delegations, including those of China, the U.S., the G-77 nations and the European Union, all acknowledged that the process of the negotiations had moved forward, but there were and still are serious disagreements on many issues. Chief among the concerns was the nature of the cuts in GHG emissions that the richer countries had promised to make, because there was widespread criticism that the cuts would be too small to keep climate change in check. Five countries (Australia, Costa Rica, Japan, Tuvalu and the United States) presented proposals for a new Protocol to the UN Climate Convention. But developing countries maintain that the Kyoto Protocol, which virtually all countries have signed—except for the US—is still good international law and will remain so, and did not wish to negotiate a new Protocol at all. It might be noted that some nations counted as developing in 1997 had thriving economies by 2009, and now make major contributions to the emissions of GHGs. Yet these nations are not required to take mitigating steps according to the provisions of the Kyoto Protocol, so that they are not keen for their status to be reassigned in a new Protocol. In the event, COP15 was a disappointment to climate scientists who had hoped for a statutory undertaking to cut carbon emissions by 50 per cent by 2050, but it was not an unmitigated failure. In a last-minute development, an 'Accord' was adopted in Copenhagen that calls for "deep cuts in global emissions...so as to hold the increase in global temperature below 2 °C". The undoubted problem is that there is a large gap between this target and the commitments that individual nations were willing to offer. National self-interest and short-term thinking dominate the responses of the majority of politicians, and somehow provision must be made for the enormous financial costs of effective action.

Several observers note that the evident disagreements and diverse objectives could only have led to a rather weak Protocol had one been formulated at the end of the COP15 meeting, and that adoption of a weak treaty would be worse than not having one at all, so long as that is only a temporary situation. Any effective Protocol must now attempt to involve all nations—regardless of their status as industrial, emerging, or more slowly developing—in common solutions. On the positive side, at COP15 the world's poorest nations (and those most likely to be the worst affected by climate change) made their voices heard plainly for the first time at such an international gathering. It seems that this meeting in Copenhagen has forced the whole world to engage with the issue of climate change.

The next step envisaged is for a further major conference (COP16) to be convened in Mexico City in late 2010, with signatories to the Copenhagen Accord required to submit their proposed national actions for emission reduction and mitigation by the end of January 2010.

Inadvertent chemical modification by Man of the troposphere (Section 11.1) or stratosphere (Sections 11.2 and 11.3) seems likely to be relatively unimportant compared to climate change resulting from the burning of fossil fuels and the consequent release of CO_2. The cause is an international activity, the effect is global in character, and the remedy will require the concerted action of the countries of the world. A major alteration of the global environment seems possible, or even probable, unless we decide to change the course of events. Man's greed for energy will either have to be satisfied by non-fossil sources, or we shall have to prepare for the climatic change that is in store for us.

The sad message with which this book must end, therefore, is that Man has many, many ways in which he can pollute his environment, and they are by no means all dependent on industrial activity. The encouraging balancing message is that scientific understanding of the problems is becoming well enough advanced that sensible policy decisions and control strategies can be formulated. A new, rigorous and demanding Protocol for GHGs will need to be adopted within a few years at most to have a chance of preventing catastrophe. We can hope that the findings that come from the enormous efforts of atmospheric scientists carrying out excellent research may persuade the legislators in Copenhagen to adopt long-term, if unpalatable, solutions. And, of course, we hope that this book has helped its readers to understand the many issues of basic science and practical implementation that will lie behind the negotiations.

Further Reading

For the reader who wishes to find out more about a topic, or to have it explained in a different way, we have listed here some suitable sources. Some of the texts delve much more deeply into certain aspects than we have been able to, while others introduce subjects that are missing from our book altogether. We have tried to group the texts and reviews according to their purpose. The larger books address the whole range of subjects that we have introduced, although some are more specialized. Listed for Chapter 11 are the most recent publications from WMO on ozone depletion (Chapter 9) and the IPCC on climate change (Chapter 11). These are major sources of information for this entire book, although they are perhaps rather heavy going for someone just starting out on the study of atmospheric chemistry.

General

C.N. Hewitt and A.V. Jackson (eds.), *Atmospheric Science for Environmental Scientists*, Wiley-Blackwell, Chichester, 2009.

B.J. Finlayson-Pitts and J.N. Pitts Jr., *Chemistry of the Upper and Lower Atmosphere: Theory, Experiments, and Applications*, Academic Press, London and San Diego, 1999.

D. Jacob, *Introduction to Atmospheric Chemistry*, Princeton University Press, Princeton, NJ, 1999.

J.H. Seinfeld and S.N. Pandis, *Atmospheric Chemistry and Physics: From Air Pollution to Climate Change*, 2nd edn, Wiley-Interscience, Hoboken, NJ, 2006.

J.M. Wallace and P.V. Hobbs, *Atmospheric Science: An Introductory Survey*, 2nd edn, Academic Press, London and San Diego, 2006.

R.P.Wayne, *Chemistry of Atmospheres: An Introduction to the Chemistry of the Atmospheres of Earth, the Planets, and their Satellites*, 3rd edn, OUP, Oxford, 2000.

Chapter 2

J.T. Houghton, *The Physics of Atmospheres*, 3rd edn, Cambridge University Press, Cambridge, 2002.

Atmospheric Chemistry
By Ann M. Holloway and Richard P. Wayne

Published by the Royal Society of Chemistry, www.rsc.org

D.G. Andrews, *An Introduction to Atmospheric Physics*, Cambridge University Press, Cambridge, 2008.

Richard Goody, *Principles of Atmospheric Physics and Chemistry*, OUP, New York, 1996.

R.B. Stull, *Meteorology for Scientists and Engineers*, 2nd edn, Brooks Cole, Cenage Learning, London, 1999.

Chapter 4

M. Z. Jacobson, *Fundamentals of Atmospheric Modeling*, 2nd edn, Cambridge University Press, New York, 2005.

Chapters 6 and 7

T.E. Graedel and P.J. Crutzen, *Atmospheric Change: An Earth System Perspective*, W.H. Freeman, New York, 1993.

C.N. Hewitt, *Reactive Hydrocarbons in the Atmosphere*, Academic Press, London and San Diego, 1998.

R. Reynolds, K. Warr, N. Dise and R. Hodgkins, *Extreme Weather, Atmospheric Chemistry and Pollution, Wetlands and the Carbon Cycle, Cryosphere*, 3rd edn, The Open University Press, Milton Keynes, 2009.

J. Lovelock, *Gaia: A New Look at Life on Earth*, New Ed., Oxford Paperbacks, Oxford, 2000.

Chapter 8

J.G. Calvert, R. Atkinson, J.A. Kerr, S. Madronich, G.K. Moortgat, T.J. Wallington and G. Yarwood, *The Mechanisms of Atmospheric Oxidation of the Alkenes*, OUP, New York, 2000.

J.G. Calvert, R.G Derwent, J.J. Orlando and G.S. Tyndall, *Mechanisms of Atmospheric Oxidation of the Alkanes*, OUP, New York, 2008.

Chapter 9

Scientific Assessment of Ozone Depletion: 2006, UNEP and WMO, Geneva, 2007.

Chapter 10

G.G. Shepherd, *Spectral Imaging of the Atmosphere*, Academic Press, London and San Diego, 2002.

V.Y. Khomich, A.I. Semenov and N.N. Shefov, *Airglow as an Indicator of Upper Atmospheric Structure and Dynamics*, Springer-Verlag, Berlin and Heidelberg, 2008.

J.M.C. Plane, Atmospheric Chemistry of Meteoric Metals, *Chem. Rev.* 2003, **103**, 4963.

R. Schunk and A. Nagy, *Ionospheres: Physics, Plasma Physics, and Chemistry*, 2nd edn, Cambridge University Press, Cambridge, 2009.

J.K. Hargreaves, *The Solar-Terrestrial Environment: An Introduction to Geospace – The Science of the Terrestrial Upper Atmosphere, Ionosphere, and Magnetosphere*, Cambridge University Press, Cambridge, 1995.

Chapter 11

M.Z. Jacobson, *Atmospheric Pollution: History, Science, and Regulation*, Cambridge University Press, New York, 2002.

P. Brimblecombe, H. Hara, D. Houle, and M. Novak (eds.), *Acid Rain—Deposition to Recovery*, Springer, Dordreeht, 2007.

S.O. Andersen and K. Madhava Sarma. *Protecting the Ozone Layer: The United Nations A History*, Earthscan Kogan Page, London, 2002.

S.H. Bakker (ed.), *Ozone Depletion, Chemistry, and Impacts*, Nova Science Publishers Inc., Hauppage NY, 2009.

Scientific assessment of ozone depletion: 2006, UNEP and WMO, Geneva, 2007

Climate change 2007 – The Physical Science Basis; Contribution of Working Group I to the Fourth Assessment Report of the IPCC, Cambridge University Press, Cambrigde, 2007.*

J.C. Goodman (ed.), *Aviation and the Environment*, Nova Publishers, Hauppage NY, 2009.

N. Stern, *Blueprint for a Safer Planet: How to Manage Climate Change and Create a New Era of Progress and Prosperity*, The Bodley Head, London, 2009.

T. Tin, *A European Update of Climate Science: An Overview of the Climate Science Published Since the UN IPCC Fourth Assessment Report*, WWF, Brussels, 2008.

UK Climate Projections 2009, Commissioned by the Department of Environment, Food and Rural Affairs (Defra), London, 2009.

J. Lovelock, *The Vanishing Face of Gaia: A Final Warning*, Allen Lane, London, 2009.

M.J. Hoffmann, *Ozone Depletion and Climate Change: Constructing a Global Response*, SUNY Press, New York, 2005.

D.J. Kennedy, *Science Magazine's State of the Planet 2008–2009: with a Special Section on Energy and Sustainability*, Island Press, Washington, DC, 2008.

* The IPCC is currently starting to outline its Fifth Assessment Report (AR5) that will be finalized in 2014. As it has been the case in the past, the outline of the AR5 will be developed through a scoping process that involves climate change experts from all relevant disciplines and users of IPCC reports, in particular representatives from governments.

Subject Index